Springer Theses

Recognizing Outstanding Ph.D. Research

Aims and Scope

The series "Springer Theses" brings together a selection of the very best Ph.D. theses from around the world and across the physical sciences. Nominated and endorsed by two recognized specialists, each published volume has been selected for its scientific excellence and the high impact of its contents for the pertinent field of research. For greater accessibility to non-specialists, the published versions include an extended introduction, as well as a foreword by the student's supervisor explaining the special relevance of the work for the field. As a whole, the series will provide a valuable resource both for newcomers to the research fields described, and for other scientists seeking detailed background information on special questions. Finally, it provides an accredited documentation of the valuable contributions made by today's younger generation of scientists.

Theses may be nominated for publication in this series by heads of department at internationally leading universities or institutes and should fulfill all of the following criteria

- They must be written in good English.
- The topic should fall within the confines of Chemistry, Physics, Earth Sciences, Engineering and related interdisciplinary fields such as Materials, Nanoscience, Chemical Engineering, Complex Systems and Biophysics.
- The work reported in the thesis must represent a significant scientific advance.
- If the thesis includes previously published material, permission to reproduce this must be gained from the respective copyright holder (a maximum 30% of the thesis should be a verbatim reproduction from the author's previous publications).
- They must have been examined and passed during the 12 months prior to nomination.
- Each thesis should include a foreword by the supervisor outlining the significance of its content.
- The theses should have a clearly defined structure including an introduction accessible to new PhD students and scientists not expert in the relevant field.

Indexed by zbMATH.

Matteo Bonanomi

Response of the High Granularity Calorimeter HGCAL and Characterisation of the Higgs Boson

With the CMS Experiment at the LHC

Doctoral Thesis accepted by
École Polytechnique, Palaiseau, France

Springer

Author
Matteo Bonanomi
Institut für Experimentalphysik
Universität Hamburg
Hamburg, Germany

Supervisor
Roberto Salerno
Palaiseau, France

ISSN 2190-5053 ISSN 2190-5061 (electronic)
Springer Theses
ISBN 978-3-031-26832-8 ISBN 978-3-031-26833-5 (eBook)
https://doi.org/10.1007/978-3-031-26833-5

This Springer imprint is published by the registered company Springer Nature Switzerland AG
The registered company address is: Gewerbestrasse 11, 6330 Cham, Switzerland

Si l'on considérait une théorie comme parfaite et si l'on cessait de la vérifier par l'expérience scientifique, elle deviendrait une doctrine.

Claude Bernard

Preface

The desire for knowledge is intrinsic to human nature and humans have always been looking for the key to disclosing the Universe's hidden secrets and the laws that regulate it... The primordial quest for the fundamental constituents of the Universe dates back to the ancient Greeks and to the philosophy of *atomism*, which postulates that ordinary matter is composed of *atoms* (from the Greek work ατομον), i.e., "uncuttable, indivisible" components.

In the nineteenth century, a remarkable series of experimental observations clarified that the "infinitely small" world is not regulated by the same laws describing ordinary and macroscopic matter. A real turning point in understanding the fundamental laws that regulate the subatomic world was possible only at the beginning of the following century, with the advent of quantum mechanics and special relativity.

These theories laid the basis for the definition of the standard model (SM) of particle physics, a quantum field theory with gauge symmetries formalized in the late 60s, when incorporating the concept of spontaneous symmetry breaking through the physical vacuum. More than 50 years later, nowadays the SM is still the most reliable theory that provides a comprehensive description of the world of the "infinitely small". However, notwithstanding the success of the SM in predicting experimental observations, it presents some severe limitations. For example, it does not include a description of quantum gravity, nor does it explain where the observed abundance of dark matter in the Universe comes from and why there is an asymmetry between ordinary matter and antimatter. The SM also presents more limitations directly related to the context of the work presented in these pages.

As detailed below, the cornerstone of the SM is the Bourt-Englert-Higgs (BEH) mechanism, which postulates the existence of a scalar field, referred to as the Higgs (H) boson, responsible of the spontaneous electroweak symmetry breaking (EWSB), ensuring that vector bosons acquire their masses, defining the Yukawa interaction of fermions, and guaranteeing the unitarity of the theory. However, the mass of the H boson itself is a free parameter of the theory and it is not protected by any symmetry. As a direct implication of this, the theory becomes sensitive to quantum fluctuations involving masses up to the grand unification energy scale or even the Planck Scale. In addition, the mass of the H boson is directly related to one of the most intriguing couplings predicted by the SM, that is the self-coupling. In fact, while the SM predicts a self-interaction of the H boson, with a strength proportional to the squared value of this particle's mass, the experimental evidence of this coupling has not been proven yet, thus increasing the experimental interest in the study of the SM scalar sector. Eventually, also the origin of the couplings of the H boson with fermions is not explained by the SM, similarly to the reason why three fermion families are observed.

Hence, while the SM contributed to unveiling many of the secrets of the subatomic world and the laws that regulate it, the quest for the fundamental constituents of the Universe initiated by the ancient Greeks is not entirely accomplished yet.

All the subatomic particles observed throughout the second century can be explained in the context of the SM, either as *elementary particles*, that is, subatomic particles with no substructure, or as their aggregate states. The experimental proof of the weak neutral currents by the Gargamelle Collaboration in 1973 and the subsequent discoveries of the $W^{\pm}$ and Z bosons at the CERN Super Proton Synchrotron (SPS) in 1983 experimentally confirmed the predictions of the SM. These discoveries have established fundamental milestones on the way toward understanding the origin of the electroweak symmetry breaking, which lead to the appearance of the known electromagnetic and weak forces. The observation of the top quark at the FNAL Tevatron in 1995 eventually demonstrated the predictive power of the SM over a wide range of energy scales.

The ultimate success of the SM came in 2012 when the ATLAS and CMS Collaborations at the CERN Large Hadron Collider (LHC) announced the observation of a scalar particle compatible with the SM elementary Higgs (H) boson. The existence of a physical H boson is presumed to be a consequence of the spontaneous EWSB caused by the Higgs field in the early Universe through the Bourt-Englert-Higgs (BEH) mechanism. This mechanism gave rise to the masses of the $W^{\pm}$ and Z bosons, and allowed the elementary fermions to gain mass, via their Yukawa interactions with the H field. The discovery was made possible by analyzing the data set collected by the experiments during the Run-I (2009–2013) period of the LHC. The quest for the truth intrinsic in human nature is well summarized by the words of Joe Incandela, spokesperson of the CMS Collaboration at the time of the discovery:

> *This boson is a very profound thing we have found. We're reaching into the fabric of the universe at a level we've never done before. We've kind of completed one particle's story [...] We're on the frontier now, on the edge of a new exploration. This could be the only part of the story that's left, or we could open a whole new realm of discoveries.*

The main hurdle in the hunt for the Higgs boson is its lifetime, which is not large enough to detect the particle directly but demands its indirect identification starting from its decay products. The experiments combine a series of sub-detectors to accomplish this task, mounted in a concentric structure around the decay point and each designed to identify and reconstruct a specific kind of particle. The discovery of this long-sought particle, almost 50 years after its postulation, proved that the fundamental principles incorporated in the SM are consistent with observations. In particular, it indicates that the BEH electroweak symmetry breaking mechanism is likely to be at the origin of the masses of the $W^{\pm}$ and Z bosons. However, the discovery raised many other questions. How does the Higgs boson couple with all the other particles of the SM? What are the properties of this long-sought particle? Are they all compatible with the predictions of the SM?

The Run-II operations of the LHC (2015-2018) addressed answering these questions by collecting more data with respect to the previous Run-I, thus giving access to a more significant number of Higgs bosons candidates to be studied. The predictive power of the SM in explaining the relative production and decays rates in various channels, once the H boson was known with high enough precision, has proven that the H field is also at the origin of fermion masses.

One of the channels that played a major role in the understanding of the scalar sector at the LHC is the Higgs boson decay into two Z vector bosons, which consequently decay into four leptons ($H \rightarrow ZZ \rightarrow 4\ell$), also known as the "golden channel" due to its many virtues, such as a large signal-to-background ratio, a completely resolved final state, and an excellent resolution. At the time of the discovery, these properties were fundamental to enhance the probability of detecting the rare Higgs boson events over the overwhelming background typical of the LHC collisions. However, due to the small branching fraction of this decay channel, only about 10 Higgs boson signal events could be identified in the dataset used for the discovery and, while they were enough to claim the first observation of this particle, they could not be used at that time to perform precision measurements of the properties of this fundamental brick of the SM. With the extensive dataset collected during the Run-II, around 400 signal events from the decay of the Higgs boson in two Z bosons were reconstructed and the "golden channel" became one of the standard candles to study in detail the properties of the Higgs boson and to look for possible deviations from the SM.

I had the opportunity to work on the full Run-II analysis of the $H \rightarrow ZZ \rightarrow 4\ell$ decay channel, aiming at precision measurement of the properties of the production mechanisms of the Higgs boson. The presence of additional objects in the final state, such as extra leptons and jets, and a set of matrix element discriminants are employed to define a total of 22 event categories to probe fine granular phase space regions of each production mode. My work also included the definition of the statistical methods used for the analysis, comprising thorough studies of the systematic uncertainties, the modeling of signal and background processes, and the development of the various fit routines employed to perform the measurements. The analysis results lead to a comprehensive characterization of the Higgs boson properties in the four-lepton final state and they are presented in a paper, of which I am co-editor, published in the European Physical Journal C.

To date, this is the most detailed study of the Higgs boson in the "golden channel" ever performed within the CMS Collaboration, and it leads to the conclusion that all the measurements are in agreement with the SM predictions within their uncertainties. Given the remarkable predictive power of the SM when including the H boson, possible new physics will require even more extensive studies at higher statistics.

To further stress the SM theory and probe its detailed predictions, CERN intends to commence the High-Luminosity LHC (HL-LHC) operations by the end of 2027.

The HL-LHC will significantly increase the statistics collected by the experiments, thus opening the doors to previously unexplored realms of physics. In the scalar sector, it will allow to observe H boson self-interaction and thus access for the first time the Higgs potential directly responsible for EWSB. It will moreover permit measurements of the coupling of the H boson to second-generation fermions, search additional scalar boson or for rare phenomena. On the other hand, the operative conditions of the HL-LHC will pose severe technical challenges and therefore a massive upgrade of the LHC infrastructure and detectors will be necessary to maintain the current physics performance and to exploit properly the HL-LHC's new physics reach. Among the many upgrades foreseen for HL-LHC, the CMS Collaboration will replace the current endcap calorimeters with a High Granularity Calorimeter (HGCAL): a detector unique in its genre, capable of measuring with high precision the energy of the particles as well as their spatial and timing positions. Because of these features the HGCAL is often referred to as "imaging", or 5D, calorimeter and it will be the first large-scale silicon-based calorimeter ever employed in a high energy physics experiment. It is not surprising that it was once described as "perhaps the most challenging engineering project ever undertaken in particle physics".[1] The validation of the detector design and the complete characterization of its physics performance are crucial aspects for the success of the project. For this reason, a series of beam tests were performed in 2016 and 2017, increasing the complexity of the tested systems incrementally and reaching a thorough understanding of the basic properties of this detector. In October 2018 the first large-scale prototype of the HGCAL was exposed to hadrons and positrons beams with energies ranging from 20 to 300 GeV at the CERN North Area. Part of the work of this thesis was dedicated to the assessment of the performance of the electromagnetic compartment of the HGCAL prototype used in this test beam. I contributed to the improvement of the data reconstruction performance and to the development of the analysis strategy, including the definition of the events' selection and of the observables to be studied. In addition to the measurement of the energy resolution and linearity, the high granularity of the prototype was exploited to study the longitudinal and transverse containment of electromagnetic cascades, as well as to measure the position and angular resolution of the detector, thus confirming the "imaging" power of the detector and its capabilities for particles identification and discrimination. The results of this analysis are summarized in a dedicated paper, of which I am co-editor, published in Journal of Instrumentation. Overall, no major concern about the HGCAL design nor about its capability of meeting the expected physics requirements resulted from the test beam data analysis, thus supporting the future application of the detector for the HL-LHC physics.

Chapters 1 and 2 set the reference frame for this work, presenting the theoretical framework of the SM and the experimental apparatus used for the analysis, respectively. The following chapters present my contribution to validating the HGCAL detector design and studying the Higgs boson properties in the "golden channel."

The results of the HGCAL test beam data analysis are presented in Chap. 3. The identification and isolation requirements imposed on leptons and jets, starting

[1] D. Barney -"The High Granularity Calorimeter upgrade project for CMS"

from the objects reconstructed with the CMS detector, are described in Chap. 4. These objects are exploited to reconstruct events with Higgs boson candidates, which are categorized by exploiting matrix element discriminants as detailed in Chap. 5. Chapter 6 describes the analytical parametrization used for the modeling of signal and background processes and the statistical methods employed for the extraction of the results of the analysis. The measurements of the Higgs boson properties in the "golden channel", along with their interpretation in the context of physics beyond the standard model, are presented in Chap. 7.

Hamburg, Germany Matteo Bonanomi

Acknowledgments

To Sergio and Giuseppe, always with me

This work is the result of a 3 years long journey, but it is also the cornerstone of a learning process that will continue for many years to come. All of this would not have been possible without several people to whom I would like to express my deepest gratitude.

I would like to thank the referees, Paolo Francavilla and Giovanni Marchiori, and the other members of the Jury, Sara Bolognesi, Tiziano Camporesi, María Luisa Cepeda, and Lucia Di Ciaccio, for having accepted to review the manuscript and for their useful comments and the following discussions that have helped me to improve the quality of these pages.

My most sincere gratitude goes to my supervisors, Roberto and Yves, who have believed in me from the beginning of this journey. Thank you for your thorough guidance throughout these years, for the uncountable discussions and opportunities you gave me, and for conveying me your passion for the beautiful job we do. I am particularly grateful to Artur for having been a fundamental guide and a constant inspiration during the first year of my Ph.D. I owe you a lot and I am looking forward to working with you again in the near future.

I had the opportunity to meet brilliant minds and researchers among all the members of the CMS group at LLR: your expertise and wisdom have helped my professional growth over these years. A particular mention goes to Claude and Florian for the interesting and pleasant discussions we have had about the world of the Higgs boson and the HGCAL.

I must express my gratitude to Christophe and Andrei for their availability and the several discussions we had in the context of the four-lepton analysis. I am glad I had the opportunity to work with two great researchers as you are and to have learnt from you most of the concepts that fill these pages. I am thankful to Elisa and Toni for making enjoyable the many hours spent working on our analysis, for teaching me your well-organized mindset, and for your never-ending enthusiasm toward what we were accomplishing. A special credit goes to Giacomo and Meng for having shed the light on the many questions I had about the statistical procedures of the analysis and all the other technical issues they helped me to solve. I also have to thank all the other members of the CJLST working group for the countless discussions and the food for thoughts that characterized our weekly meetings and which I extremely profited from.

I am grateful to Catherine, Stathes, Shilpi, and Thorben for the time spent together working on the analysis of the HGCAL test beam data. Thank you for the precious advice and guidance you gave me.

I want to thank Alex for the opportunity he gave me being on his side in teaching at TREX. You are an amazing teacher: I admire and share your passion for teaching and I will always bring with me this experience. It goes without saying that I will always be in debt with all the students I had the luck to supervise. Your curiosity was a constant stimulus for me to do better and to learn more.

Thank you Alessandro for the uncountable laughs together and for being such a great colleague. Thank you Jona for the several kilometers we ran together. You guys are such curious and cultured people and constantly motivated me to be a better person. I am lucky to have met you and to account you as friends. I wish you all the best for your Ph.D. journeys. Thank you Cristina and Chiara for the discussions we had together that helped me to take with more ease and peacefulness these 3 years. Thank you Jonas for the good times spent together as office mates and for your help and precious insights about several topics. Eventually, thank all my Ph.D. and postdoc fellows at LLR. It has been a pleasure to share with you the joys and sorrows of this Ph.D. journey!

Thank you to all the members of the administration at LLR for their support and prompt action every time there was the need.

A successful Ph.D. journey requires a good balance between working activity and private life. Hence, I must express my gratitude to all the people I met outside the lab that made this possible, with special regard to all my friends. Thank you for having made me feel home during these 3 years in Paris!

Grazie a tutti i miei amici in Italia per esserci sempre stati, per non avermi mai fatto sentire lontano da casa e per essere riusciti ad annullare le distanze anche quando una pandemia mondiale ci ha tenuti distanti, chiusi in casa, per un anno. Parte di questo successo è senza dubbio grazie a voi!

Infine, grazie alla mia famiglia per aver sempre creduto in me e per aver sempre assecondato le mie decisioni. Se non fosse per voi, oggi non sarei la persona che sono e probabilmente non avrei mai raggiunto questo traguardo. Grazie!

Contents

Chapter 1
The Higgs Boson and the Standard Model of Particle Physics

The Standard Model of particle physics (SM) [1] is the pillar on which the current understanding of the subatomic world is based. It describes the elementary particles and their fundamental interactions in the context of a Lorentz-invariant and renormalizable quantum field theory. The theoretical framework of the SM is corroborated by an extensive set of experimental measurements, of which the last is the discovery of a particle compatible with the Higgs boson by the ATLAS and CMS Collaborations at the CERN LHC in June 2012 [2–4]. A brief theoretical introduction to the SM is given in Sect. 1.1, emphasizing the electroweak symmetry breaking mechanism (EWSB) in Sect. 1.1.2. This process ensures that elementary particles acquire mass while respecting the gauge symmetries of the SM by introducing an additional scalar field to the SM Lagrangian: the Higgs boson, a long-sought particle originally postulated in the second half of the XX century, but experimentally detected only 50 years later. More details about the properties and the phenomenology of this particle at hadron colliders are given in Sect. 1.2. The chapter closes in Sect. 1.3, with the description of the Higgs boson decay into four leptons, being it the channel studied in this work.

1.1 The Standard Model of Particle Physics

1.1.1 Particles and Fields

The SM describes three of the four fundamental forces in the context of a renormalizable quantum field theory. The strong, weak, and electromagnetic interactions are represented by a $SU(3)_C \times SU(2)_L \times U(1)_Y$ local gauge symmetry. The $SU(3)_C$ symmetry is manifested in the existence of *gluons*: the mediators of the strong force

M. Bonanomi, *Response of the High Granularity Calorimeter HGCAL and Characterisation of the Higgs Boson*, Springer Theses,
https://doi.org/10.1007/978-3-031-26833-5_1

described with Quantum Chromodynamics (QCD). The $SU(2)_L \times U(1)_Y$ symmetry represents the unification of the weak and electromagnetic forces, mediated by the $W^{\pm}$ and Z bosons and the photon (γ). While the gravitational interaction is not included in the formulation of the SM, it is also true that the strength of this interaction is negligible at the subatomic level.

The SM comprises of a total of twelve elementary spin $s = 1$ *bosons*: eight[1] gluons, the mediators of the strong force, and the $W^{\pm}$, Z, and γ, mediators of the weak and electromagnetic forces. The second set of elementary particles present in the SM are the *fermions*, which comprise the ordinary matter and interact via the exchange of virtual bosons. Fermions are $s = \frac{1}{2}$ particles, divided into *quarks* and *leptons*, for a total of twelve fermion fields included in the SM formulation. These particles are divided into three *families*, each comprising of two quarks with electric charge $+\frac{2}{3}$ and $-\frac{1}{3}$, and two leptons with charge -1 and 0. As predicted by the solutions of Dirac's equation, for each fermion, there is a corresponding *antiparticle*, that is, a particle with the same properties but opposite quantum numbers.

The Higgs boson completes the SM picture. This particle is the only $s = 0$ boson in the SM and gives mass to all the other particles in the SM via the so-called *electroweak symmetry breaking* (EWSB) mechanism, of which a more extensive description is given in Sect. 1.1.2.

A review of all the particles included in the SM formulation is presented in Fig. 1.1 and detailed in the following.

Quarks

The SM formulation, supported by the current state-of-art experimental observations, includes six quarks, divided into three doublets: the *up* (u) and *down* (d) quarks, that compose "everyday matter," as protons and neutrons, the *charm* (c) and *strange* (s), and the *top* (t) and *bottom* (b). These particles have masses ranging from a few MeV up to 172 GeV, and the first quark of each doublet has an electric charge of $Q = +\frac{2}{3}$, while the second has $Q = -\frac{1}{3}$. Quarks are subject to all the three interactions of the SM and have a quantum number associated with each of them. Besides the charge, each quark has a *flavor*, which is conserved in the electromagnetic and strong interactions, but not in the weak one. Quarks also have a *color*, from which the QCD name, that regulates their behavior under the strong interaction. As a result of the QCD color confinement, quarks are experimentally observed only as bound states, referred to as *hadrons*. Hadrons composed of a quark-antiquark pair are called *mesons*, while aggregated states of three quarks take the name of *baryons*. This is true for all the quarks but the top, which has a lifetime so short[2] that decays[3] before forming any bound state.

---

[1] Gluons correspond to the generators of the $SU(3)_C$ group. A $SU(N)$ group has $(N^2 - 1)$ generators, resulting in eight gluons in the SM.

[2] The lifetime of the top quark is: $\tau_t = 0.5 \times 10^{-24}$ s.

[3] Via weak interaction into a W boson and a lighter quark, often a b [6].

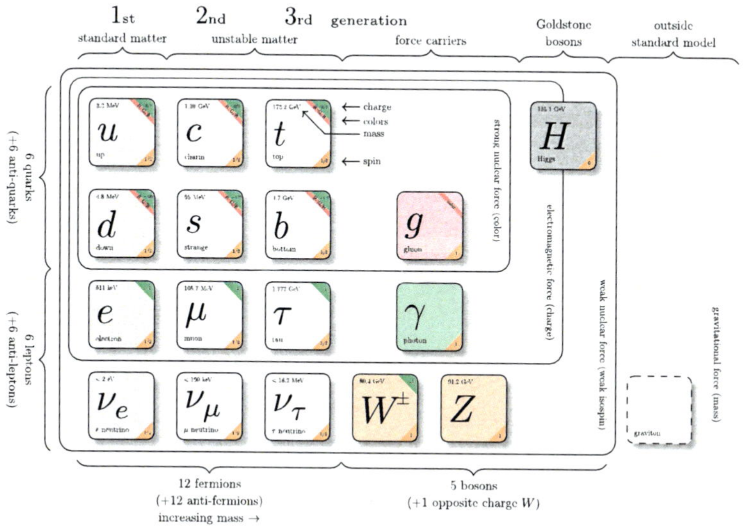

Fig. 1.1 Illustration of the elementary particles of the Standard Model (SM) and their properties. Outside the SM box, the graviton is depicted: this particle is unobserved as of today and should represent a bosonic field responsible of the gravitational force. Figure taken from [5]

As it will be explained in the following, left-handed fermions are identified as members of doublets, while fermions of right chirality are identified as singlets of the $SU(2)_L \times U(1)_Y$ electroweak gauge group. Consequently, the mass eigenstates of down-type quarks do not coincide with their weak eigenstates. The Cabibbo-Kobayashi-Maskawa matrix regulates the mixing between these two bases [6].

Leptons

The second kind of fermions in the SM are the *leptons*. Conversely from quarks, these particles interact only via the electromagnetic and weak forces. There are three charged leptons in the SM: the electron (e), the muon (μ), and the tau (τ), all with electric charge $Q = -1$. While these particles have similar properties, their masses cover a wide range: going from 511 keV for the electron to 105.7 MeV for the muon, up to 1.8 GeV for the tau. The electron and the muon are considered stable particles at the LHC experiments: the former is actually stable, and the latter has a sufficiently large lifetime to be detected before decaying. Conversely, the tau has a short lifetime of 2.9×10^{-13} s, which makes it detectable only via its decay products. Each charged lepton is associated with a neutrino (ν_e, ν_μ, and ν_τ): these are neutral particles that interact only via the weak force. In the original formulation of the SM neutrinos are massless. However, in 2016, neutrino oscillations were observed [7], proving that also these particles are massive ($m_\nu < 2$ eV [6]). Similarly to quarks, also for neutrinos, the weak and mass eigenstates are subject to mixing, regulated by the Pontecorvo-Maki-Nakagawa-Sakata (PMNS) matrix [8].

Strong Interaction

Quantum Chromodynamics (QCD) is the underlying theory for the strong interaction. QCD is a non-abelian gauge theory based on the local $SU(3)_C$ symmetry, where the subscript C refers to the color charge.

The Dirac Lagrangian of a free massive fermionic field ψ, associated with a particle of spin $s = 1/2$, can be written as:

$$\mathcal{L}_f = \bar{\psi}(x)(i\partial\!\!\!/ - m)\psi(x), \tag{1.1}$$

where $\partial\!\!\!/ = \gamma_\mu \partial^\mu$ and γ are the Dirac matrices.

Such a lagrangian is invariant under the "global" $SU(3)$ transformation:

$$\psi'(x) = U\psi(x), \tag{1.2}$$

where U is a 3×3 unitary matrix of the form:

$$U = e^{ig_s\theta^a \frac{\lambda^a}{2}}, \tag{1.3}$$

where $\frac{\lambda^a}{2}$ denote the 8 Gell-Mann matrices, corresponding to the generators of the $SU(3)$ group, and g_s is a constant. Such a global invariance is promoted to a "local" one, i.e. dependent on the spacetime coordinate x, by introducing gauge fields in the Lagrangian of Eq. (1.1). More precisely, this is ensured by defining the covariant derivative (D_μ):

$$D_\mu = \partial_\mu + ig_s\frac{\lambda_a}{2}A^a_\mu, \tag{1.4}$$

where g_s is the arbitrary coupling constant introduced above and the vector bosons fields A^a_μ, that can be identified as the eight gluon fields[4] transform as:

$$A^a_\mu \rightarrow A^a_\mu + g_s\alpha^b(x)f^{abc}A^c_\mu + \partial_\mu\alpha^a(x), \tag{1.5}$$

where f^{abc} are the structure constants of the group and must satisfy the commutation rules $\left[\frac{\lambda^a}{2}, \frac{\lambda^b}{2}\right] = if^{abc}\frac{\lambda^c}{2}$. Substituting these terms into the Dirac Lagrangian of Eq. (1.1), one gets a QCD Lagrangian invariant under local $SU(3)_C$ transformations:

$$\mathcal{L}_{QCD} = \bar{\psi}(i\partial\!\!\!/ - m)\psi - g_s\bar{\psi}\gamma^\mu\frac{\lambda_a}{2}\psi A^a_\mu - \frac{1}{4}F^{\mu\nu}_a F^a_{\mu\nu}, \tag{1.6}$$

---

[4] As described above, eight gluon fields are expected in the SM as the $SU(3)$ group has 8 generators.

where

$$F^{\mu\nu} = \partial_\mu A^a_\nu - \partial_\nu A^a_\mu + g_s f^{abc} A^b_\mu A^c_\nu$$

is the field strength tensor. The corresponding term in Eq. (1.6) is required for the propagation of the gluon field and, not being linear in terms of the gluon potential, introduces trilinear and quadrilinear terms in the Lagrangian, corresponding to the self-interactions of gluons. The coupling constant g_s determines the strength of the interaction and it is often redefined as $\alpha_s = g_s^2/4\pi$.

Electroweak Interaction

The $SU(2)_L \times U(1)_Y$ gauge group jointly describes the electromagnetic and weak interactions and is therefore referred to as electroweak (EW) group.

The electromagnetic interaction can be represented by the Quantum Electrodynamics (QED) Lagrangian, invariant under the local $U(1)_{em}$ symmetry:

$$\mathcal{L}_{QED} = \bar{\psi}(i\not{D} - m)\psi - \frac{1}{4}F^{\mu\nu}F_{\mu\nu}. \tag{1.7}$$

In an analogous way to the QCD Lagrangian, a kinetic term,[5] proportional to $F^{\mu\nu}$, accounts for the propagation of the gauge boson (here the photon), while the covariant derivative ($D^\mu = \partial - ieA^\mu$) introduces the interaction term:

$$\mathcal{L}^{int}_{QED} = e\bar{\psi}A^\mu\gamma_\mu\psi.$$

A theoretical description of the weak interaction is more challenging, as it has to reflect the experimental evidence of the parity violation. This is taken into account assigning different interaction terms to left- and right-handed fermion fields. Theoretically this is achieved using the gamma matrix $\gamma^5 \equiv i\gamma^0\gamma^1\gamma^2\gamma^3$ to identify the left and right chirality[6] projection operators:

$$P_L = \frac{1-\gamma^5}{2}; \quad P_R = \frac{1+\gamma^5}{2} \tag{1.8}$$

Ultimately, the weak interaction is characterized by two quantum numbers: the weak isospin (T_3) and hypercharge (Y). The weak isospin (T_3) quantum number is associated with the non-abelian group $SU(2)_L$ and corresponds to one of the group generators. More precisely, the group has three generators $T_i = \frac{\sigma_i}{2}$, where σ_i are the Pauli's matrices, and the gauge invariance is reflected by the presence of the 3 gauge fields W^i_μ. In the $SU(2)_L$ group, left-handed fermions are represented by a doublet,

---

[5] In the QED Lagrangian $F^{\mu\nu} = \partial^\mu A^\nu - \partial^\nu A^\mu$.

[6] For massless particles the chirality corresponds to the projection of the spin vector $\vec{s}$ over the momentum vector $\vec{p}$.

whereas right-handed particles are $SU(2)_L$ singlets that do not interact with the W^i_μ fields. Both left- and right-handed fields are obtained from the chirality projection operators defined above (cf. Eq. (1.8)):

$$\begin{aligned}\Psi_L &= \frac{1-\gamma^5}{2}\begin{pmatrix}\psi\\ \psi'\end{pmatrix} = \begin{pmatrix}\psi_L\\ \psi'_L\end{pmatrix}\\ \psi_R &= \frac{1+\gamma^5}{2}\psi \\ \psi'_R &= \frac{1+\gamma^5}{2}\psi',\end{aligned} \tag{1.9}$$

where the ψ and ψ' fields represent either the neutrino and charged lepton or the up- and down-type quark fields. The weak hypercharge (Y) quantum number is associated with the $U(1)_Y$ abelian group. This group has a single generator $\frac{Y}{2}$, with the $U(1)_Y$ gauge invariance reflected by the presence of the B_μ gauge field. The Gell-Mann-Nishijima formula relates the weak hypercharge and isospin to the electric charge associated with the $U(1)_{em}$ group:

$$Q = T_3 + \frac{Y}{2}. \tag{1.10}$$

Hence, the EW Lagrangian density can be written as:

$$\mathcal{L} = \overline{\Psi}_L(i\not{D})\Psi_L + \overline{\psi}_R(i\not{D})\psi_R + \overline{\psi}'_R(i\not{D})\psi'_R, \tag{1.11}$$

where the mass terms are omitted to preserve the $SU(2)_L$ invariance. Following the same approach adopted with the QED and QCD Lagrangians, the introduction of the covariant derivative (D^μ) ensures a $SU(2)_C \times U(1)_Y$ local symmetry:

$$D^\mu = \partial^\mu - ig\overrightarrow{T}\overrightarrow{W}^\mu - ig'\frac{Y}{2}B^\mu, \tag{1.12}$$

with g and g' regulating the strength of the interactions.

Equation (1.11) can be rewritten separating the kinetic term from those responsible for the charged and neutral currents:

$$\mathcal{L} = \mathcal{L}_{\text{kin}} + \mathcal{L}_{\text{CC}} + \mathcal{L}_{\text{NC}}, \tag{1.13}$$

where:

$$
\begin{aligned}
\mathcal{L}_{\text{kin}} &= \overline{\Psi}_L (i\partial\!\!\!/)\Psi_L + \overline{\psi}_R (i\partial\!\!\!/)\psi_R + \overline{\psi}'_R (i\partial\!\!\!/)\psi'_R \\
\mathcal{L}_{\text{CC}} &= \frac{g}{\sqrt{2}} W^+_\mu \overline{\Psi}_L \gamma^\mu \sigma^+ \Psi_L + \frac{g}{\sqrt{2}} W^-_\mu \overline{\Psi}_L \gamma^\mu \sigma^- \Psi_L = \\
&= \frac{g}{\sqrt{2}} W^+ \left(\overline{\psi}_L \gamma^\mu \psi'_L\right) + \frac{g}{\sqrt{2}} W^- \left(\overline{\psi}'_L \gamma^\mu \psi_L\right) \\
\mathcal{L}_{\text{NC}} &= \frac{g}{\sqrt{2}} W^3_\mu \left[\overline{\psi}_L \gamma^\mu \psi_L - \overline{\psi}'_L \gamma^\mu \psi'_L\right] + \\
&+ \frac{g'}{\sqrt{2}} B_\mu \left[Y\left(\overline{\psi}_L \gamma^\mu \psi_L + \overline{\psi}'_L \gamma^\mu \psi'_L\right) + Y \overline{\psi}_R \gamma^\mu \psi_R + Y \overline{\psi}'_R \gamma^\mu \psi'_R\right]
\end{aligned}
\tag{1.14}
$$

where the $W^\pm$ charged bosons can be expressed as linear combinations of the W_1 and W_2 gauge bosons:

$$
W^\pm_\mu = \frac{1}{\sqrt{2}}\left(W^1_\mu \mp i W^2_\mu\right) \tag{1.15}
$$

and similarly for the Pauli's matrices:

$$
\sigma^\pm = \frac{1}{\sqrt{2}}\left(\sigma^1 \pm i\sigma^2\right). \tag{1.16}
$$

The neutral current Lagrangian is expressed as a function of W^3_μ and B_μ, but none of them can be interpreted as a physical neutral boson at this stage of the formulation. For this purpose, one introduces the Z_μ and A_μ fields that directly represent the neutral Z boson and the photon by means of a Weinberg's angle (θ_W) rotation of the W^3_μ and B_μ fields:

$$
\begin{pmatrix} A_\mu \\ Z_\mu \end{pmatrix} = \begin{pmatrix} \cos\theta_w & \sin\theta_w \\ -\sin\theta_w & \cos\theta_w \end{pmatrix} \begin{pmatrix} B_\mu \\ W_\mu \end{pmatrix}, \tag{1.17}
$$

which allows to rewrite the NC Lagrangian of Eq. (1.14) separating the terms associated with the photon and the Z boson fields:

$$
\mathcal{L}^Z_{\text{NC}} = \overline{\psi}_L \gamma^\mu Z_\mu \left(g\frac{\sigma_3}{2}\cos\theta_w - g'\frac{Y}{2}\sin\theta_w\right)\psi_L \tag{1.18}
$$

$$
\mathcal{L}^\gamma_{\text{NC}} = \overline{\psi}_L \gamma^\mu A_\mu \left(g\frac{\sigma_3}{2}\sin\theta_w + g'\frac{Y}{2}\cos\theta_w\right)\psi_L, \tag{1.19}
$$

where the two neutral currents associated with the Z_μ and A_μ fields manifest explicitly as a function of the weak isospin ($T_3 = \frac{\sigma_3}{2}$) and hypercharge (Y). In addition, as a consequence of the Gell-Mann-Nishijima formula (cf. Eq. (1.10)), the following relation holds:

$$
e = g'\sin\theta_w = g\cos\theta_w \tag{1.20}
$$

and it relates the value of the fundamental electric charge e to the coupling constants g, g' as well as to the Weinberg's angle θ_w.

To complete the description of the Electroweak Lagrangian (cf. Eq. (1.11)), one has to introduce a kinetic term for the gauge fields:

$$\mathcal{L}_{EW} = \overline{\Psi}_L(i\not{D})\Psi_L + \overline{\psi}_R(i\not{D})\psi_R + \overline{\psi}'_R(i\not{D})\psi'_R - \frac{1}{4}W_i^{\mu\nu}W^i_{\mu\nu} - \frac{1}{4}B^{\mu\nu}B_{\nu\mu}, \tag{1.21}$$

with the fields strength tensors $W^i_{\mu\nu}$ and $B_{\nu\mu}$ defined as:

$$\begin{aligned} B_{\nu\mu} &= \partial_\mu B_\nu - \partial_\nu B_\mu \\ W^i_{\mu\nu} &= \partial_\mu W^i_\nu - \partial_\nu W^i_\mu + g\varepsilon^{ijk} W^j_\mu W^k_\nu, \end{aligned} \tag{1.22}$$

where ε^{ijk} is the Levi-Civita tensor. This additional kinetic term gives rise to the self-interaction of the gauge bosons, predicting the existence of trilinear (ZW^+W^-, γW^+W^-) and quadrilinear (ZZW^+W^-, $\gamma\gamma W^+W^-$, $Z\gamma W^+W^-$, $W^+W^-W^+W^-$) self-couplings of the Z, $W^\pm$, and γ bosons.

Hence, the electroweak and strong interactions can be represented by a Lagrangian comprising of the two terms $\mathcal{L}_{QCD}$ and $\mathcal{L}_{EW}$, resulting in a gauge theory with a $SU(3)_C \times SU(2)_L \times U(1)_Y$ symmetry. However, neither $\mathcal{L}_{QCD}$ nor $\mathcal{L}_{EW}$ contain explicit mass terms of either the gauge or the fermion fields, as these would break the gauge invariance. While this is consistent with the evidence of massless photons and gluons, it does not reflect the existence of massive gauge bosons, such as the $W^\pm$ and Z, and fermions.

In order to maintain the gauge invariance of the theory, whilst at the same time reflecting the experimental evidence of massive gauge bosons and fermions, the electroweak symmetry must be broken. In the SM this happens via the so-called Brout-Englert-Higgs (BEH) mechanism [9–12], detailed in the next section.

1.1.2 *Spontaneous Symmetry Breaking: The Brout-Englert-Higgs Mechanism*

The BEH mechanism is based on the introduction of a complex scalar field $\phi(x)$ in the SM Lagrangian to solve the issue of massless particles arising from the theoretical formulation presented above. The potential of such a field ($V(\phi)$) must be invariant under the symmetry of the system, but when a specific ground state is selected, among a continuum of possible states, the symmetry of the system is said to be *spontaneously broken*. When this happens, a scalar field corresponding to a massless spin 0 boson is introduced in the Lagrangian density for each generator of the group

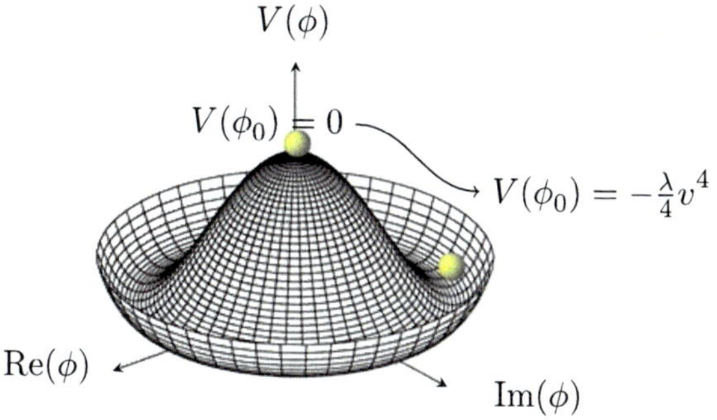

Fig. 1.2 Schematic illustration of the Higgs potential $V(\phi)$ (cf. Eq. (1.25))

associated with the broken symmetry, as a consequence of the Goldstone theorem[7] [13].

Hence, introducing a complex scalar doubled field with hypercharge $Y_\phi = 1$:

$$\phi = \begin{pmatrix} \phi^+ \\ \phi^0 \end{pmatrix} = \frac{1}{\sqrt{2}} \begin{pmatrix} \phi^1 + i\phi^2 \\ \phi^3 + i\phi^4 \end{pmatrix} \tag{1.23}$$

the SM Lagrangian density is supplemented with the term:

$$\mathcal{L}_{\text{BEH}} = (D_\mu \phi)^\dagger (D^\mu \phi) - V(\phi^\dagger \phi)^2, \tag{1.24}$$

where:

$$V(\phi^\dagger \phi) = -\mu^2 \phi^\dagger \phi + \lambda (\phi^\dagger \phi)^2, \quad (\mu^2, \lambda > 0) \tag{1.25}$$

is the scalar field potential and D_μ is the covariant derivative introduced in Eq. (1.12). The functional form of the potential is represented in Fig. 1.2.

The peculiar expression of this potential is such that at $\phi = 0$ it presents an unstable local maximum, whereas a continuum of ground states is found at

$$| \phi^\dagger \phi | = \frac{\mu^2}{2\lambda} \equiv \frac{v^2}{2}, \tag{1.26}$$

where the quantity v is referred to as vacuum expectation value (VEV). The choice of a ground state is said to "spontaneously break" the symmetry, however leaving the Lagrangian gauge invariant. The choice of the ground state breaks both the $SU(2)_L$ and the $U(1)_Y$ symmetries, but it preserves the $U(1)_{em}$ symmetry when the ground state is parallel to the ϕ^0 component of the doublet (cf. Eq. (1.23)).

The small perturbations around the minimum can be parametrized as:

$$\phi(x) = \frac{1}{\sqrt{2}} e^{i\sigma^i \theta^i(x)} \begin{pmatrix} 0 \\ v + H(x) \end{pmatrix}, \tag{1.27}$$

[7] The massless spin 0 boson(s) take the name(s) of "Goldstone boson(s)".

where θ_i and $H(x)$ are the massless and massive spin 0 fields predicted by the Goldstone theorem. While the massive $H(x)$ field can be associated with the Higgs boson (H), needed to complete the SM picture presented above, the massless θ_i Goldstone bosons are unphysical.

The dependence from the Goldsone bosons is removed from Eq. (1.27) by means of a $SU(2)_L$ local transformation, also referred to as "unitary gauge", under which the Lagrangian is invariant:

$$\phi(x) \rightarrow e^{-i\alpha(x)\frac{g}{2}}\phi(x) = \phi'(x) = \frac{1}{\sqrt{2}}\begin{pmatrix} 0 \\ v + H(x) \end{pmatrix}, \tag{1.28}$$

where $\alpha_i(x) = 2\theta^i(x)$, and $\phi(x)$ now only depends on the VEV value and on $H(x)$.

Eventually, one can rewrite the $\mathcal{L}_{\mathrm{BEH}}$ Lagrangian explicitly as:

$$\begin{aligned}\mathcal{L}_{\mathrm{BEH}} = &\frac{1}{2}\partial_\mu H \partial^\mu H + \mu^2 H^2 + \\ &+ \frac{g^2 v^2}{4} W_\mu^+ W^{-\mu} + \frac{1}{2}\frac{(g^2 + g'^2)}{4} v^2 Z_\mu Z^\mu + \\ &+ \frac{g^2}{2} v H W_\mu^+ W^{-\mu} + \frac{g^2}{4} H^2 W_\mu^+ W^{-\mu} + \frac{g'^2}{2} v H Z_\mu Z^\mu + \frac{g^2}{4} H^2 Z_\mu Z^\mu + \\ &+ \frac{\mu^2}{v} H^3 + \frac{\mu^2}{4v^2} H^4\end{aligned} \tag{1.29}$$

From the first line one gets the mass of the scalar Higgs field:

$$m_H^2 = 2\lambda v^2 = 2\mu^2, \tag{1.30}$$

which is a free parameter of the theory.

The second line describes the massive $W_\mu^\pm$ and Z_μ fields, resulting in:

$$\begin{aligned} m_Z &= \frac{\sqrt{g^2 + g'^2}}{2} v \\ m_{W^\pm} &= \frac{gv}{2} = m_Z \cos\theta_w \end{aligned} \tag{1.31}$$

The A_μ field is not present in Eq. (1.29), reflecting the $U(1)_{em}$ invariance of the Lagrangian and the fact that the photon is massless.

The third line introduces trilinear and quadrilinear couplings of the Higgs boson to the weak vector bosons Z and $W^\pm$ (HW^+W^-, HZZ, HHW^+W^-, $HHZZ$), of which the HZZ is of particular interest for the analysis presented in the context of this thesis. Eventually, the fourth line describes the trilinear and quadrilinear self-couplings (H^3 and H^4) of the Higgs boson, from which the scalar potential can be written as:

$$V(\phi) = \frac{1}{2} m_{\mathrm{H}}^{2} \mathrm{H}^{s} + \lambda_{\mathrm{HHH}} v \mathrm{H}^{3} + \frac{1}{4} \lambda_{\mathrm{HHHH}} \mathrm{H}^{4} - \frac{\lambda}{4} v^{4}, \tag{1.32}$$

where the H boson self-couplings are defined as:

$$\lambda_{\mathrm{HHH}} = \lambda_{\mathrm{HHHH}} = \lambda = \frac{m_{\mathrm{H}}^{2}}{2v^{2}}. \tag{1.33}$$

This strong relation of the self-couplings to the Higgs boson mass and the VEV value is one of the cornerstones of the SM. Precision measurements of these quantities provide one of the most stringent tests of the SM, giving direct access to the Higgs boson potential and probing directly the spontaneous symmetry breaking mechanism detailed here.

The VEV value is one of the two free parameters of the BEH mechanism, but it can be measured precisely from the charged current interaction in the $\mu \rightarrow e\overline{\nu}_{e}\nu_{\mu}$ decay, noticing the relation between the W boson mass and the Fermi's constant:

$$\frac{G_F}{\sqrt{2}} = \frac{g^{2}}{8m_{W}^{2}} = \frac{1}{2v^{2}}, \tag{1.34}$$

from which one obtains:

$$v = \frac{1}{\left(\sqrt{2}G_F\right)^{1/2}} = 246\ \mathrm{GeV}. \tag{1.35}$$

In addition to vector bosons, also fermions acquire mass via the BEH mechanism. More precisely, the mass terms are generated by the Yukawa interaction of the H field with left- and right-handed fields[8]:

$$\mathcal{L}_{\text{Yukawa}} = -y_{f'} \left(\overline{\Psi}_L \phi \psi'_R + \overline{\psi}'_R \phi^{\dagger} \Psi_L\right) - i y_f \left(\overline{\Psi}_L \sigma_2 \phi^{*} \psi'_R + \overline{\psi}'_R \sigma_2 \phi^{*\dagger} \Psi_L\right), \tag{1.36}$$

where y_f and $y_{f'}$ are the coupling constants for the up- ($T_3 = 1/2$) and down-type ($T_3 = -1/2$) fermions, respectively. After electroweak symmetry breaking one gets:

$$\mathcal{L}_{\text{Yukawa}} = -\sum_{f} m_f \overline{\psi}\psi \left(1 + \frac{H}{v}\right) - \sum_{f'} m_f \overline{\psi}' \psi' \left(1 + \frac{H}{v}\right), \tag{1.37}$$

where the mass terms for the up- and down-type fermions are:

$$m_{f^{(\prime)}} = y_{f^{(\prime)}} \frac{v}{\sqrt{2}}. \tag{1.38}$$

[8] Defined as of Eq. (1.9).

Hence, the mass terms of the fermion fields originate from their interaction with the Higgs field and are proportional to the strength of the Yukawa couplings $y_{f^{(\prime)}}$ and to the VEV value v.

1.2 Phenomenology of the Higgs Boson at the LHC

In July 2012 the ATLAS and CMS Collaborations announced the observation of a new spin 0 scalar boson [2–4] with a mass of about 125 GeV. The sensitivity of the analysis, performed exploiting proton-proton collision data collected by the experiments at a centre-of-mass energy of $\sqrt{s} = 8$ TeV, was driven mainly by the $H \rightarrow ZZ \rightarrow 4\ell$ and $H \rightarrow \gamma\gamma$ decay channels. The former is also known as the "golden channel" because of its many virtues and is the target of the analysis presented in this manuscript. It is remarkable to note here that the analysis of the $H \rightarrow ZZ \rightarrow 4\ell$ decay channel, initially developed to identify a significant signal excess over a continuous background, nowadays has become a precision physics analysis, where many of the properties of the H boson can be measured to probe in detail the predictions of the SM.

After the claim of the discovery, the analyses focused on the measurement of the properties of this long-sought particle [14–17]: its spin-parity is measured to be $J^P = 0^+$ [14, 15] and the first measurement of m_H yields to $m_H = 125.09 \pm 0.21$ (stat) $\pm\, 0.11$ (syst) GeV [16], both in agreement with the SM predictions.

The discovery of the Higgs boson at the LHC, along with the first measurements of its properties, set a significant breakthrough in the study of the BEH mechanism.

Conversely from the VEV value, which can be measured precisely exploiting m_W and G_F, a measurement of m_H requires studying exclusive decay channels and production mechanisms in detail. The first precision measurement of this parameter was performed combining the Run-I results obtained by the ATLAS and CMS Collaborations [16]. Nowadays, the most precise measurement of m_H results from a combination of the CMS $H \rightarrow ZZ \rightarrow 4\ell$ and $H \rightarrow \gamma\gamma$ decay channels using the full Run-I statistics and the 2016 Run-II data [18]:

$$m_H = 125.38 \pm 0.11\ (\text{stat}) \pm 0.08\ (\text{syst})\ \text{GeV}. \tag{1.39}$$

The precision measurement of m_H is fundamental as, being a free parameter of the SM, its value is not fixed *a priori*. Setting the value of this parameter enables predictions on the production and decay rates of the SM Higgs boson. Both these two quantities have a dependence on m_H, as shown in Fig. 1.3, where the production cross sections of the Higgs boson are shown as a function of m_H. The orange line in the plot shows the value of $m_H = 125.38 \pm 0.14$ GeV [18], corresponding to value used for the extraction of the results presented in this manuscript.

At hadron colliders, the H boson is produced predominantly in four processes, of which the leading order (LO) Feynman diagrams are illustrated in Fig. 1.4.

Their expected cross section at $m_H = 125$ GeV are shown in the top plot of Fig. 1.5, as a function of the centre-of-mass energy of the proton-proton collisions.

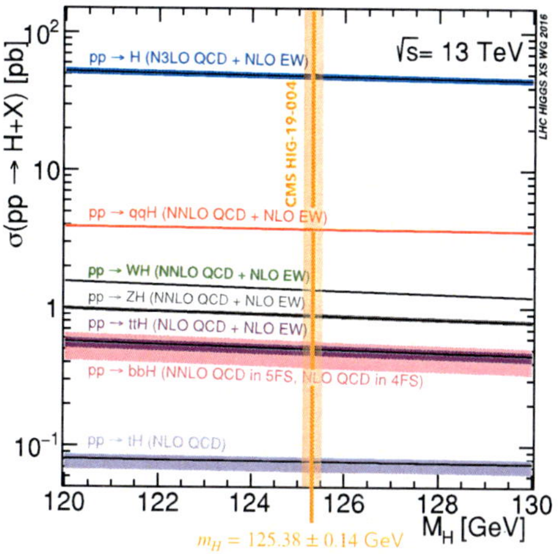

Fig. 1.3 Production cross sections of the SM Higgs boson at hadron colliders as a function of the Higgs boson mass m_H. Figure taken and edited from [19]

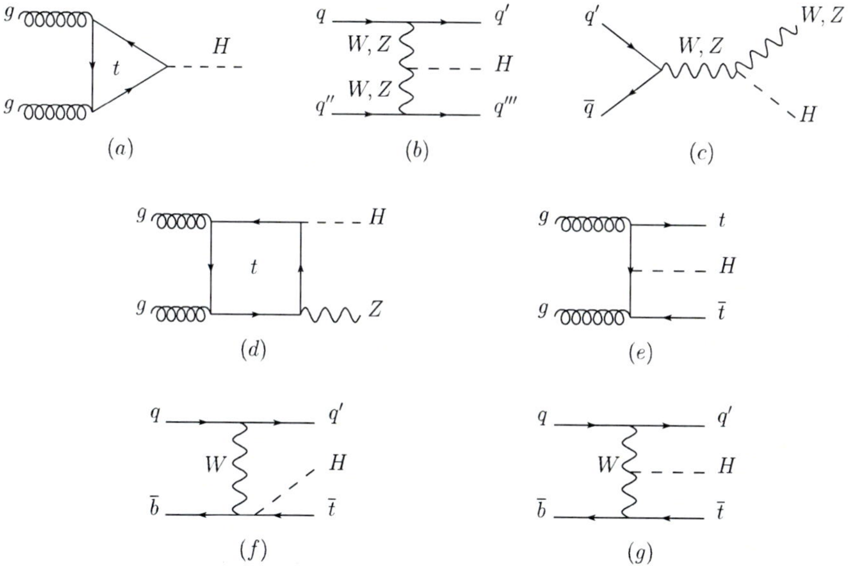

Fig. 1.4 Leading order Feynman diagrams for the main production mechanisms of the Higgs boson: **a** gluon fusion, **b** Vector-boson fusion, **c–d** associated production with a gauge boson, **e** associated production with a pair of top (or bottom) quarks, **f–g** production in association with a single top quark. Figure taken from [6]

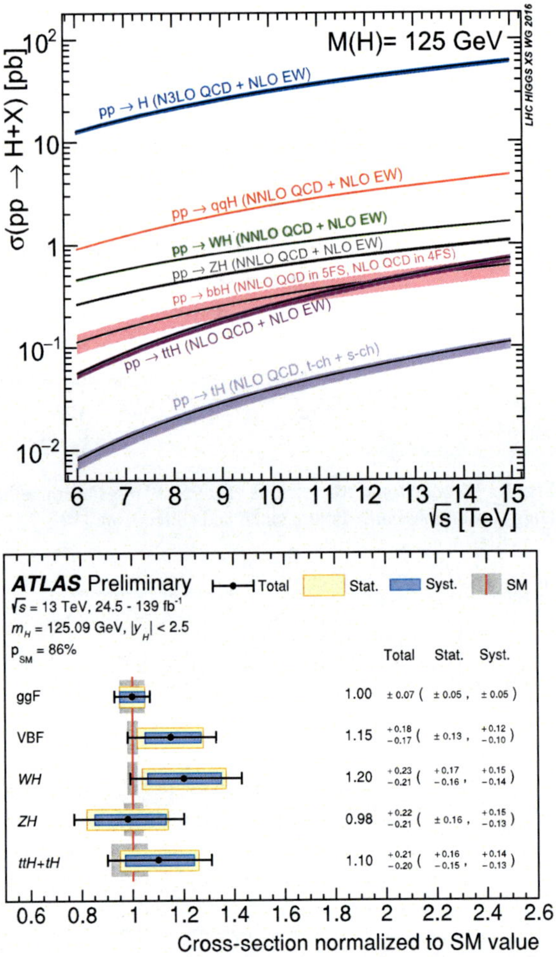

Fig. 1.5 *Top*: Production cross sections of the SM Higgs boson at hadron colliders as a function of the centre-of-mass energy ($\sqrt{s}$) at $m_{\mathrm{H}} = 125$ GeV. The different orders in perturbative QCD (pQCD) at which each cross section is computed are reported in parentheses. *Bottom*: Cross sections of the main Higgs boson production mechanisms measured by the ATLAS Collaboration using the data collected in Run-II at $\sqrt{s} = 13$ TeV. Figures taken from [19] and [20]

The dominant production mode is the **gluon-gluon fusion** (ggH), with an expected cross section of 49 pb. The **vector boson fusion** (VBF) and **Higgs-strahlung** (VH) processes come immediately after, with similar cross sections of 3.8 pb and 2.3 pb, respectively. The productions in association with heavy quarks ($\mathrm{t\bar{t}H}$, $\mathrm{b\bar{b}H}$, and tH) are the rarest and more challenging to probe. The latest measurement of the Higgs boson production cross sections, performed by the ATLAS Collaboration exploiting the full Run-II statistics [20], is shown in the bottom plot of Fig. 1.5.

The cross sections of the main production processes (ggH, VBF, WH, ZH) are found to be in agreement with the SM predictions. As of today, all the production mechanisms but the $\mathrm{b\bar{b}H}$ and tH have been observed at the LHC. In particular, it is worth mentioning the recent observation of the $\mathrm{t\bar{t}H}$ production mode by the ATLAS

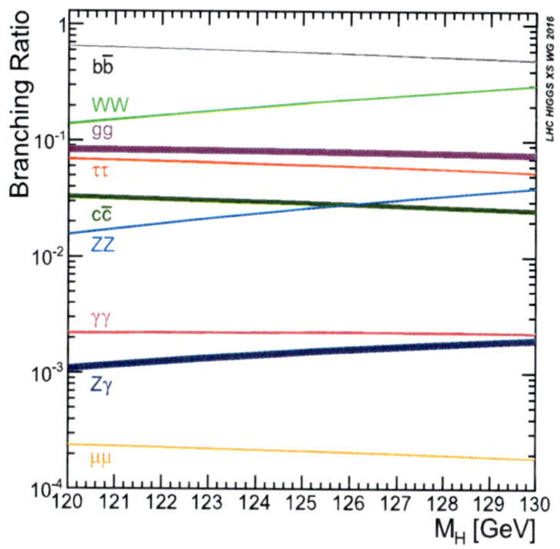

Fig. 1.6 Branching fractions of the main H boson decay modes as a function of m_H. Figure taken from [19]

and CMS Collaborations in 2018 [21, 22], as this result gives direct access to the tree-level coupling of the Higgs boson to the top quark.

The Higgs boson branching fractions (or branching ratios, BR) present a dependence on m_H too, as shown in Fig. 1.6, and while a rather flat dependence on m_H is observed, it is worth to remark the wide range between the most and the least probable decays of the Higgs boson, covering about four orders of magnitude.

The specific value of m_H, set by the discovery of this particle and by the more recent measurements performed by the ATLAS and CMS, fixes the values of the various BR, thus determining which decay modes can be experimentally probed at the LHC. A large plethora of decay channels is therefore accessible for a H boson of about 125 GeV: these allow to probe both the coupling to vector bosons (H → ZZ, $H \to \gamma\gamma$, $H \to W^{\pm}W^{\mp}$, but also the rarer $H \to Z\gamma$) and the coupling to fermions (e.g. $H \to b\bar{b}$, $H \to c\bar{c}$, and $H \to \mu\mu$). The complete list of the expected BR for the main decay channels of the Higgs boson is given in Table 1.1.

The searches for some of these decay channels, such as $H \to b\bar{b}$ and $H \to c\bar{c}$, are limited by the large QCD backgrounds that characterize the dominant production mechanisms, such as the gluon- and vector boson-fusion processes, while others suffer from a low branching fractions, such as $H \to \mu\mu$, thus making their searches more challenging and requiring sophisticated analysis techniques. For example, the recent precision measurement of the $H \to \mu^+\mu^-$ process by the CMS Collaboration [23] sets an important milestone in the understanding of the SM, confirming the expected proportionality between the fermionic couplings of the Higgs and the fermion masses. Figure 1.7 summarizes the current status of the results obtained by the CMS Collaboration for the measurement of the Higgs boson couplings to fermions and vector bosons, where a remarkable agreement with the SM predictions is observed.

Table 1.1 Branching fraction of the principal H boson decays at $m_H = 125.09$ GeV. Theoretical uncertainties are obtained from the sum in quadrature of the uncertainties on the H boson partial width, the value of α_s, and the quark masses [6]

Decay mode	Branching fraction [%]
$H \to b\bar{b}$	$58.09^{+0.72}_{-0.73}$
$H \to W^{\pm}W^{\pm *}$	21.52 ± 0.33
$H \to gg$	8.18 ± 0.42
$H \to \tau^+\tau^-$	6.27 ± 0.10
$H \to c\bar{c}$	$2.88^{+0.16}_{-0.06}$
$\mathrm{H} \to \mathrm{ZZ}$	2.641 ± 0.040
$H \to \gamma\gamma$	0.2270 ± 0.0047
$H \to Z\gamma$	0.1541 ± 0.0090
$H \to \mu^+\mu^-$	$0.02171^{+0.00036}_{-0.00037}$

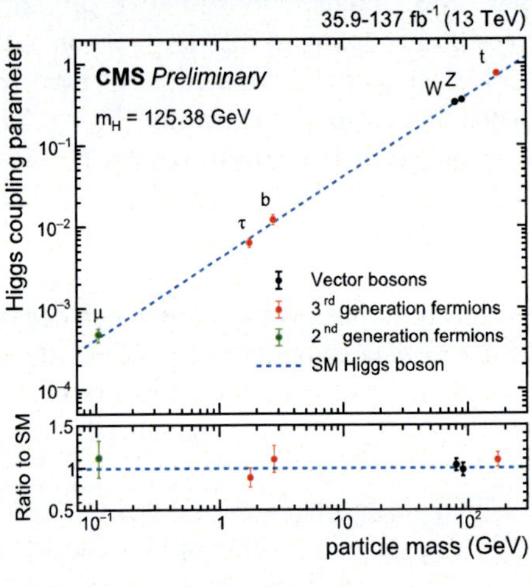

Fig. 1.7 Couplings of the SM H boson measured from the combination of the proton-proton collisions data collected by the CMS experiment at a centre-of-mass energy of $\sqrt{s} = 13$ TeV. Figure taken from [23]

With access to a precise value of m_H, a further and more stringent test of the SM is possible. In fact, Fig. 1.7 does not comprise the self-coupling of the Higgs boson. Combining Eqs. (1.39) and (1.33) one gets:

$$\lambda_{HHH} = \left(\frac{m_H}{2v}\right)^2 \simeq 0.13. \tag{1.40}$$

Now, a precision measurement of this quantity is extremely appealing, as it would provide a closure test of the SM, nailing down its only free parameter and ultimately proving that this Higgs boson is the SM Higgs boson. With the large statistics collected by the LHC experiments during the Run-II phase, the sensitivity of the analyses to λ_{HHH} has substantially increased. The most natural way to measure this observable is to exploit the double Higgs boson production cross section, which depends on λ_{HHH} at LO.

The most stringent value on λ_{HHH} is derived from the CMS and ATLAS $b\bar{b}\gamma\gamma$ di-Higgs analysis [24, 25] that exclude at 95 % CL values of $\lambda_{HHH} < -3.3\,\lambda_{HHH}^{SM}$ and $\lambda_{HHH} > 8.5\,\lambda_{HHH}^{SM}$, and $\lambda_{HHH} < -1.5\,\lambda_{HHH}^{SM}$ and $\lambda_{HHH} > 6.7\,\lambda_{HHH}^{SM}$, respectively. More stringent limits on λ_{HHH} are expected from the combination of different di-Higgs channels using the 137 fb^{-1} of data collected by the experiments over the Run-II phase of the LHC. More recently, theoretical calculations have also shown the possibility of setting constraints to λ_{HHH} via single-Higgs decay channels [26, 27], exploiting electroweak corrections to the loops of the Higgs boson production and decay processes. Although at the current stage this approach is limited by the statistics, it provides complementary information on the self-coupling of the Higgs, with exclusion limits that can become competitive with the direct ones from the di-Higgs measurements. In the context of this thesis, the $H \to ZZ \to 4\ell$ decay channel is used to infer constraints on the trilinear self-coupling of the Higgs boson, as further detailed in Chap. 7.

1.3 The "Golden Channel": Past, Present, and Future

The analysis presented in this manuscript consists of the study of the Higgs boson properties in the $H \to ZZ \to 4\ell$ decay channel, often referred to as the "golden channel".

The Higgs boson decays in pairs of vector bosons can have very different topologies: on one side the decay to a pair of $W^{\pm}$ bosons, which features a large branching fraction, despite a poor mass resolution due to missing energy in the final state; on the other hand the decay to two Z bosons, where the limited statistics due to the low branching fraction (cf. Table 1.1) is compensated by a large signal-to-background ratio, an excellent mass resolution, and a completely resolved final state.

- A large signal-to-background ratio: in the $H \to ZZ \to 4\ell$ decay, the Higgs boson can be identified starting from the four leptons in the final state. The typical resonant shape of the invariant mass of the four-lepton system represent a clear signature to distinguish between the signal and the background processes, as the latter has a rather flat shape under the Higgs peak;
- An excellent mass resolution: such a great separation between signal and background is also the result of the reconstruction power of the CMS detector. The precise identification of electrons and muons, to which CMS owes its name, allows to probe with great precision this decay channel, making it one of the driving channels in the determination of the Higgs boson properties, such as its mass and width [15];

- A completely resolved final state: where Higgs boson candidates can be identified relatively easily by imposing selection cuts on the leptons reconstructed in the final state. Once the four leptons in the final state are identified and reconstructed, the information can be used to characterize completely the $H \to ZZ \to 4\ell$ decay and to construct dedicated kinematic discriminants that the analysis can exploit to enhance the background rejection.

Hence, this channel dominates the scene from the early stages of Run-I, where it was fundamental for the announcement of the discovery in 2012 [2–4], and for the subsequent measurements of the Higgs boson mass, width, and spin-parity [14–17]. All these results turned out to be in agreement with the SM predictions, which began to be probed in more detail with the measurement of signal strength modifiers [15, 28] and fiducial cross sections [29, 30], in order to assess the properties of the different production mechanisms and the compatibility with the SM in specific phase space regions. At the same time, the sought for anomalies in the SM predictions and the possible existence of BSM scenarios were probed with measurements of anomalous couplings to vector bosons and fermions [31–33].

While all the results were found to be in agreement with the SM predictions, their ultimate precision was essentially limited by the statistics available for the analysis, as a consequence of the low branching ratio that characterizes the $H \to ZZ$ decay. The measurements of signal strength modifiers, fiducial cross sections, and anomalous couplings were repeated at a centre of mass of $\sqrt{s} = 13$ TeV [34], exploiting the data collected in the early stages of the Run-II. The agreement with the SM expectations was confirmed with more strength and an overall reduction of the global uncertainty was observed. However, most of the measurements were still dominated by the statistical uncertainty. Exploiting the full statistics collected in the first year of the Run-II, the properties of the Higgs boson in the four lepton decay channel were measured again [35], setting a clear turning point in the history of this analysis. First of all, some of the results started to have comparable statistical and systematical components of the uncertainties; in addition, the increased statistics available for the analysis made possible the exploration of more fine-granular phase space regions. Hence, the $H \to ZZ \to 4\ell$ analysis entered the realm of precision physics, targeting the test of the SM in a more "production-oriented" way. With the analysis of the combined 2016 and 2017 data collected at $\sqrt{s} = 13$ TeV some of the inclusive measurements started to be limited by the systematic uncertainties, and the production mechanisms of the Higgs boson could be probed individually [36]. The apex was reached in 2018, when the analysis could profit from the entire statistics collected during the Run-II phase. The results of this study have recently been published in the European Physics Journal C [37] and are extensively detailed in this manuscript. Exploiting the full Run-II data set, the $H \to ZZ \to 4\ell$ decay channel could be used to measure the signal strength modifiers for each production mode and to measure fine granular cross sections in the so-called Simplified Template Cross Section (STXS) framework [19]. The large statistics available, corresponding to 137 fb^{-1}, also contributed to the improvement of the measurement of fiducial cross sections in differential bins of certain kinematic observables. The full Run-II statistics will also allow measurements of the Higgs boson's mass and width, improving the current constraints on these two observables. More precisely, the width of the Higgs boson is currently constrained to be $\Gamma_H = 3.2^{+2.8}_{-2.2}$ MeV, under the assumption of couplings with structure similar to that of the SM, for an expectation measured to be $\Gamma_H = 4.1^{+5.0}_{-4.0}$ MeV, using the data collected by the CMS experiment during the Run-I and combined with a portion of Run-II data [38]. This result was obtained from the combination of the on- and off-shell productions of the Higgs boson and led to the first confidence interval on Γ_H, which is expected to improve even further when the full Run-II statistics will be used for the analysis. The combination of the on-shell and off-shell productions of the Higgs boson will also permit studies of CP-violation and anomalous couplings of the Higgs boson. Eventually, the future Runs of the LHC will change the fate of these measurements, providing substantially more data and enhancing the sensitivity to possible BSM effects.

Before diving into this "golden journey" and present the measurements of the Higgs boson properties in the four lepton decay channel, it is useful to concentrate for a moment on the description of the detector that made this analysis possible: the CMS Experiment at the CERN LHC.

References

1. Herrero M (1999) The standard model. NATO Sci Ser C 534:1–59
2. ATLAS Collaboration (2012) Observation of a new particle in the search for the standard model Higgs boson with the ATLAS detector at the LHC. Phys Lett B 716:1
3. Collaboration CMS (2012) Observation of a new boson at a mass of 125 GeV with the CMS experiment at the LHC. Phys Lett B 716:30
4. Collaboration CMS (2013) Observation of a new boson with mass near 125 GeV in pp collisions at $\sqrt{s} = 7$ and 8 TeV. JHEP 06:081
5. Illustration of the standard model (SM) of particle physics. https://texample.net/tikz/examples/model-physics/
6. Zyla PA et al (2020) Review of particle physics. PTEP 2020(8):083C01
7. Kajita T (2016) Nobel lecture: discovery of atmospheric neutrino oscillations. Rev Mod Phys 88(3)
8. Maki Z, Nakagawa M, Sakata S (1962) Remarks on the unified model of elementary particles. Prog Theor Phys 28:870–880
9. Higgs PW (1964) Broken symmetries, massless particles and gauge fields. Phys Lett 12:132
10. Higgs PW (1964) Broken symmetries and the masses of gauge bosons. Phys Rev Lett 13:508
11. Higgs PW (1966) Spontaneous symmetry breakdown without massless bosons. Phys Rev 145:1156
12. Englert F, Brout R (1964) Broken symmetry and the mass of gauge vector mesons. Phys Rev Lett 13:321
13. Goldstone J (1961) Field theories with «?superconductor?» solutions. Il Nuovo Cimento 19(1):154–164
14. ATLAS Collaboration (2016) Measurements of the Higgs boson production and decay rates and coupling strengths using pp collision data at $\sqrt{s} = 7$ and 8 TeV in the ATLAS experiment. Eur Phys J C 76
15. Collaboration CMS (2015) Precise determination of the mass of the Higgs boson and tests of compatibility of its couplings with the standard model predictions using proton collisions at 7 and 8 TeV. Eur Phys J C 75:212

16. ATLAS and CMS Collaborations (2015) Combined measurement of the Higgs boson mass in pp collisions at $\sqrt{s} = 7$ and 8 TeV with the ATLAS and CMS experiments. Phys Rev Lett 114:191803
17. ATLAS and CMS Collaborations (2016) Measurements of the Higgs boson production and decay rates and constraints on its couplings from a combined ATLAS and CMS analysis of the LHC pp collision data at $\sqrt{s} = 7$ and 8 TeV. JHEP 08:45
18. Collaboration CMS (2020) A measurement of the Higgs boson mass in the diphoton decay channel. Phys Lett B 805:135425
19. de Florian et al (2016) Handbook of LHC Higgs cross sections: 4. Deciphering the Nature of the Higgs Sector. 2/2017, 10
20. ATLAS Collaboration (2020) A combination of measurements of Higgs boson production and decay using up to 139 fb^{-1} of proton–proton collision data at $\sqrt{s} = 13$ TeV collected with the ATLAS experiment. CERN CDS, 8
21. ATLAS Collaboration (2018) Observation of Higgs boson production in association with a top quark pair at the LHC with the ATLAS detector. Phys Lett B 784:173–191
22. Collaboration CMS (2018) Observation of $t\bar{t}H$ production. Phys Rev Lett 120(23):231801
23. CMS Collaboration (2021) Evidence for Higgs boson decay to a pair of muons. JHEP 01(148)
24. Collaboration CMS (2021) Search for nonresonant Higgs boson pair production in final states with two bottom quarks and two photons in proton-proton collisions at $\sqrt{s} = 13$ TeV. JHEP 03:257
25. ATLAS Collaboration (2021) Search for Higgs boson pair production in the two bottom quarks plus two photons final state in pp collisions at $\sqrt{s} = 13$ TeV with the ATLAS detector. CERN CDS, 3
26. Degrassi G, Giardino PP, Maltoni F, Pagani D (2016) Probing the Higgs self coupling via single Higgs production at the LHC. JHEP 12:080
27. Maltoni F, Pagani D, Shivaji A, Zhao X (2017) Trilinear Higgs coupling determination via single-Higgs differential measurements at the LHC. Eur Phys J C 77(12):887
28. ATLAS Collaboration (2015) Measurements of Higgs boson production and couplings in the four-lepton channel in pp collisions at center-of-mass energies of 7 and 8 TeV with the ATLAS detector. Phys Rev D 91(1):012006
29. ATLAS Collaboration (2014) Fiducial and differential cross sections of Higgs boson production measured in the four-lepton decay channel in pp collisions at $\sqrt{s}$=8 TeV with the ATLAS detector. Phys Lett B 738:234–253
30. Collaboration CMS (2016) Measurement of differential and integrated fiducial cross sections for Higgs boson production in the four-lepton decay channel in pp collisions at $\sqrt{s} = 7$ and 8 TeV. JHEP 04:005
31. Collaboration CMS (2015) Constraints on the spin-parity and anomalous HVV couplings of the Higgs boson in proton collisions at 7 and 8 TeV. Phys Rev D 92(1):012004
32. Collaboration CMS (2015) Limits on the Higgs boson lifetime and width from its decay to four charged leptons. Phys Rev D 92(7):072010
33. ATLAS Collaboration (2015) Study of the spin and parity of the Higgs boson in diboson decays with the ATLAS detector. Eur Phys J C 75(10):476. [Erratum: Eur Phys J C 76:152 (2016)]
34. CMS Collaboration (2016) Measurements of properties of the Higgs boson and search for an additional resonance in the four-lepton final state at sqrt(s) = 13 TeV. CERN CDS
35. CMS Collaboration (2017) Measurements of properties of the Higgs boson decaying into the four-lepton final state in pp collisions at $\sqrt{s} = 13$ TeV. JHEP 11(047)
36. CMS Collaboration (2018) Measurements of properties of the Higgs boson in the four-lepton final state at $\sqrt{s} = 13$ TeV. CERN CDS
37. Collaboration CMS (2021) Measurements of production cross sections of the Higgs boson in the four-lepton final state in proton-proton collisions at $\sqrt{s} = 13$ TeV. Eur Phys J C 81:488
38. Collaboration CMS (2019) Measurements of the Higgs boson width and anomalous HVV couplings from on-shell and off-shell production in the four-lepton final state. Phys Rev D 99(11):112003

Chapter 2
The CMS Detector at the LHC

Founded in 1954 and located near the Franco-Swiss border west of Geneva, the *Conseil Européen pour la Recherche Nucléaire* (CERN), or European Council for Nuclear Research, is nowadays the largest particle physics laboratory worldwide. Counting more than 10000 researchers from over 100 nationalities, CERN is at the forefront of fundamental research, innovation, and knowledge transfer.

Scientists at CERN analyze data from proton-proton collisions in the Large Hadron Collider (LHC) at a design centre-of-mass energy of 14 TeV. The LHC is the largest particle accelerator ever built, and is located in an underground tunnel of 27 km in length. This collider was constructed between 1998 and 2008 in the same tunnel that hosted its predecessor, the Large Electron-Positron (LEP) collider, posing stringent requirements on the design of the LHC infrastructure and its installation. The tunnel is placed underground, at a depth ranging between 45 m and 170 m, and herein are located the two vacuum beam-pipes where proton beams circulate in opposite directions before colliding at four interaction points, where the main experiments are installed. Given the vastness of the LHC scientific programme, each experiment was designed to probe specific physics scenarios. The ALICE and LHCb detectors study heavy ions and b-quark physics, respectively. The ATLAS and CMS are two general-purpose detectors, initially designed to hunt for the Higgs boson and to explore possible Beyond Standard Model (BSM) scenarios at the TeV energy scale. The results presented in this thesis are obtained from the analysis of proton-proton collisions collected by the CMS experiment at a center-of-mass energy of $\sqrt{s}$ =13 TeV, corresponding to the total 137 fb^{-1} recorded during the so-called Run-II phase of the LHC.

This chapter is structured as follows: an overview of the LHC accelerator, with particular attention to its design and operations, is given in Sect. 2.1; the CMS experiment, along with the description of its various sub-detectors, is presented in Sect. 2.2. Section 2.3 presents the algorithms used for the offline objects reconstruction and identification.

M. Bonanomi, *Response of the High Granularity Calorimeter HGCAL and Characterisation of the Higgs Boson*, Springer Theses,
https://doi.org/10.1007/978-3-031-26833-5_2

2.1 The Large Hadron Collider at CERN

The LHC inherits its location from the LEP e^+e^- collider and it was designed to deliver proton-proton (pp) collisions at a center-of-mass energy of $\sqrt{s} = 14$ TeV, with an instantaneous luminosity of $\mathcal{L} = 10^{34}$ cm^{-2}s^{-1} [1, 2]. The LHC was primarily conceived to probe the spontaneous symmetry breaking mechanism via the search of the Higgs boson, observed by the ATLAS and CMS experiments in 2012 [3–5], and to investigate for any hint of BSM physics at the TeV energy scale. In addition, the LHC features a nuclear physics programme, from which CERN owes its name, based on lead ions collisions at 2.76 GeV/nucleon at a luminosity of $\mathcal{L} = 10^{27}$ cm^{-2}s^{-1}, with the ultimate goal of understanding the behaviour of quarks and gluons in plasma.

2.1.1 *Design*

The LHC [6] is the last element of the CERN accelerator complex: herein, proton beams are produced, accelerated, in subsequent steps, up to 7 TeV, and injected into two separate beam pipes that cross at dedicated interaction points (IP), where the detectors are placed. The position of the four main experiments along the ring and the different elements of the accelerator complex are illustrated in Fig. 2.1.

The first step of the injection chain consists of the extraction of protons from H atoms exploiting strong electric fields. Bunches of protons are formed using the magnetic field generated in a Radio Frequency Quadrupole (RFQ) and are accelerated up to an energy of 750 keV. In a subsequent step, the bunches are supplied to the Linear Accelerator (LINAC 2), which increases their energy up to 50 MeV and sends them to the Proton Synchrotron Booster (PSB), a 150 m ring where their intensity is increased, and the beam is accelerated up to 1.4 GeV. Two more steps are required before injecting the beams in the LHC: the Proton Synchrotron (PS) and the Super Proton Synchrotron (SPS) accelerate the beams up to 25 GeV and 450 GeV, respectively. When ready to be sent to the LHC, the beam is split into two parallel beamlines with fast kicker magnets and accelerated to their nominal energy in high frequency (RF) cavities.

The beam optics in the LHC ring is controlled with superconducting NbTi magnets. The beams are kept on their circular trajectory exploiting 1232 dipoles, disposed into eight arcs of 2.45 km each. A current of about 11 kA flows in the dipoles and generates the 8.3 T magnetic field that keeps the beams on their circular trajectory. One of the most stringent requirement for the accelerator is the narrow size of the bunches, which is kept constant by using additional magnets, also known as quadrupoles. Superfluid He-4 is used to cool down the magnets to 1.9 K, thus ensuring their superconducting state.

The beams are made to collide at four interaction points (IP), where the main experiments are located, after a further focusing obtained with additional quadrupoles placed before each IP.

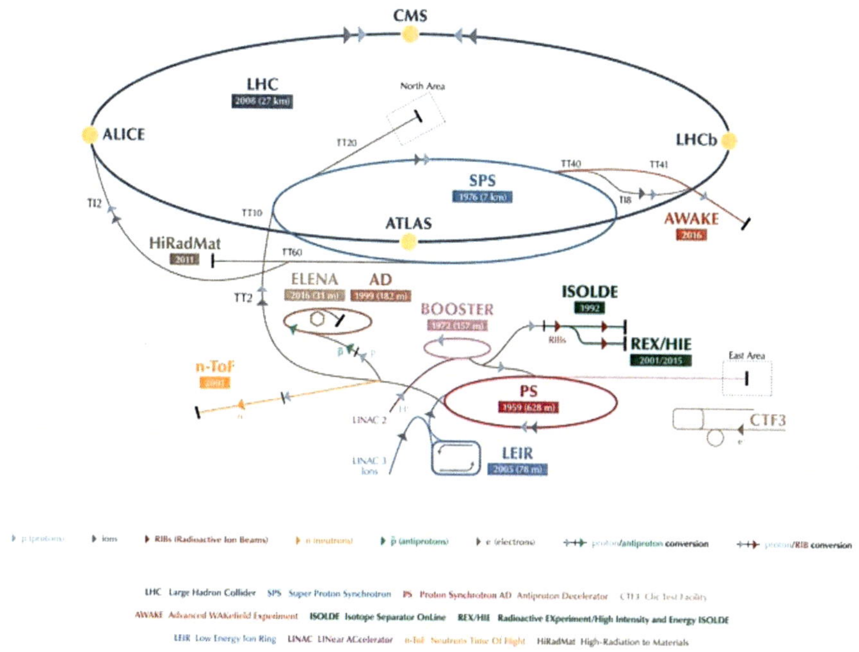

Fig. 2.1 Schematic representation of the CERN accelerator complex [7]. The LHC (dark blue line) is only the last part of a vast acceleration chain that begins at the LINAC 2 (pink line)

The number of events per unit time $\partial N/\partial t$ for a process with given cross section σ is determined by the *instantaneous luminosity* $\mathcal{L}$ [6]:

$$\frac{\partial N}{\partial t} = \mathcal{L} \times \sigma. \tag{2.1}$$

Hence, the only way to observe processes with low cross section at the LHC is to have a machine with a sizeable instantaneous luminosity. However, this parameter can not be increased indefinitely, as it conflicts with the performance of the data acquisition system, as described in Sect. 2.2.4. Often one reports the integral of $\mathcal{L}$ over time, referred to as integrated luminosity, $L = \int \mathcal{L}dt$, which defines the amount of data recorded over a given period of time, usually a year of data taking. The instantaneous luminosity is measured in units of $\text{cm}^{-2}\text{s}^{-1}$, and its design value for the LHC machine is $\mathcal{L} = 10^{34}\ \text{cm}^{-2}\text{s}^{-1}$, while the integrated luminosity is usually expressed in inverse femtobarns (fb^{-1}) or picobarns (pb^{-1}).

The operations of the accelerator are often quantified quoting the instantaneous luminosity, which can be defined from the beam parameters as:

$$L = \frac{N_p^2 n_b f_{\text{rev}} \gamma_r}{4\pi \epsilon_n \beta^\star} F\,, \tag{2.2}$$

Table 2.1 Nominal parameters of the LHC in proton-proton collisions

Parameter	Meaning	Nominal value
$\sqrt{s}$	Center-of-mass energy	14 TeV
Δt	Bunch separation	25 ns
n_b	Number of bunches	2808
N_p	Number of protons per bunches	$1.15 \cdot 10^{11}$
f_{rev}	Revolution frequency	11245 Hz
σ_z	Transverse bunch r.m.s. at the IP	16.7 μm
σ_{xy}	Longitudinal bunch r.m.s.	7.55 cm
$\beta^\star$	Beta function at the IP	0.55 m
θ_c	Crossing angle at the IP	285 μrad
ϵ_n	Transverse emmittance	3.75 μm

where N_p is the number of particles in each of the n_b bunches present in the beam, f_{rev} is its revolution frequency, and γ_r is the relativistic factor. The beam optics enters in the $\epsilon_n \beta^\star$ factor, where the first term represents the emittance and the latter the so called *beta function*, i.e. the beam focusing at the IP [6]. The F factor accounts for the reduction of instantaneous luminosity due to the crossing angle (θ_c) of the beams and the transverse and longitudinal r.m.s. bunch sizes (σ_{xy} and σ_z):

$$F = \left(1 + \frac{\theta_c \sigma_z}{2\sigma_{xy}}\right)^{-1/2} \tag{2.3}$$

The nominal values of the different parameters defined above are reported in Table 2.1.

As mentioned above, the large luminosity of the LHC is fundamental to have the possibility to observe processes with small cross sections. However, this also means that several interactions can happen within the same bunch crossing. This phenomenon is referred to as *pileup* and its average value is given by:

$$\langle \text{P.U.} \rangle = \frac{L\sigma_{pp}^{\text{inel}}}{n_b \cdot f_{\text{rev}}}\,, \tag{2.4}$$

where $\sigma_{pp}^{\text{inel}}$ is the inelastic pp cross section. At the design value of the LHC one has $\sqrt{s} = 14$ TeV and $\sigma_{pp}^{\text{inel}} = 69$ mb, for a resulting pileup rate of about 22 interactions [8]. As further detailed in the next section, the LHC has been increasing constantly its operational values over the years, reaching the current pileup rate of about 60 interactions per bunch crossing. This number will become even larger with the High Luminosity phase of the LHC (HL-LHC), foreseen to start in 2027, where an average pileup rate of $\langle \text{P.U.} \rangle = 140/200$ events is expected. These constantly increasing values of pileup rate have a direct consequence on the detector's performance, as a higher detector occupancy means a reduction in the particle reconstruction's efficiency and resolution. In order to cope with such a harsh environment expected for the HL-LHC operations, all the LHC experiments are currently undergoing an upgrade programme [9]. The mitigation of the pileup rate and of the high radiation dose caused by such a harsh environment are two of the most stringent requirements the design of the new sub-detectors has to meet. For example, the upgrade of the CMS endcap calorimeters with a High Granularity Calorimeter (HGCAL) [10] goes in this direction and it has been one of the topics studied in this thesis. A more detailed description of the CMS upgrade campaign in view of the HL-LHC programme, with a particular focus on the HGCAL and its design validation using test beam data, is given in Chap. 3.

2.1.2 Operations

After more than ten years of commissioning and installation, the first proton beams were injected into the LHC in September 2008. The inaugural run was stopped shortly after, and intervention was needed to repair a severe mechanical damage induced by a large loss of liquid He-4 due to a faulty electrical connection between two magnets. The LHC underwent a one-year technical stop until November 2009, when 450 GeV proton beams were injected into the machine for the first time after the incident. A few weeks after, the beam energy was ramped up to 1.18 TeV per beam, and the LHC officially became the highest-energy particle accelerator in the world.

In 2010 the beam energy was constantly increased until reaching 3.5 TeV per beam. In March 2010, the first high-energy proton-proton collision with a center-of-mass energy of 7 TeV took place. The data collected over the years 2011 and 2012 comprise the Run-I phase of the LHC: in the first year 6 fb^{-1} were delivered at a center-of-mass energy of $\sqrt{s} = 7$ TeV, while in 2012, a total of 23 fb^{-1} were delivered at $\sqrt{s} = 8$ TeV. The former dataset allowed the ATLAS and CMS collaborations to set stringent upper limits to the permitted mass range of the Higgs boson. The latter featured a larger luminosity, and its partial combination with the $\sqrt{s} = 7$ TeV dataset permitted to claim the Higgs boson discovery in July 2012. During the following years, the luminosity of the machine has been increased constantly to reach its nominal design value. After the beginning of Run-II in 2015, the LHC ran at a centre-of-mass of $\sqrt{s} = 13$ TeV, for a total luminosity of 139 fb^{-1} collected by the end of 2018. Figure 2.2 shows the evolution of the integrated and peak luminosities recorded by the CMS detector over the last eight years of LHC operations.

Figure 2.3 shows a summary of the past LHC operations and a schedule for future runs. After the Run-I phase, the LHC underwent two years of a technical stop, also referred to as the first *long shutdown* (LS1), with the goal of preparing the machine for the beginning of the Run-II. A substantial upgrade of the entire infrastructure

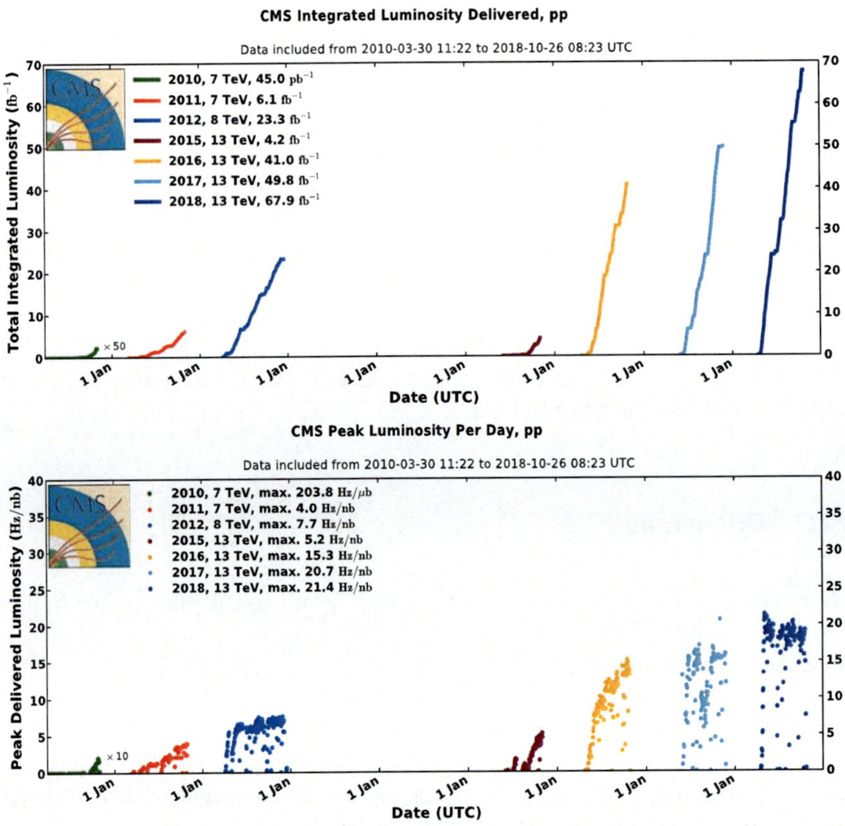

Fig. 2.2 Total integrated luminosity (top) and peak instantaneous luminosity (bottom) delivered to CMS as a function of the date. The plots comprise both the Run-I (2010–2012) and the Run-II (2015–2018) operations of the LHC [11]

was needed, with particular attention to the magnets and some of the experiments' sub-detectors.

With the luminosity steadily growing over the years (cf. Fig. 2.2), the average number of collisions per bunch crossing is expected to rise. Figure 2.4 shows the pileup rates for the different years of the Run-II data taking phase: the average number of collisions per bunch crossing almost tripled going from $\langle \text{P.U.} \rangle = 13$ in 2015, up to $\langle \text{P.U.} \rangle = 40$ in 2018. The operations in the period ranging from 2011 to 2024 are referred to as *Phase 1*, while *Phase 2* refers to the period between 2027 and 2037, and it will correspond to the High-Luminosity era of the LHC (HL-LHC).

The LHC resumed its operations after the LS1 in 2015 and it kept running until the end of 2018, for a period that is commonly referred to as Run-II. The 2015 runs were intended for the commissioning of the machine at the new center-of-mass energy of $\sqrt{s} = 13$ TeV, a record value that opened the doors to new physics scenarios for the experiments. In this same year, 4.2 fb^{-1} were delivered to CMS at $\mathcal{L} =$

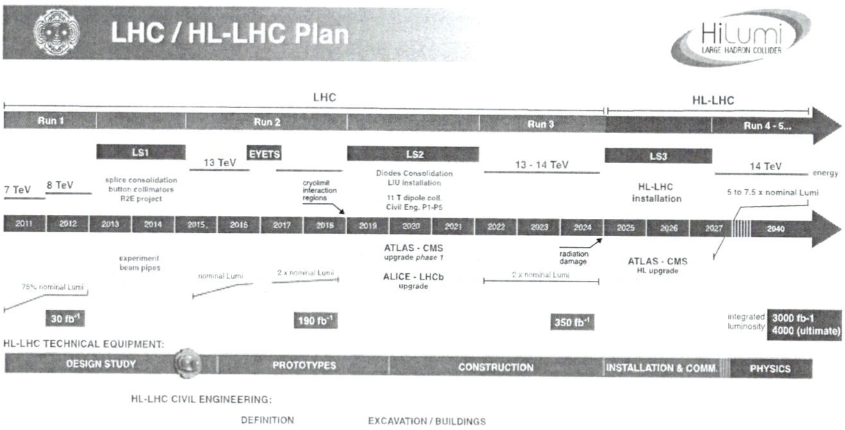

Fig. 2.3 Subdivision of the past LHC activities and plans for the future of the LHC and HL-LHC programmes [12]

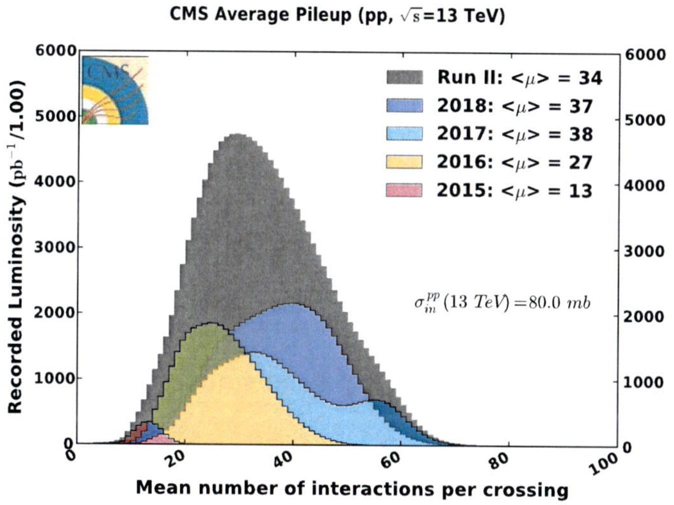

Fig. 2.4 Average pileup profile recorded by the CMS detector over the Run-II (2015–2018) period. The average pileup value $\langle\mu\rangle$ is computed as a function of the σ_{pp}^{inel} values indicated in the figure for each centre-of-mass energy [11]

5×10^{33} cm^{-2}s^{-1}, but in June 2016 the design instantaneous luminosity was reached and later on exceeded until reaching the current value of $\mathcal{L} = 1.5 \times 10^{34}$ cm^{-2}s^{-1}. The CMS detector recorded integrated luminosities of 41.0 fb^{-1}, 49.8 fb^{-1}, and 67.9 fb^{-1} in the years from 2016 to 2018, as shown in Fig. 2.2.

After discovering of the Higgs boson in Run-I, one of the main goals of the Run-II physics programme has been the precision measurement of the properties of this long sought particle: from its production and decay rates, up to the characterization of its couplings with the other particles of the SM.

Up to now, the CMS Run-II analyses have not revealed any hint of new physics. BSM processes are expected to have lower cross sections and affect primarily the distributions' tails, where the statistics available is still limited. Hence, to probe any BSM scenario or use the Higgs boson as a portal for new physics, more data are required, alongside more precise techniques for events selection, identification, and reconstruction.

At the moment of writing this thesis, the LHC machine is in its second *long shutdown* (LS2). LS2 aims to upgrade and maintain the accelerator complex and infrastructure in preparation for the Run-III, when the LHC is expected to run at unprecedented center-of-mass energy of $\sqrt{s} = 14$ TeV, finally reaching its original design value. The end of Run-III in 2024 will mark the closure of the LHC Phase 1, and by then, the experiments are expected to have recorded an integrated luminosity of about 350 fb^{-1}: a record-breaking value that could allow the study of rare phenomena and would help to nail down even more precisely the properties of the Higgs boson. However, by this time, the detectors of the various experiments are expected to have suffered severe radiation damage, and a substantial upgrade will be needed to maintain the current physics performance and enhance the machine's discovery potential. The LHC is preparing its High Luminosity phase (HL-LHC), which foresees a total integrated luminosity of about 3000 fb^{-1} and a peak instantaneous luminosity of $5 \times 10^{34}\ \text{cm}^{-2}\text{s}^{-1}$, leading to unprecedented radiation levels and pile up rates of about 140/200 events per bunch crossing, but also opening the doors to physics scenarios previously inaccessible. The HL-LHC is expected to produce about 15 million Higgs bosons per year, substantially exceeding the three millions produced in 2017, thus allowing detailed studies of the Higgs boson properties and possibly unveiling new physics not observed before.

2.1.3 Experiments

As shown in Fig. 2.1, the CERN accelerator complex is supplied with several experimental facilities placed along the acceleration chain. The PS serves as low-energy proton source for the *East Area*, while the SPS is used to stream protons up to 450 GeV in the *North Area*. These two locations are commonly used for test beam purposes and also host other detectors used for non-Higgs physics, such as dark matter studies, crystallography and irradiation studies.

Proton beams circulate in counter-wise directions in two separate beam pipes and are made to collide at four IPs, where the main experiments are placed. While for the ALICE and LHCb detectors two of the LEP IPs were reused, for the two multi-purpose detectors ATLAS and CMS larger caverns were obtained from the old ones. At each IP the beam pipes are made to cross for collisions to take place and be detected by the corresponding experiment:

- **Interaction point IP1**
  The IP1 hosts **A Toroidal LHC Apparatus (ATLAS)** [13], one of the two general-purpose detectors designed to hunt for the SM Higgs boson and to study its properties, but also to perform QCD analyses and detect possible BSM physics.
  The ATLAS apparatus comprises many sub-detectors, each dedicated to identifying and reconstructing a given type of particle or signature. The ATLAS experiment has a cylindrical structure around the interaction point: after an inner tracker and a Transition Radiation Tracker (TRT), that feature a mixed silicon-based and gaseous technology, the core of the detector is its sampling calorimeter, consisting of an inner liquid Argon electromagnetic component and an outer steel-tile calorimeter. This design choice represents one of the main differences between the ATLAS and CMS detectors, as the latter features a homogeneous $PbWO_4$ electromagnetic calorimeter. Muon chambers are located in the experiment's outer region and cover it completely, resulting in the huge 25 m diameter and 46 m length of the ATLAS detector and in the second major difference with respect to the CMS design. A solenoid magnet surrounding the IP provides an axial field of 2 T, while several toroidal coils are used to reach up to 4 T fields around the solenoid.
  The detectors of the **Large Hadron Collider forward (LHCf)** [14] experiment are placed in the same cavern of the ATLAS detector, and they are located about 140 m away from the IP. The LHCf is made of two detectors, each weighing 40 kg for a total volume of 30 cm $\times$ 80 cm $\times$ 10 cm. The physics goal of the LHCf experiment is the characterization of the ultra-high-energy cosmic rays.
- **Interaction point IP2**
  A Large Ion Collider Experiment (ALICE) [15] is located at IP2, one of the former collision points of LEP. The ALICE experiment is designed to study heavy-ion physics with the final aim of unveiling the nature of the quark-gluon plasma (QGP), the state of the matter expected to be present in the primordial Universe. The ALICE detector is unique in its genre and it features many sub-detectors not present in the other LHC experiments. The golden point of ALICE is its Time Projection Chamber (TPC) that complements the inner tracking system and allows precise particle identification and discrimination. The former is enhanced by the presence of several calorimeters and Time of Flight (TOF) and Ring-imaging Cherenkov (RICH) detectors. In the very forward region the detector is complemented with a muon spectrometer, an additional handle when studying QGP with the decay products of heavy quarkonium states.
- **Interaction point IP5**
  The **Compact Muon Solenoid (CMS)** [16] is located at IP5, diametrically opposed to the ATLAS detector. Both are general-purpose detectors, designed to study a large variety of physics phenomena, even though their structure is substantially different, resulting in complementary measurements. A more detailed description of this detector is given in Sect. 2.2.
  Sharing the same IP as CMS and partially placed within it there is the **TOtal Elastic and diffractive cross section Measurement (TOTEM)** [17]. The TOTEM experiment owes its name to its physics goal, that is the measurement of total cross section, elastic scattering, and diffractive processes in *pp* collisions.

The particles coming out from the IP are tracked using cathode strip chambers (CSC) and Gas Electron Multipliers (GEM) placed over multiple stations and forming two *telescopes*. The other main detector of the TOTEM experiment are *Roman Pots*, i.e. silicon sensors used to measure scattered protons.

- **Interaction point IP8**
  The IP8 hosts the **LHC beauty (LHCb)** [18] experiment which, alike the ALICE one, is located within one of the original LEP caverns. LHCb is specialized in the analysis of b-quark physics with the aim of unveiling any new BSM process or leading to direct hints of CP violation.
  The LHCb detector is built in a different fashion than the ATLAS or CMS to reflect the different physics studied: as b-hadrons are mostly produced in the forward region, the detector is a single-arm forward spectrometer. The first sub-detector is placed close to the IP, with the rest disposed over a length of 20 m.
  In order to measure efficiently b-quarks, the LHCb detector features a Vertex Locator (VELO), i.e. a movable tracking detector close to the beam-pipe. The VELO is an array of semi-circular silicon-based detectors which allows to detect secondary vertices of the b-hadrons decay.
  Sharing the same location as LHCb the **Monopole and Exotics Detector At the LHC (MoEDAL)** [19] experiment aims at a direct measurement of the magnetic monopole, a hypothetical particle with a magnetic charge. The MoEDAL detector is the seventh experiment approved and installed at the LHC.

Not all the LHC experiments aforementioned can sustain the same instantaneous luminosities and pile up rates. More precisely, the ALICE and LHCb detectors have limiting factors that prevent them to collect data at the typical LHC rates. For the former the constraint is set by the TPC dead-time, while the large particle flux traversing the latter almost saturates its data acquisition capabilities. Hence, a *luminosity leveling* is performed: the beams are automatically separated and delivered to the various detectors when the target luminosity is reached. ALICE works with a stable instantaneous luminosity of about $\mathcal{L} = 10^{27}\ \text{cm}^{-2}\text{s}^{-1}$, while a larger value of $\mathcal{L} = 10^{32}\ \text{cm}^{-2}\text{s}^{-1}$ is used for LHCb.

2.2 The CMS Detector

Located in an experimental cavern 100 m deep underground at Interaction Point 5, in the French village of Cessy, the CMS detector [16] is one of the two multi-purpose experiments installed at the CERN LHC to hunt for the Higgs boson and to explore a large variety of physics scenarios at the TeV energy scale. The detector has a cylindrical structure, hermetically surrounding the IP5, for a total diameter of 15 m and a length of 21.5 m. The CMS detector owes the "compact" of its acronym to the fact that it is almost twice as small as the ATLAS experiment but with a twice larger weight of 12500 t.

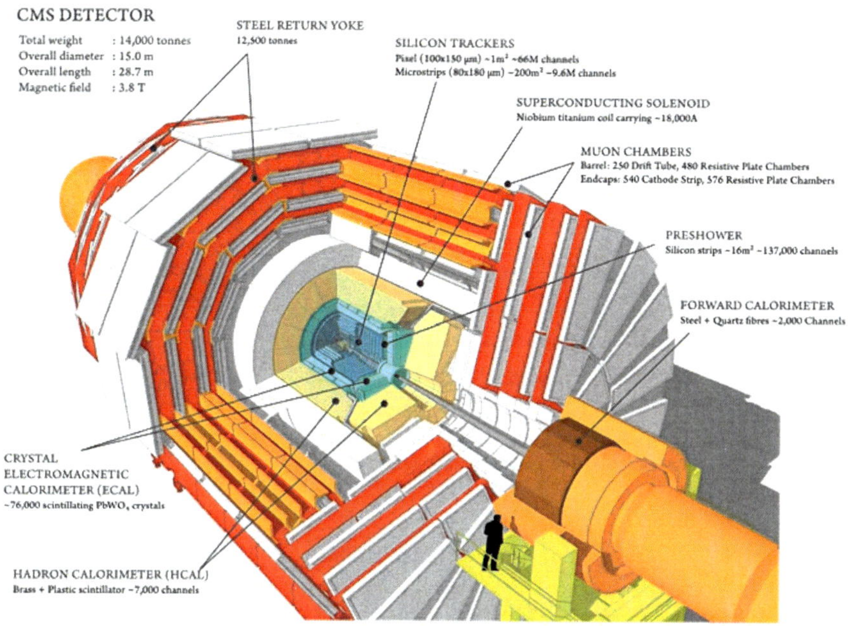

Fig. 2.5 Schematic illustration of the CMS detector at the CERN LHC [20]

The central component of the detector is a superconducting 4 T solenoid magnet, which encloses the several sub-detectors that build up the CMS experiment. Each of these has a specific role in detecting a particular type of particle or in the measurement of a given observable. Immediately after the IP, 6 m within the coil, the inner tracking system is surrounded by electromagnetic and hadronic calorimeters. The large muons' detection subsystem is located around the magnet for an additional radius of 7 m, and it composes about 80% of the detector's total volume, from which the "muon" of the CMS acronym. Figure 2.5 depicts a cross-section of the CMS experiment, where one can identify all the subsystems briefly mentioned above. A more detailed description of all the CMS components is given in Sect. 2.2.2.

2.2.1 Coordinate System

The CMS experiment exploits a right-handed coordinate system, with the origin placed at the nominal interaction point. The x-axis points towards the geometrical center of the LHC, the y-axis points upwards, and the z-axis points in the anticlockwise proton beam direction.

Given the cylindrical structure of the detector, a polar coordinate system is also used. The azimuthal angle ϕ is defined as the angle formed in the transverse plane with respect to the x-axis, and r as radial coordinate.

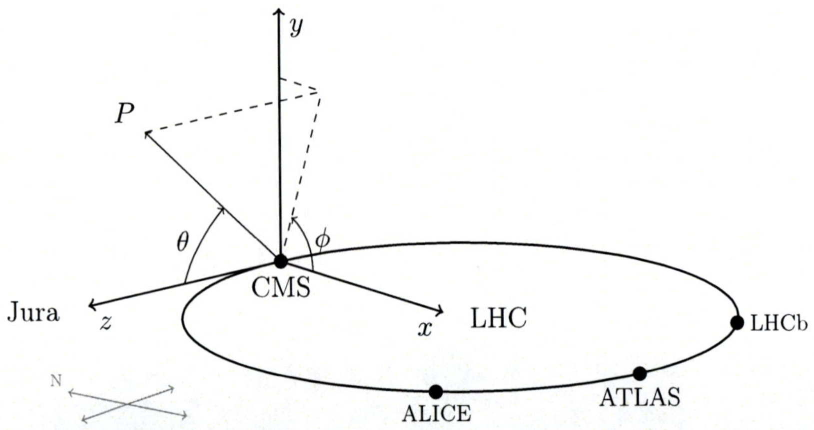

Fig. 2.6 Schematic representation of the CMS coordinate systems [21]

A schematic illustration of the CMS Cartesian and polar coordinate systems is given in Fig. 2.6.

While this coordinate system is well suited to describe the orientation of the detector and macroscopic observables, it is not used to identify proton-proton collisions, where the interaction takes place at parton level. Here the momentum fraction carried by each parton along the z-axis and the longitudinal boost of the collision's rest frame are unknown observables. Hence, the event is preferably described by means of Lorentz boost-invariant quantities, such as the projections of observables on the transverse plane.

Therefore one can define the *transverse momentum* (p_T) and the *transverse mass* (m_T):

$$p_T^2 = p_x^2 + p_y^2$$

$$m_T^2 = m^2 + p_x^2 + p_y^2 = E^2 - p_z^2$$

The rapidity of a particle is expressed as:

$$y = \frac{1}{2} \ln \left(\frac{E + p_z}{E - p_z} \right) \tag{2.5}$$

The polar angle is usually converted into *pseudorapidity*, i.e. the rapidity for ultra-relativistic ($m/E \ll 1$) particles:

$$\eta = \frac{1}{2} \ln \left(\frac{p + p_z}{p - p_z} \right) = -\ln \tan \left(\frac{\theta}{2} \right) \tag{2.6}$$

which varies from 0 at $\theta = \frac{\pi}{2}$ to $\pm\infty$ at $\theta = 0\,(\pi)$, as shown in Fig. 2.7.

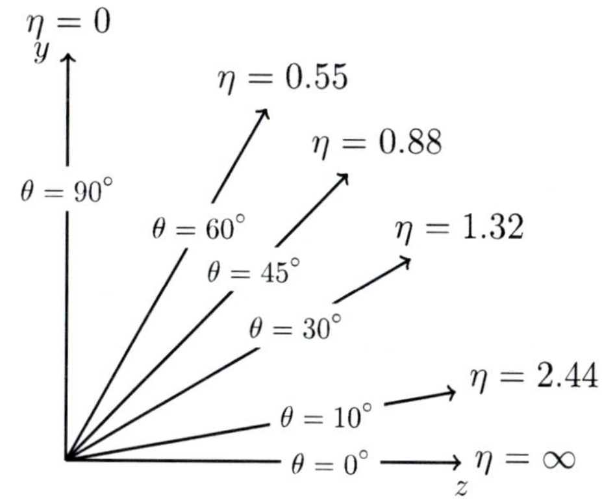

Fig. 2.7 Relation between the pseudorapidity η and the polar angle θ

The spatial separation between two particles is expressed, in a Lorentz boost-invariant frame, as a function of their angular distance as:

$$\Delta R = \sqrt{(\Delta\eta)^2 + (\Delta\phi)^2}. \tag{2.7}$$

An additional handle on the event's topology comes from the negative sum of the transverse momenta of all the reconstructed particles, which goes under the name of missing transverse momentum p_T^{miss}. Being the detector symmetric around the interaction point and hermetic, this quantity is usually interpreted as the total transverse momentum of neutrinos, or other hypothetical non-interacting particles, escaping the detector without leaving any signature whatsoever.

2.2.2 *Detector Structure*

The CMS detector is built around a 4 T superconducting solenoid, used to bend the tracks of the particles traversing the different sub-modules of which CMS consists of. In the following the various components are detailed, with particular attention to their role in the detection of a specific kind of particle. This same information is also exploited for the offline particles' identification and reconstruction, as discussed in Sect. 2.3.

Superconducting Magnet

The most essential component of the CMS experiment is the niobium-titanium (NbTi) superconducting solenoid magnet [22].

With a 6 m diameter and an operational temperature of about 4.5 K, the magnet dictates the design of the CMS detector. The magnet volume hosts the tracking and

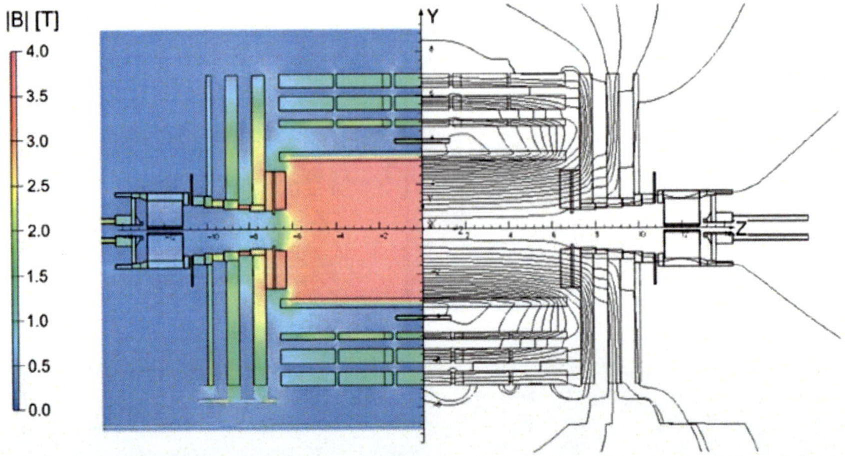

Fig. 2.8 Longitudinal view of the CMS detector showing the magnetic | B | field (left) and field lines (right). Central magnetic flux density of 3.8 T. Each field line represents an increment of 6 Wb of the total flux. Figure taken from [23]

calorimetric systems, and it features a constant magnetic field of 3.8 T. The iron return yoke, used to confine the high magnetic field, is interspersed with the muon detection system and is placed outside the NbTi magnet. Hence, the muon chambers are immersed in a 2 T magnetic field, thus resulting in muons tracks having two different bendings before and after crossing the magnet.

The CMS collaboration decided to operate the magnet at a nominal value of 3.8 T, slightly lower than the 4 T design value, because of the unique design of the magnet and its unknown aging properties. The magnetic field strength and lines estimated by means of Monte Carlo simulation are shown in Fig. 2.8.

Inner Tracking System

Placed directly around the interaction point, the tracking system [24, 25] is the innermost part of the CMS detector. The CMS tracker covers a volume of 5.6 m in length and 2.4 m in diameter, and it is instrumented with silicon sensors, sensitive to the passage of charged particles originating from the interaction vertex. A uniform magnetic field of 3.8 T permeates the tracking volume, and it is used as an additional handle in the particles identification and reconstruction. The tracker also has to identify the in-flight decays and discriminate the hard scattering interactions vertex, also referred to as *primary vertex*, from additional pile-up collisions. As a result, the tracker is the sub-detector that has to cope with the most stringent requirements in terms of design, detection efficiency, and radiation hardness. The tracking system is subdivided into two components to cope with these challenges: the pixel detectors and the silicon strip tracker, shown in Fig. 2.9.

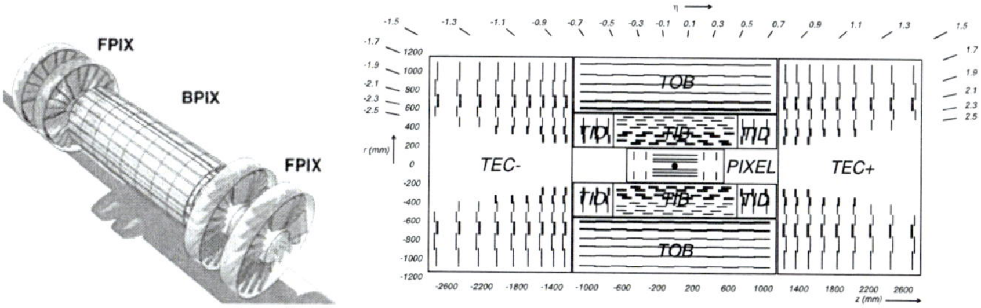

Fig. 2.9 *Left*: Illustration of the CMS pixel tracking sub-systems. Three cylindrical layers are disposed in the barrel (BPix), while the endcaps are covered by two disks (FPix) on each side. *Right*: Schematic representation of the CMS tracking systems' cross section in the $r - z$ plane. Figures taken from [26, 27]

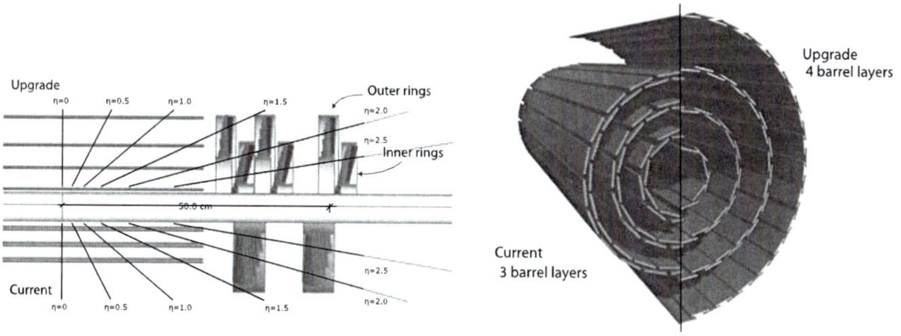

Fig. 2.10 *Left*: Side view of the CMS tracking systems. In the top part of the figure the conceptual layout of the upgraded tracker is shown, while the old one is depicted in the bottom part of the figure. *Right*: Transverse view of the pixel barrel layers in the two designs described in the text. Figures taken from [28]

- **Pixel:** The CMS pixel tracking detector surrounds the IP, in a region of $| \eta |< 2.5$, and it comprises 65 millions silicon pixels of $100 \times 150\ \mu m^2$, disposed in three cylindrical layers in the barrel (BPix) and in two endcap disks (FPix), one on each side of the IP, as shown in the left picture in Fig. 2.9.
  The BPix layers are located at radii of 4.4, 7.3 and 10.2 cm, while the FPix disks start at 6 cm from the beam pipe center and extend up to 15 cm. Each pixel features a spatial resolution of 10 μm in the (r, ϕ) plane and of 20 μm along the z axis.
  The downside of having such a finely segmented detector is the increased need for readout electronics and power supply components, which demands a large cooling capacity. This infrastructure results in passive material encountered by the particles while traversing the detector. Therefore it increases the probability of parasitic phenomena such as multiple scattering, bremsstrahlung, pair production, and nuclear interactions, which may impact the reconstruction of the particles' trajectory.

In order to improve the physics reconstruction performance and to cope with the exceedingly large luminosity registered by 2017, the pixel system has been upgraded in the technical stop between the 2016 and 2017 data taking periods [28]. The new detector features four layers in the barrel and three disks in the endcap regions, thus providing an additional tracking point in both directions. The major challenge of such an upgrade is clearly the reduction of the total budget material, notwithstanding the increased number of tracking station. A complete redesign of the support and CO_2 cooling systems, along with the displacement of readout boards and connections outside the tracking volume, allowed a consistent reduction of the dead material, with the new barrel and endcap detectors weighing 40% and 80% of the old ones, respectively. The sensitive thickness of the BPix sensors of 285 μm remained unchanged after the upgrade, whereas the FPix sensors thickness was increased from 270 μm to 300 μm. The layout of the upgraded pixel detector is depicted in Fig. 2.10.

- **Strips:** The outermost region is equipped with silicon strip sensors, covering the same η range as of the pixels, a radial occupancy between 20 and 116 cm, and an extension along the z-axis of ±282 cm. Micro-strip sensors can be used instead of pixels because of the increased distance from the nominal IP, resulting in a lower flux of particles.
  The micro-strip tracking detector is divided into three subsystems, each with a different strip pitch and thickness according to the location of the modules. The innermost part is organised in four barrel layers (TIB) and three disks (TID) of up to 55 cm of radius on each side. These two subsystems combined provide up to 4 (r, ϕ) measurements and they are composed of strips with varying pitch from 80 to 120 μm, a sensor thickness of 320 μm, and a single point spatial resolution from 23 to 25 μm.
  The Tracker Outer Barrel (TOB) encloses the TIB and TID systems in a outer radius of 116 cm and 236 cm length. It embeds 500 μm thick sensors with a strip pitch of 122 to 183 μm and a single point resolution ranging from 35 to 53 μm for the six measurements of (r, ϕ) provided.
  The forward region is covered by the Tracker Endcaps (TEC), placed within a radius of 22.5 cm to 113.5 cm and located at 124 cm $<| z |<$ 282 cm. The TEC comprises nine disks covered by 320–500 μm thick radial micro-strip sensors, with a pitch ranging from 97 μm to 184 μm which adds up to nine measurements in ϕ.
  As shown in the right picture of Fig. 2.9, several layers and disks are supplemented with a second micro-strip sensor mounted back-to-back and tilted by 100 mrad to allow the measurement of an additional coordinate, being it longitudinal (z) in the barrel or radial (r) in the endcaps.
  With an area of 198 m^2 and a total of 9.3×10^3 strips, the CMS micro-strip tracking system is the largest silicon-based detector worldwide.

The operational temperature is one of the critical aspects of silicon-based detectors, as it is necessary to reduce the ionizing radiation damage on the sensors. The CMS tracker is designed for dry gas cooling at -20°C. Nevertheless, during the Run-

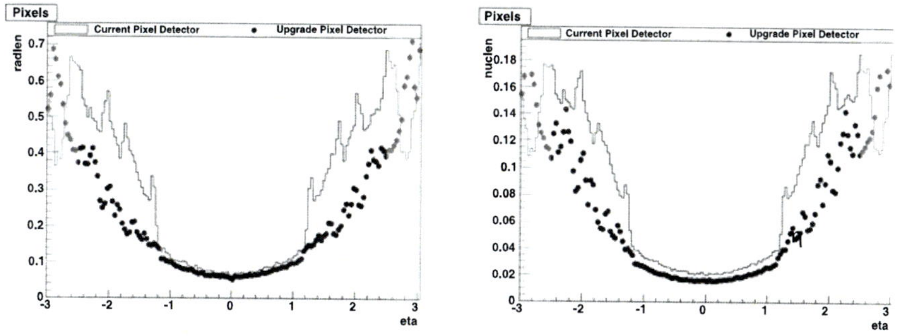

Fig. 2.11 Material budget in the tracker pixels, as a function of the pseudorapidity, in radiation length (left) and nuclear interaction length (right). The histograms show the material budget of the pixel detector before the upgrade, while the black dots represent the reduced material budged after the 2017 upgrade. Figures taken from [29]

I operations, a working temperature of only 4°C could be achieved. The tracker's cooling and thermal isolation systems were upgraded during the LS1 of the LHC [29] to prevent from further radiation damage. From the beginning of Run-II, the tracker pixel sensors have been operated at a temperature of -10°C, while the rest of the tracker system has been working at -15°C. These working conditions lasted until the end of the Run-II operations, after which the pixel system, along with several other sub-detectors, is going to be replaced during the CMS Phase-II Upgrade programme (cf. Chap. 3).

The design of the CMS tracking system is a tradeoff between the best detector performance and the least possible inactive material. In this regard, the reduction of dead material is a crucial issue in minimizing the parasitic interaction of particles originating from the primary vertex. Limiting multiple scattering, bremsstrahlung, nuclear interactions, and photon conversion as much as possible is fundamental for a precise reconstruction of the trajectory of the particles in the tracker and for the measurement of their energy in the subsequent calorimeters. Figure 2.11 shows the simulated material budget in the CMS tracker in units of radiation length (X_0) and hadronic interaction length (λ_i) for both the original and the upgraded pixel detector. The total passive material for the tracking and the relative services (cabling, support, and cooling system) added up for a total of 1.6 λ_0 in the original design. With the pixel upgrade, this value was reduced by about 40% in the forward region and 10% in the barrel, thus improving the resolution on the interaction point by a factor 1.5 in the z-direction.

Electromagnetic Calorimeter (ECAL)

The Electromagnetic CALorimeter (ECAL) [30] surrounds the inner tracking system and it is designed to provide a precise energy measurement of the incoming electrons and photons. The ECAL is a hermetic homogeneous calorimeter consisting of almost

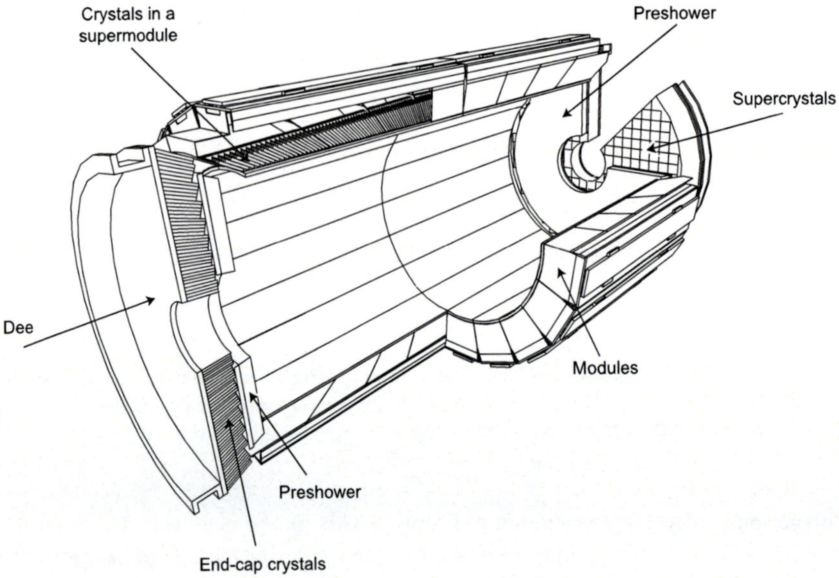

Fig. 2.12 Schematic view of the CMS ECAL mechanical structure. The $PbWO_4$ crystals in the barrel are organized into modules and super-modules. In the endcaps a preshower is followed by two half-disks, or "Dees", on each side. Figure taken from [30]

70000 lead tungstate ($PbWO_4$) crystals. The choice of $PbWO_4$ is driven by the need of a high density ($\rho = 8.29\,\text{g/cm}^3$) material, with a small radiation length ($X_0 = 0.89\,\text{cm}$) and a short Molière radius ($R_M \simeq 2.2\,\text{cm}$), to ensure a full containment of electromagnetic showers and a relatively fast response, which has to cope with the fact that about 80% of the scintillating light is emitted within the LHC bunch crossing time of 25 ns. The layout of the CMS ECAL is depicted in Fig. 2.12.

The ECAL Barrel (EB) is divided into two half-cylinders enclosed by two end-cap discs (EE). The two EB halves consists of 18 super-modules, each weighing approximately 1.5 t and grouping about 1700 crystals: 20 in ϕ and 85 in η. Each super-module has a coverage of $|\eta| < 1.479$ and 20° in ϕ. The EB groups in total 62000 crystals, each covering 0.0174×0.0174 in $\eta \times \phi$, or a $22 \times 22\,\text{mm}^2$ front face. The EE disks are composed by groups of 5×5 $PbWO_4$ crystals, also referred to as super-crystals, for a total coverage of $1.479 < |\eta| < 3.0$. For a hermetic coverage the ECAL crystals are oriented with a tilt of up to 3° with respect to the nominal IP direction. Nevertheless, some gaps, referred to as *cracks*, remain between the modules and the energy reconstruction, especially in the $\eta = 0$ region and in the transition between EB and EE (Fig. 2.13).

An additional electromagnetic preshower detector (ES) is placed in front of the two EE disks in order to enhance the spatial resolution in the $1.65 < |\eta| < 2.6$ region. The ES is a sampling calorimeter made of two lead absorbers followed by a silicon sensors plane made of 22 m long strips, for a total of around 1 X_0. In addition to the

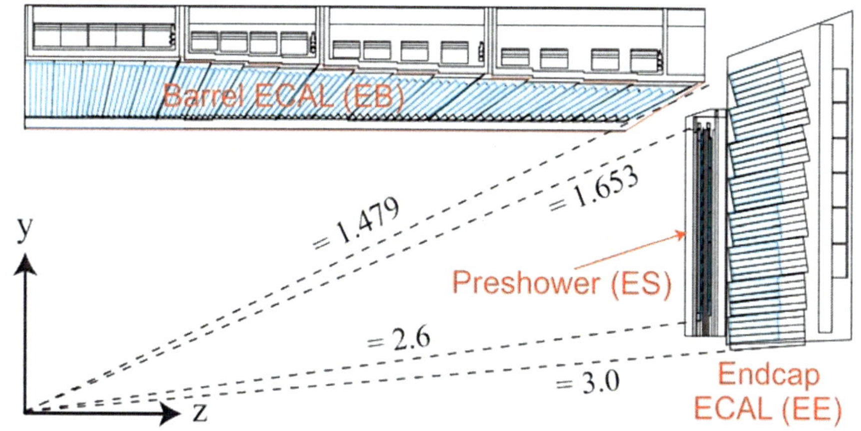

Fig. 2.13 Schematic illustration of the spatial coverage of an ECAL quarter. The three sub-components of the ECAL are shown: barrel, preshower and endcap systems. $PbWO_4$ crystals are depicted in blue. Figure taken from [31]

2 X_0 of the tracker, this depth ensures that most of the single-incident photons start showering within the sensors. The main purpose of the ES is to distinguish between di-photon pairs coming from the $\pi^0 \rightarrow \gamma\gamma$ decay and single high-energy forward photons. The choice of silicon is motivated by the larger radiation dose the detector has to sustain in this η region. As for the CMS tracker, one of the most critical aspects of silicon detectors is the operational temperature. The optimal performance of the ES is achieved at a working temperature between -15 °C and -10°C. The ES provides a fundamental handle in the reconstruction of electromagnetic showers in the event, as approximately 6% to 8% of the shower energy is deposited in this sub-detector.

The CMS ECAL intrinsic energy resolution was measured on a 3×3 crystals matrix in a test beam environment [32] and it is parametrized as

$$\left(\frac{\sigma_E}{E}\right)^2 = \left(\frac{2.8\%}{\sqrt{E}}\right)^2 + \left(\frac{12\%}{E}\right)^2 + (0.3\%)^2$$

where the first term, also called *stochastic*, represents the statistical fluctuations related to the physical development of the shower. The second term represents the electronic *noise* of the readout chain and it strictly depends on the DAQ system. Being it inversely proportional to the shower energy, it may become dominant at energies lower than the GeV scale. The *constant* term is energy independent and represents an intrinsic property of the calorimeter due to non-uniformities in the detector response, but also energy leakage(s) and residual miscalibrations.

Being the CMS ECAL a homogeneous calorimeter, it has the advantage of providing a more precise and complete energy reconstruction with respect to a sampling calorimeter. On the other hand, such a calorimeter does not allow for precise shower shape measurements, nor does it provide additional tracking points to those obtained from the pixel and strips in the tracker. Besides this design choice, the main limitation of the CMS ECAL is the fact that it was conceived to sustain luminosities up to 500 fb^{-1}, after which the crystals will suffer severe radiation damage. Since the CMS detector is expected to integrate about 350 fb^{-1} by the end of the LHC Run-III (cf. Fig. 2.3), it is clear that the CMS ECAL will have to be upgraded in view of the High-Luminosity LHC (HL-LHC) project. The endcap calorimeters are intended to be replaced with a High Granularity Calorimeter (HGCAL) during the HL-LHC upgrade campaign. The HGCAL will be a sampling calorimeter, thus allowing transverse and longitudinal shower shape measurements. In addition, it will be the first calorimeter providing a precise timing information. Due to these many features, the HGCAL is often referred to as 5D, or *imaging calorimeter*. A detailed description of the HGCAL, along with the study of its expected performance in a test beam environment, is given in Chap. 3.

Hadronic Calorimeter (HCAL)

Hadronic showers deposit at most 30% of their energy in the ECAL. The CMS hadronic calorimeter (HCAL) [33] is placed immediately after the ECAL and it is designed to fully absorb hadronic-induced cascades. The HCAL plays a fundamental role in the reconstruction of events with hadronic jets, especially in the measurement of their neutral components, which require the usage of a huge amount of material to be completely absorbed. For this reason, the HCAL is mostly a sampling detector with alternating layers of brass absorber and plastic scintillator tiles. Directly after the ECAL and within the volume of the solenoid there is the HCAL barrel (HB), which extends the EB in the barrel region, as of the EE is followed by the HCAL endcap (HE) system. The magnet sets an important constraint in the design of the HCAL, as the total depth of interaction lengths of about 10 λ_i up to the HB is not enough to contain completely hadronic showers. Hence, the barrel component of the HCAL is instrumented with an additional outer hadron calorimeter (HO), placed just outside the solenoid, which adds up for a total depth of 11 λ_i, as shown in Fig. 2.14, thus ensuring an better absorption of hadronic jets.

The HCAL is supplemented with a very forward calorimeter (HF), placed just outside the CMS detector at 11.2 m from the interaction point, thus ensuring a high-η coverage. A schematic view of the CMS HCAL is shown in Fig. 2.15.

- **HB and HE:** The HB and HE are sampling calorimeters consisting of brass absorbers and plastic scintillator tiles. The former has a coverage of $|\eta|<1.3$ and adds up with the previous sub-detectors for a total depth of about 7 λ_i, while the latter covers the region $1.3 <|\eta|< 3$ and reaches a maximum depth of 10 λ_i. The HB is organized into two halves, each composed of 18 identical wedges where plastic scintillator trays are placed in between absorber layers and divided into 16 η sectors. A total coverage of 72 sectors in ϕ is achieved with an additional split of the wedges into four parts, for a high granularity of 0.087×0.087 in $\eta - \phi$. The

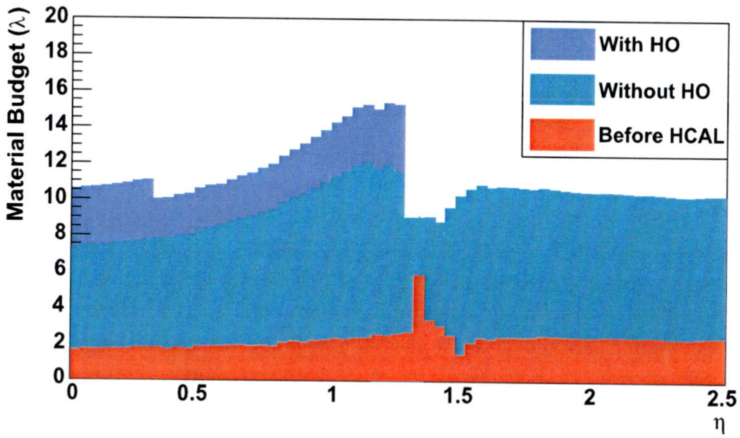

Fig. 2.14 Material budget of the CMS HCAL with and without the HO. The introduction of the HO brings the material budget up to 11 λ_i everywhere in η, thus ensuring the complete absorption of hadronic showers. Figure taken from [33]

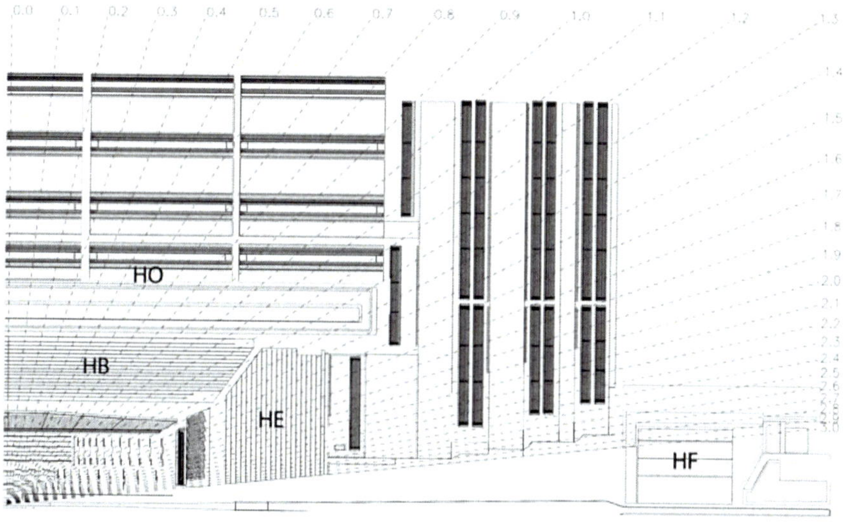

Fig. 2.15 Illustration of a HCAL quarter: the sub-modules that compose the detector are indicated, along with their coverage in pseudorapidity. Figure taken from [33]

scintillator tiles of different layers, located at the same $\eta - \phi$ coordinates, are grouped into stacks pointing towards the IP, also referred to as HCAL *projective* towers. The scintillating light produced by the hadronic interactions is extracted using thin wavelength shifting fibers (WLS) of 1 mm in diameter and it is carried out to the readout electronics to be digitized and read by means of hybrid photodiodes (HPD).
The HE follows a design close to the one of the barrel system, it extends the EE in the endcaps and it is shaped to fit perfectly in the space left by the HB, partially overlapping its last tower to avoid any gap. It is organised into 18 brass layers matching the HB ϕ segments and it covers about 0.17×0.17 in the $\eta - \phi$ plane.

- **HO:** As mentioned above, the total budget material from the interaction point up to the solenoid is about 10 λ_i and it does not guarantee a full containment of the hadronic showers. Hence, the HCAL is supplemented with an additional outer hadron calorimeter (HO), consisting of a single plastic scintillator in the outer barrel region ($0.4<|\eta|<1.26$) and a two modules in the central one. The HO uses the solenoid coil as dead material and ensures a full shower containment, reaching a depth of $\sim$11 λ_i (cf. Fig. 2.14).
The scintillating light readout, in the original design similar to the one of HB and HE, was updated during the LS1 when the old HPDs were replaced by silicon multipliers (SiPM) to profit from a faster response and to improve the physics performance [34].
- **HF:** The HCAL sub-system is completed by the presence of a very forward calorimeter (HF) placed at $z = \pm 11.2$ m from the interaction point and ensuring a coverage up to $|\eta| = 5.2$. The HF is a sampling calorimeter made of steel absorber alternating with quartz fibers as active material. The choice of this radiation-hard materials is dictated by the large radiation doses and neutron fluxes that the HF has to sustain at such extreme η regions. The scintillating light is produced in the quartz fibers by means of Cherenkov effect and it is collected by photomultiplier tubes (PMTs) placed just behind the absorber.

The measurement of hadronic jets is an extremely challenging task, which can not be accomplished by the HCAL alone. In fact, the presence of non-Poissonian events in the development of hadronic showers, the possible production of undetectable particles, and electromagnetic components in the showers are limiting factors on the HCAL performance. The HCAL and ECAL combined energy resolution is measured to be

$$\left(\frac{\sigma_E}{E}\right)^2 = \left(\frac{84.7\%}{\sqrt{E}}\right)^2 + (7.4\%)^2,$$

and it can be enhanced only combining the HCAL information with the one measured in the other sub-detectors, by means of particle flow reconstruction techniques, as further discussed in Sect. 2.3.

In order to improve the HCAL performance many of its components were upgraded at the end of 2017, similarly to what was done with the HO readout system. SiPM were found to be ideal replacements for many of the HPDs and to allow a more granular depth segmentation of the HE and HB calorimeters. In addition to the finer longitudinal segmentation, an upgrade of the front-end electronics allowed precision timing measurement, thus providing an additional handle for the particle-flow reconstruction of hadronic jets.

Muon Detectors

The CMS detector is designed to identify and reconstruct muons precisely, as it goes by its name. The typical interactions at the LHC can produce muons with energies ranging from few up to hundreds of GeV, which travel the detector leaving only a mild signature in the tracker and the following sub-systems. However, identifying them and measuring their charge and momentum with high precision is a key handle for the analysis of events where W, Z, or H bosons are involved, as well as for the study of new physics signatures.

Hence, the muons detection system [35] is placed in the outermost region of the CMS detector, it is organized into four stations embedded in the iron yokes, also called *wheels*, and it exploits the 2 T return magnetic field to infer particle momenta from their curvature. The detector design is dictated by a compromise between the large surface to be covered and the costs to be minimized. The CMS muons detection system is composed of three different technologies of gaseous detectors, for a total of 1400 chambers placed according to the expected background rates and the magnetic field intensity: Drift Tubes (DTs) cover the central region up to $|\eta|<1.2$, while Cathode Strip Chambers (CSCs) in the endcaps region $0.9<|\eta|<2.4$ are supplemented with Resistive Plate Chambers (RPCs) covering up to $|\eta|<1.6$ for triggering and redundancy purposes. All these elements add up for a total of about $25000\,m^2$ detection planes and reach a depth of 20 λ_i at low values of pseudorapidity. A schematic view of the CMS muons detection system is depicted in Fig. 2.16.

- **DT:** 250 rectangular ($2\times2.5\,m^2$) Drift Tubes (DTs) are arranged into groups of four stations and divided into five *wheels* along the entire barrel region, three of which shown in Fig. 2.16. They are located in the central region of the detector, providing a coverage up to $|\eta|<1.2$. The low muons rate and the magnetic field in this part of the detector ensure an adequate performance even with the use of standard rectangular drift cells. A cell is a 2.4 m long drift tube of $1.3\times4.2\,cm^2$ surface, consisting of an anode wire and two cathode strips, immersed in a mixture of Ar (85%) and CO_2 (15%). DTs provide a two dimensional measurement by taking the drift time of knocked-off electrons to the anode within a cell. Besides giving a precise information of the angle and position of traversing muons, DTs are also used to fire muon triggers without the need of any information from the tracker. The DTs layers are stacked one on top of the others with an offset of half cell width, resulting in a time resolution lower than 3 ns, a single cell efficiency of 99.8%, and a spatial resolution per-cell of about 180 μm, corresponding to an overall position resolution of 80–120 μm.

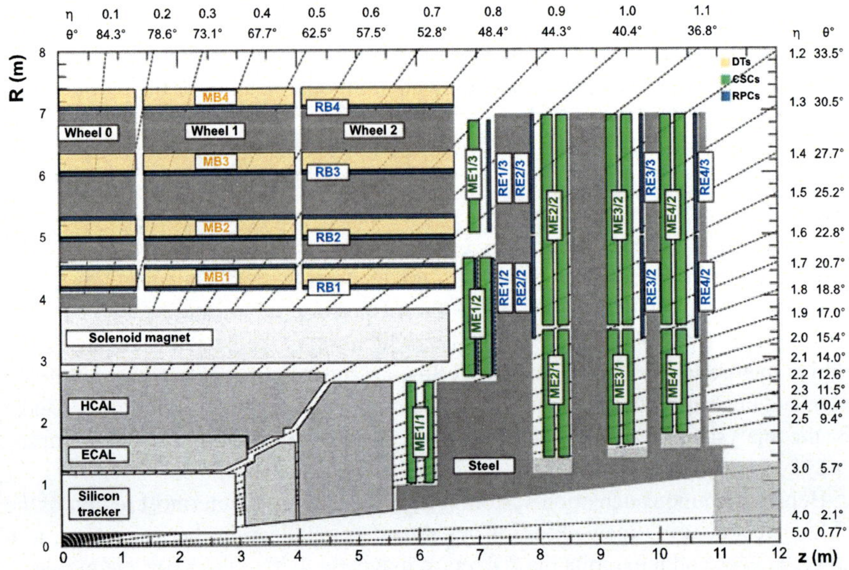

Fig. 2.16 Schematic longitudinal view of a quarter of the CMS detector with the layout of the muon detection systems. DTs, CSCs and RPCs are represented in orange, green, and blue, respectively. Figure taken from [35]

- **CSC:** The endcap region features a higher background rate and a stronger, non-uniform, magnetic field with respect to the barrel, as shown in Fig. 2.8. Hence the requirement of a fast and radiation-hard detector, accommodated by the choice of finely segmented multi wire proportional cathode strip chambers (MWP CSCs) to detect muons in the endcaps at $0.9 <| \eta |< 2.4$. The CSCs have a trapezoidal shape, covering an angle of 10° or 20°, and are grouped into four stations placed between the iron yoke. They are disposed into alternating layers of anode wires and cathode strips: the former provides a measurement in η, while the r, ϕ coordinates are extracted from the latter. The CSCs alone cover an area of about $5000\,\text{m}^2$ with a volume of $50\,\text{m}^3$, composed by a mixture of Ar (45%), CO_2 (50%) and CF_4 (10%). The choice of CSCs comes with several advantages: the closely spaced wires provide a fine segmentation, which allows to get down to a spatial resolution of 40–150 μm and a precise timing resolution of about 3 ns. The fast response of CSCs also makes this detector suitable for triggering purposes.
- **RPC:** Resistive plate chambers (RPCs) support both the DTs and the CSCs with a standalone triggering system and are capable of resolving tracking ambiguities in events with multiple hits in a certain cell. RPCs are fast detectors, mounted both in the barrel and the endcaps, that provide additional redundancy to the measurements of the two other sub-systems. In the barrel RPCs are mounted on the inner and outer side of the first two DT stations and in front of the drift tube chamber in the last two stations. This arrangement ensures a good triggering performance also for

low p_T muons. In the endcaps we find one RPC plane per station. Initially both the endcap's RPC and CSC were mounted on three stations, but as of LS1 both were upgraded to their baseline design with four stations.

RPCs consist of a double-gap chamber operated in avalanche mode and filled with a mixture of $C_2H_2F_4$ (96.2%), i-C_4H_10 (3.5%) and SF_6 (0.3%). Despite their relatively poor spatial resolution (~1 cm), RPCs have a faster response than DTs and CSCs, reaching a time resolution below 3 ns and being a fundamental handle in triggering muons in events with high pileup rate.

As of the end of 2017, the muons detection system has been extended with gas electron multiplier detectors (GEMs) in the region $1.6 <| \eta |< 2.2$. GEMs are ideal to increase the redundancy of the system in the endcaps, as they feature an excellent rate capability and an optimal radiation hardness. The 10 prototype chambers installed in 2017 are going to be supplemented with additional GEMs during the LS2, to meet the design value of 144 detectors in view of the HL-LHC program.

2.2.3 Luminosity Measurement

Precision measurement of both the integrated and the instantaneous luminosities is crucial for the online operations of CMS and the offline physics analyses. For example, the precise feedback on the online luminosity is used directly by the DAQ and trigger systems to regulate the beam intensity. At the same time, the final number of reconstructed events scales directly with the experimental luminosity (cf. Eq. 2.1).

The HF is used together with two dedicated luminometers for the online measurement: the Fast Beam Conditions Monitor (BCM1f) [36] and the more recent Pixel Luminosity Telescope (PLT), installed before Run-II to reduce the systematic uncertainty on the luminosity measurement [37–39]. The combined information of these three detectors provides in situ feedback about the integrated luminosity. The offline measurement is more accurate and considers the information of the pixel tracker and the DTs. These detectors are supplemented with a dedicated readout system to speed up the operations.

The measurement of the integrated luminosity has been performed for each data taking period separately, and it has an associated systematic uncertainty that ranges between 2.3% and 2.5%, depending on the year. The impact of this systematic uncertainty on the results of the analysis and the correlations associated to the luminosity uncertainties are discussed in detail in Chap. 6.

2.2.4 Trigger and Data Acquisition Systems

When a collision occurs at the interaction point, one has to quickly decide whether or not to save the resulting event for the offline analysis. With a pp interaction rate of

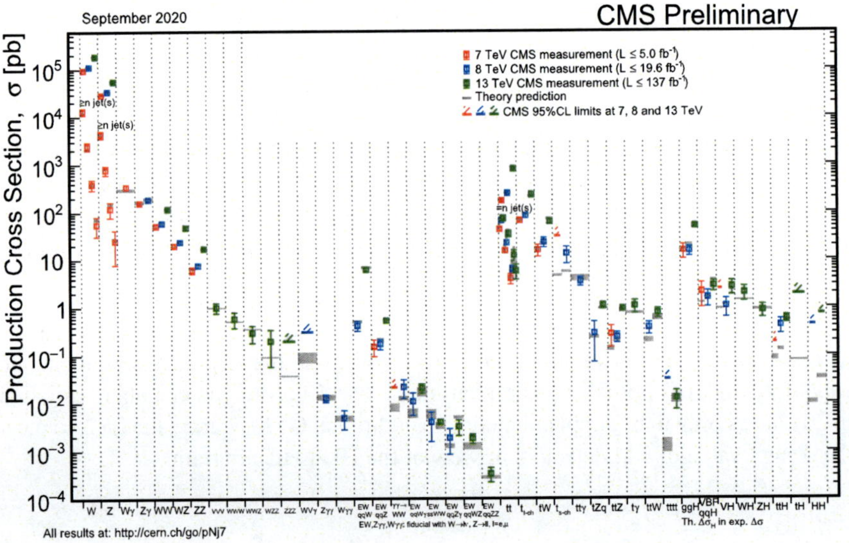

Fig. 2.17 Summary of the CMS the SM cross sections measurements at $\sqrt{s} = 7,\ 8$, and 13 TeV and their corresponding theoretical predictions

40 MHz and full detector information taking about 1 Mb of disk space per event, one of the main challenges for the LHC experiments is the permanent storage of such a huge amount of data. Most of the events will be discarded and, in a fraction of a second, a decision about the importance of a given event is taken. The CMS detector demands these tasks to the trigger and data acquisition systems (TriDAS) [40], which have to reduce the information rate of a factor 10^6: from the MHz produced within a pp collision, down to the Hz that can be saved on disk for the subsequent reconstruction processes and for the offline analyses. To make an idea, Fig. 2.17 presents a summary of the SM production cross sections measured at CMS. The total proton-proton interaction cross section is $\sigma_{\mathrm{pp}} = 10^{11}$ pb [41]: at least six orders of magnitude larger than any of the processes reported in Fig. 2.17.

Hence, the TriDAS system serves as a "bridge" between the online data acquisition and the offline data analysis. Decisions are taken with a two-level trigger architecture:

- **L1 trigger:** The first skim of the events is demanded to the *Level-1 trigger*, which has to reduce the relevant information up to 100 kHz, within a *latency* time of 3.8 μs. Having to take fast decisions in such a short time, the L1 trigger is mounted on low level hardware such as Field Programmable Gate Arrays (FPGAs) and Application Specific Integrated Circuits (ASICs) to access the information of the various CMS sub-detectors directly. Coarsely segmented and low-resolution physics objects from the calorimeters and the muon chambers are used to produce the so-called "L1 candidates." The ECAL and HCAL information is clustered, as a function of the energy deposit, into groups, also referred to as "trigger towers", representing electrons and photons, jets, and taus. On the other hand, muons are

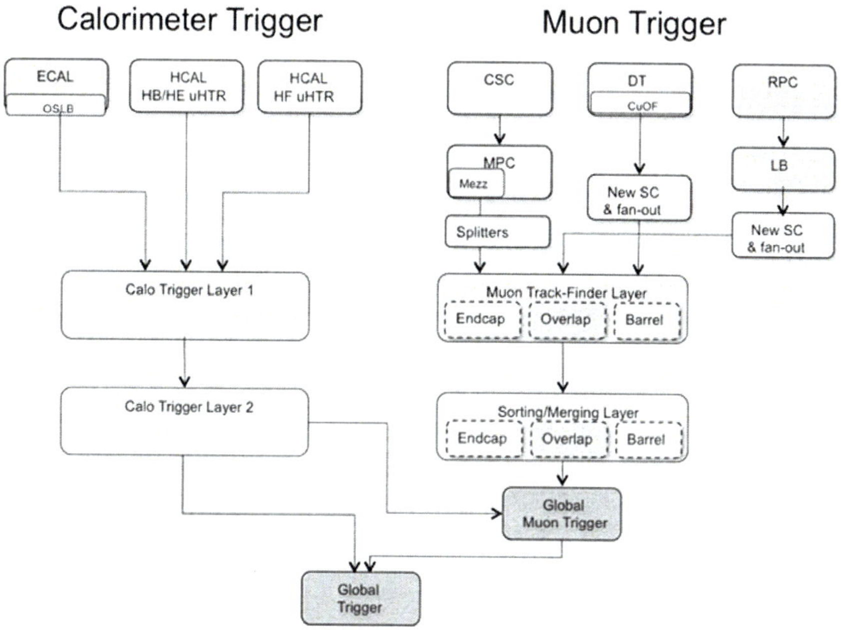

Fig. 2.18 Data flow of the upgraded L1 trigger system. The muon and calorimeter triggers work independently: the former creates L1 muon candidates combining the information from CSC, RPC, and DT, while the latter creates L1 objects for electrons and photons, as well as for jets and taus, as explained in the text. Figure taken from [40]

identified by reconstructing the hits in the DTs, CSCs, and RPCs. The structure of the L1 trigger system allows the muon and calorimeter parts to exchange information with each other to enhance the objects' identification. The overall data flow, along with the interconnections of the various components, is shown in Fig. 2.18. The Calo and the Global Muon Triggers are eventually combined into a Global Trigger, used by the L1 trigger system to accept or reject the event. An L1 accept (L1A) signal is ultimately passed to the front-end electronics and DAQ system for the acquisition and reconstruction processes.
As for many of the sub-detectors aforementioned, also the L1 trigger system was upgraded for the Run-II phase of the LHC [40]. Both the calorimeter and muon triggers were replaced by the use of new-generation FPGAs and fast optical connections, resulting in the current architecture sketched in Fig. 2.18.

- **DAQ and HLT:** Events marked with a L1A signal are streamed to the High Level Trigger (HLT) computing farm, where a further selection of the events is performed. The HLT reduces the information rate from 100 kHz down to around 1 kHz,[1] stores, and processes it within a 32000 CPU cores computing farm located

[1] With this upper limit being set by the maximum speed of writing an event to permanent storage.

on top of Point 5. The HLT exploits the full detector information and granularity to perform an online reconstruction of the various objects present in the event, classifying them into Primary Datasets (PD) according to the trigger fired. The HLT menu comprises more than 600 "paths", i.e. combinations of reconstruction and filtering instructions with increasing complexity, each selecting a particular physics signature. Trigger paths usually accept the event by performing subsequent steps of increasing complexity: at first a limited reconstruction and a pre-filtering of the events is performed; then, only for the more interesting ones, a more detailed reconstruction, which takes into account the tracking information to match the objects, is ran and the final trigger decision is taken. Figure 2.19 shows an example of bandwidth occupancy from the various CMS analyses during the 2018 data taking period.

The crucial importance of the HLT and its impact on the physics analyses, make it one of the CMS components regarded with most care: the several HLT algorithms are constantly revised and improved to keep up with the upgrades of the different sub-detectors, as it happened with the 2017 trigger upgrade. With the beginning of the Run-II phase of the LHC, new data taking strategies have also been introduced to cope with the evolving exigencies of the analyses. The "data scouting" technique largely exploits thc close relation between the HLT and the offline objects to save directly the former at a rate of 1–5 kHz and use them directly in the analyses. A second technique goes under the name of "parking" and it consists of streaming events directly to tape without running any reconstruction, with the transcription rate being ultimately limited by the DAQ and disk storage only. The "parked" events, consisting of complete raw data, can be recovered for an *a posteriori* reconstruction and analysis, in case any deviation from the SM physics is observed and an improved sensitivity is required.

Events passing the trigger criteria are eventually streamed to the Tier-0 at CERN and prepared for the offline reconstruction and analysis.

The analysis presented in this thesis makes use of the data samples recorded during the 2016–2018 data taking periods, corresponding to the 137 fb^{-1} of data collected by the CMS experiment during the Run-II. The Run-II datasets are summarized in Appendix A.

The HLT paths used to select the events of the analysis define four different PDs: *DoubleMuon*, *MuonEG*, *EGamma*[2], and *SingleMuon*.

To avoid duplicate events from different primary datasets, events are taken:

- from EGamma if they pass the diEle or triEle or singleElectron triggers;
- from DoubleMuon if they pass the diMuon or triMuon triggers and fail the diEle and triEle triggers;
- from MuEG if they pass the MuEle or MuDiEle or DiMuEle triggers and fail the diEle, triEle, singleElectron, diMuon and triMuon triggers;
- from SingleMuon if they pass the singleMuon trigger and fail all the above triggers.

[2] For 2016 and 2016 datasets the *DoubleEG* and *SingleElectron* PDs are used.

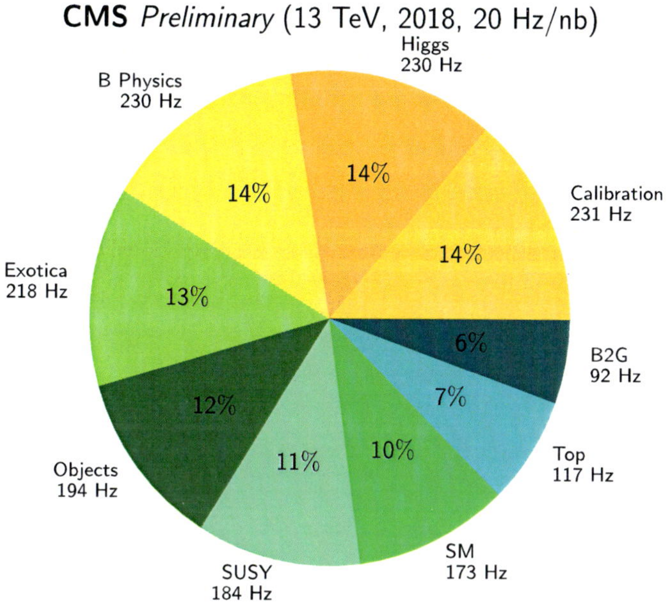

Fig. 2.19 HLT bandwidth occupancy for the different analyses in a typical configuration during the 2018 data taking period

The HLT paths used for 2016–2018 collision data are listed in Tables 2.2, 2.3 and 2.4, respectively. The efficiency in data of the combination of triggers used in the analysis with respect to the offline reconstruction and selection is measured by considering 4ℓ events triggered by single lepton triggers. An overall trigger efficiency measured on simulation is found to be larger than 99% for each final state.

2.3 Physics Objects: Identification and Reconstruction

The products of the hard scattering processes happening at the LHC are recorded by the different sub-modules of the CMS detector: particles enter the Si tracker and leave signals (*hits*) in the active layers; these are used to reconstruct their trajectories (*tracks*) and identify the position of secondary interaction origins (*vertices*). Electrons and photons continue their path and are absorbed within the ECAL volume. The direction and the energy of electromagnetic showers can be inferred from the energy deposit in neighbouring ECAL cells, often grouped into *clusters*. Charged and neutral hadrons follow a more complicated path due to the more complex nature of hadronic showers, which can initiate either in the ECAL or the HCAL volumes. Similar to what happens for electrons and photons, also in this case, the shower's energy and direction are inferred from clustered cells. Muons and neutrinos can continue their

Table 2.2 Trigger paths used in 2016 collision data

HLT path	Prescale	Primary dataset
HLT_Ele17_Ele12_CaloIdL_TrackIdL_IsoVL_DZ	1	DoubleEG
HLT_Ele23_Ele12_CaloIdL_TrackIdL_IsoVL_DZ	1	DoubleEG
HLT_DoubleEle33_CaloIdL_GsfTrkIdVL	1	DoubleEG
HLT_Ele16_Ele12_Ele8_CaloIdL_TrackIdL	1	DoubleEG
HLT_Mu17_TrkIsoVVL_Mu8_TrkIsoVVL	1	DoubleMuon
HLT_Mu17_TrkIsoVVL_TkMu8_TrkIsoVVL	1	DoubleMuon
HLT_TripleMu_12_10_5	1	DoubleMuon
HLT_Mu8_TrkIsoVVL_Ele17_CaloIdL_TrackIdL_IsoVL	1	MuonEG
HLT_Mu8_TrkIsoVVL_Ele23_CaloIdL_TrackIdL_IsoVL	1	MuonEG
HLT_Mu17_TrkIsoVVL_Ele12_CaloIdL_TrackIdL_IsoVL	1	MuonEG
HLT_Mu23_TrkIsoVVL_Ele12_CaloIdL_TrackIdL_IsoVL	1	MuonEG
HLT_Mu23_TrkIsoVVL_Ele8_CaloIdL_TrackIdL_IsoVL	1	MuonEG
HLT_Mu8_DiEle12_CaloIdL_TrackIdL	1	MuonEG
HLT_DiMu9_Ele9_CaloIdL_TrackIdL	1	MuonEG
HLT_Ele25_eta2p1_WPTight	1	SingleElectron
HLT_Ele27_WPTight	1	SingleElectron
HLT_Ele27_eta2p1_WPLoose_Gsf	1	SingleElectron
HLT_IsoMu20 OR HLT_IsoTkMu20	1	SingleMuon
HLT_IsoMu22 OR HLT_IsoTkMu22	1	SingleMuon

path beyond the superconducting solenoid: the former are detected by the muon chambers placed up to 7 m away from the IP, while the latter leave the detector with no signature whatsoever. A schematic view of the CMS detector is shown in Fig. 2.20, along with the typical signature of the various particles aforementioned.

The L1 and HLT take rapid decisions on whether to discard an event or not: selected events are streamed to the Tier-0 and saved in the form of *RAW* Data. At this stage, an event takes up to 2 Mb of space, mainly because of the triggers instructions being saved. Triggers only use coarse information from the calorimeters and the

Table 2.3 Trigger paths used in 2017 collision data

HLT path	Prescale	Primary dataset
HLT_Ele23_Ele12_CaloIdL_TrackIdL_IsoVL_*	1	DoubleEG
HLT_DoubleEle33_CaloIdL_GsfTrkIdVL	1	DoubleEG
HLT_Ele16_Ele12_Ele8_CaloIdL_TrackIdL	1	DoubleEG
HLT_Mu17_TrkIsoVVL_Mu8_TrkIsoVVL_DZ_Mass3p8	1	DoubleMuon
HLT_Mu17_TrkIsoVVL_Mu8_TrkIsoVVL_DZ_Mass8	1	DoubleMuon
HLT_TripleMu_12_10_5	1	DoubleMuon
HLT_TripleMu_10_5_5_D2	1	DoubleMuon
HLT_Mu23_TrkIsoVVL_Ele12_CaloIdL_TrackIdL_IsoVL	1	MuonEG
HLT_Mu8_TrkIsoVVL_Ele23_CaloIdL_TrackIdL_IsoVL_DZ	1	MuonEG
HLT_Mu12_TrkIsoVVL_Ele23_CaloIdL_TrackIdL_IsoVL_DZ	1	MuonEG
HLT_Mu23_TrkIsoVVL_Ele12_CaloIdL_TrackIdL_IsoVL_DZ	1	MuonEG
HLT_DiMu9_Ele9_CaloIdL_TrackIdL_DZ	1	MuonEG
HLT_Mu8_DiEle12_CaloIdL_TrackIdL	1	MuonEG
HLT_Mu8_DiEle12_CaloIdL_TrackIdL_DZ	1	MuonEG
HLT_Ele35_WPTight_Gsf_v*	1	SingleElectron
HLT_Ele38_WPTight_Gsf_v*	1	SingleElectron
HLT_Ele40_WPTight_Gsf_v*	1	SingleElectron
HLT_IsoMu27	1	SingleMuon

muon chambers without relying on the tracker. The information of the different CMS sub-modules is combined to produce a set of reconstructed objects that can be used in the physics analyses to identify and reconstruct all the final state particles. The final data format is called *miniAOD*, and it contains only the most relevant object variables in 40 times less space than the original *RECO* data. Table 2.5 shows the list of the different formats used for processing CMS data, along with the typical disk size required to save events of a given format.

This approach of merging *tracks* and *clusters* to obtain the best description of all the final state objects in the event goes under the name of *Particle-Flow* reconstruction

Table 2.4 Trigger paths used in 2018 collision data

HLT path	Prescale	Primary dataset
HLT_Ele23_Ele12_CaloIdL_TrackIdL_IsoVL_v*	1	DoubleEG
HLT_DoubleEle25_CaloIdL_MW_v*	1	DoubleEG
HLT_Mu17_TrkIsoVVL_Mu8_TrkIsoVVL_DZ_Mass3p8_v*	1	DoubleMuon
HLT_Mu23_TrkIsoVVL_Ele12_CaloIdL_TrackIdL_IsoVL_v*	1	MuonEG
HLT_Mu8_TrkIsoVVL_Ele23_CaloIdL_TrackIdL_IsoVL_DZ_v*	1	MuonEG
HLT_Mu12_TrkIsoVVL_Ele23_CaloIdL_TrackIdL_IsoVL_DZ_v*	1	MuonEG
HLT_DiMu9_Ele9_CaloIdL_TrackIdL_DZ_v*	1	MuonEG
HLT_Ele32_WPTight_Gsf_v*	1	SingleElectron
HLT_IsoMu24_v*	1	SingleMuon

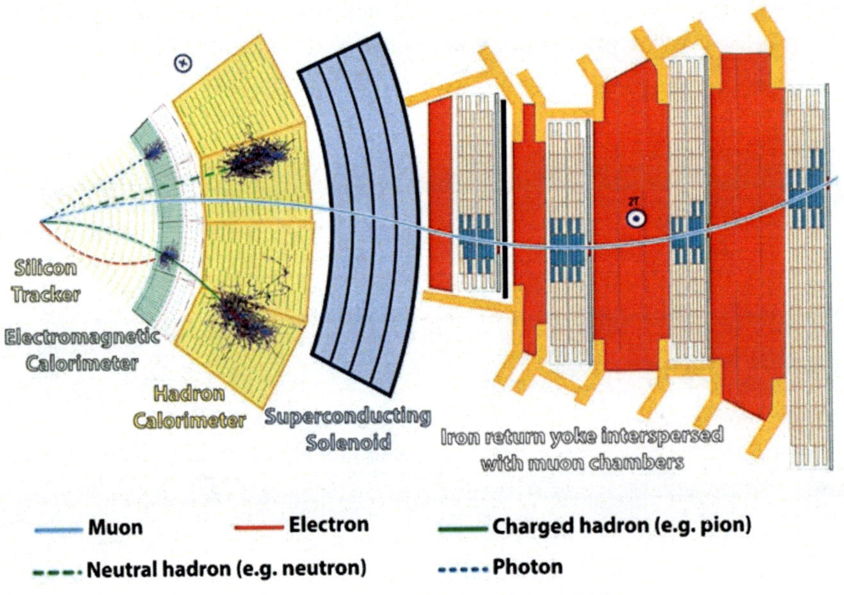

Fig. 2.20 Schematic representation of a transverse slice of the CMS detector with the typical signatures of the different kind of particles detected. Figure taken from [42]

Table 2.5 The second column shows the different data formats of the CMS preprocessing phase. The purpose of each of them is explained in the third column, while the last column indicates the typical memory disk usage needed to store an event

Application	Data format	Purpose	Mb/evt
MC	*GEN*	Output of the MC generator. Full description of the event, including weights and model parameters	O(0)
	SIM	Energy deposits from the GEANT4-based simulated response of the detector	2
	DIGI	MC level detector response, similar to the real detector output	2
Data/MC	*RAW*	Events passing L1 and HLT trigger criteria. Include the several variables used for trigger decisions	2
	RECO	Reconstructed hits and clusters: contains the information of tracks, vertices, jets, and leptons	2
	AOD	Analysis Object Data: reduced set of RECO information that can be already used for offline processing and analysis	0.5
	miniAOD	Skimmed version of the AOD format, contains only relevant information	0.05

algorithm [42]. Conventionally PF physics objects are classified based on the signals collected by a given sub-module:

- *jets*: collection of hadrons and photons. These objects can be reconstructed without needing the tracker information, as the energy of the individual jet particles can be inclusively measured by the calorimeters. The same holds for the *missing transverse momentum* ($p_{\mathrm{T}}^{\mathrm{miss}}$) reconstruction;
- *EGamma*: isolated photons and electrons (e/γ object) are reconstructed exploiting primarily the ECAL. Additional information from the tracker is also used to enhance the reconstruction efficiency, as further described in Sect. 2.3.2;
- *jets tagging*: tagging exploits the tracker information to discriminate between hadronic τ and b-quark jets;
- *muons*: primarily based on the information from the muon chambers.

The PF reconstruction algorithm was initially conceived for e^+e^- colliders: originally developed for the ALEPH experiment at LEP, is now the pillar at the basis of the design of future lepton colliders, such as ILC and CLIC, FCC-ee, and CEPC. This new generation of colliders features a fine spatial granularity: an ideal property when it comes to the application of the PF algorithm. The CMS Collaboration adopted for the first time the PF at a hadron collider [42], breaking the barriers of what was thought to be one of the limits of this approach. As shown in Fig. 2.20 and as described in detail in Sect. 2.2, the CMS detector is granular enough to provide a complete discrimination between particles of different nature. This information is combined together with the signal in all the sub-detectors to provide a comprehensive description of the final state detected. The replacement of the current CMS endcap calorimeters with a High Granularity CALorimeter (HGCAL), further discussed in Chap. 3, is designed with the goal of reducing the busy environment typical of the HL-LHC collisions exploiting a PF-based reconstruction: the high granularity of the detector make it ideal for particles discrimination and identification, providing an additional handle for pile-up mitigation and events selection in the forward region.

2.3.1 Fundamental Elements of the Particle-Flow Algorithm

The PF algorithm [42] relies on the creation of *tracks* and *clusters* used as input for the reconstruction of all the final state physics objects in the event.

- **Tracks:** An iterative approach based on the combinatorial track finder (CTF) algorithm exploits the Kalman Filtering (KF) [43] to reconstruct charged particles' tracks. Initially only tracks believed to have originated from the primary vertex are reconstructed and the corresponding hits are not considered in the subsequent stages of the tracking process [44]. The reconstruction of more complex tracks is performed in successive steps, for a total of about 12 iterations performed to ensure high efficiency and low fakes rate [45]. Tracks of charged particles with p_T as low as 0.1 GeV can be reconstructed.
- **Clusters:** Neighbouring cells in the ECAL and in the HCAL are grouped to form energy clusters. As for the tracking, a high efficiency is required for PF clusters: for this reason the algorithm is run independently on the preshower, the ECAL, and the HCAL. This approach is adopted also to have more discrimination power in disentangling overlapping showers. The local maxima of energy deposit (*cluster seeds*) are identified following the longitudinal shower profile. Neighbouring energy deposits are aggregated to form *topological clusters* if the corresponding cell signal exceeds 2σ of the electronic noise. Finally *PF clusters* are identified within a topological cluster by means of a Gaussian-mixture model[3] expectation-maximization algorithm.

Tracks and clusters are merged (*linked*) together to form *PF blocks*, in turn used to identify all the final state objects aggregating pairs of nearest neighbours, in the (η, ϕ) plane, according to their topology.

A practical example of how the PF algorithm works is given in Fig. 2.21. The image depicts a simulated jet, with transverse momentum of 65 GeV, composed of only five particles: two photons originating from a $\pi^0 \rightarrow \gamma\gamma$ decay, two charged hadrons (π^+ and π^-), and one neutral hadron (K^0_L). The two charged hadron tracks

[3] The Gaussian-mixture model states that a topological cluster composed of M individual cells can be seen as the combination of N Gaussian energy deposits, where N is the number of seeds.

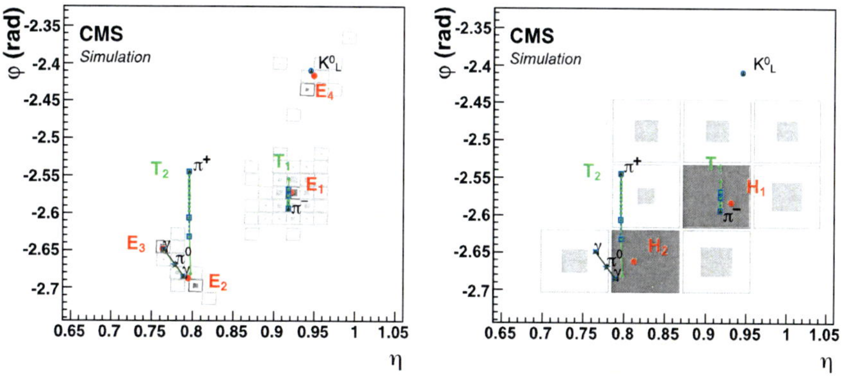

Fig. 2.21 Monte Carlo simulation of a 65 GeV jet made of five particles. Particle Flow tracks and clusters are represented in green and gray, respectively. Figure taken from [42]

are reconstructed as $T_{1,2}$, pointing towards the HCAL clusters $H_{1,2}$. The two photons, the neutral hadron, and one of the two charged π deposit their energy in the ECAL, as shown by the four well separated clusters $E_{1,2,3,4}$. One can identify a PF block corresponding to the two charged hadrons and three PF blocks in the ECAL, one for each cluster. To sum up, in each PF block the following identification and reconstruction steps are performed.

- Identification and reconstruction of **muon** tracks, using the information from the muon chambers, and consequent removal of the associated hit;
- identification and reconstruction of **electron** candidates, with energy recovery for bremsstrahlung photons, and of energetic and well **isolated photons**;
- **number of fakes** reduced by removing all those tracks with a bad χ^2;
- the presence of **muons within jets** and **fake tracks** would result in PF tracks' transverse momentum much larger than the corresponding clusters' energy sum. Well-identified muons are removed from the PF block and if the total p_T is still larger than the clusters energy sum, an additional fakes removal is performed by iteratively discarding tracks with momentum uncertainty $\sigma_{p_T} > 1$ GeV;
- the remaining tracks in the PF block are used to identify **charged hadrons**;
- **photons** and **neutral hadrons** are identified by looking at the excess of the clusters energy sums with respect to the sum of track momenta. A photon is created if the excess is smaller than the ECAL energy, a neutral hadron in the other cases;
- isolated clusters in the ECAL or HCAL, i.e. not linked to any track[4] are used to create PF photons and neutral hadrons, respectively.

The analysis presented in this thesis exploits the $H \to ZZ \to 4\ell$ decay channel and uses the full Run-II statistics collected by the CMS experiment. This decay channel has the cleanest signature among all the Higgs boson decays, as it is enough to reconstruct relatively simple physics objects: leptons. With such large statistics

[4] For example the $E_{1,2,3,4}$ clusters in Fig. 2.21.

available, the analysis can also probe the main production mechanisms of the Higgs boson, and in this case, jets bring much discrimination power. For these reasons, the following sections detail the reconstruction of these three physics objects, starting from the PF candidates built as described in Sect. 2.3.1.

2.3.2 Electrons

The reconstruction of electrons may become challenging when it comes to accounting for their interactions with the passive material encountered along their path, especially in the tracker (cf. Fig. 2.11). The primary source of energy loss is radiated bremsstrahlung photons, which can carry up to 86 % of the total energy at $\eta \simeq 1.4$, where the detector features the largest material budget. Consequently, electrons undergo deviations with respect to their original path, increasing the complexity in the reconstruction of their tracks. Eventually, some photons can convert to an e^+e^- pair, resulting in more populated showers.

Tracking

A dedicated tracking procedure has been developed to account for possible electrons' energy losses and consequent sudden changes in their path. Instead of the KF algorithm, electrons' tracking relies on a Gaussian Sum Filter (GSF) algorithm [46]. The GSF algorithm works with looser selection criteria than the KF, and it features a more significant number of components that allow sudden curvature changes in the tracks, thus enhancing the overall tracking efficiency. On the other hand, this method is limited by the computing power required, which can go up to a few hundred milliseconds per track. For this reason, the GSF tracking is used only on those seeds that most likely represent an electron.

The original electron seeding [47], referred to as the *ECAL-driven* approach, exploits energetic ECAL clusters with $E_T > 4$ GeV. To consider all the radiated energy, the most energetic ECAL cluster is merged with other nearest neighbouring cells to form a so-called *supercluster*, i.e., a group of ECAL clusters reconstructed in a window around the electron direction, covering a small η range but a larger one in ϕ. The tracks are projected from the ECAL back to the innermost layers of the trackers, and electron seeds are formed if pairs or triplets are matched.

The ECAL-driven approach can be completed to avoid missing particular topologies of electrons, such as those in jets and those at low transverse momentum. In the first case, the supercluster energy and position can be biased mainly by the presence of other particle deposits, whereas in the latter, tracks are bent significantly by the magnetic field, resulting in a comprehensive (η, ϕ) coverage that a single supercluster can not cover. A dedicated *tracking-driven* seeding was developed specifically for these electrons, in place of the less efficient ECAL-driven one. This method relies on tracks with $p_T > 2$ GeV, determined with the iterative tracking algorithm afore-

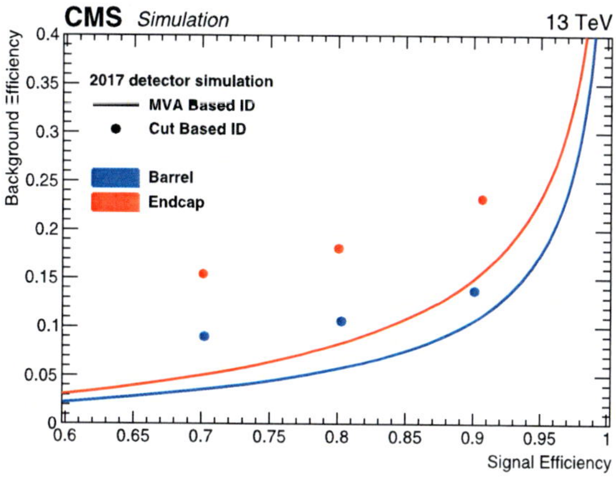

Fig. 2.22 Fakes identification efficiency as a function of the true electrons efficiency for both barrel (blue) and endcaps (red). Solid dots correspond to the three WPs derived using a cut-based approach. Figure taken from [42]

mentioned (cf. Sect. 2.3.1). Tracks are propagated to the ECAL surface and matched to the nearest ECAL cluster to form an electron seed if the ratio between the cluster energy and the track transverse momentum is compatible with unity. Electrons undergoing significant energy radiation are harder to track: the algorithm might be able to retrieve all the associated hits in the tracker, but the extrapolated track would have a sizeable χ^2 value. In these cases, after a preselection on the number of tracker hits, the tracks are fit again with a GSF algorithm with reduced a number of components. An electron seed is eventually formed based on the score of a boosted-decision-tree (BDT) classifier that takes into account the track parameters from both the KF and GSF fits, the electron's energy loss, and the $(\Delta\eta, \Delta\phi)$ distance between the ECAL cluster and the extrapolated track's position on the ECAL's inner surface. The score of the BDT is used to discriminate between actual electrons and *fakes*, i.e., misreconstructed pions. Three levels of identification (WP) are defined, similarly to what is done for muons: loose, medium, and tight, with increasing selection efficiency and purity. The receiver operating characteristic curve (ROC) of the BDT is depicted in Fig. 2.22, where the electrons identification efficiency is plotted against the fakes rate. The dots correspond to the three WP defined with a cut-based approach, now dropped in favour of the outperforming BDT.

A unique collection of electron seeds is formed by merging the ECAL- and tracker-driven seeds, yet saving the information on the origin of the seed.

The combination of the two sub-detectors brings a substantial improvement to the overall seeding efficiency, as shown in Fig. 2.23 for electrons in b quark jets.

The final electron track reconstruction is performed by a fit on the layer hits using a GSF method. The electron loss is modeled with a Bethe-Heitler function and loose requirements on the estimation of the hits position in each layer are used to take into account the effect of bremsstrahlung. Each track is described by the fit free parameters and the post-fit values define the *GSF tracks*.

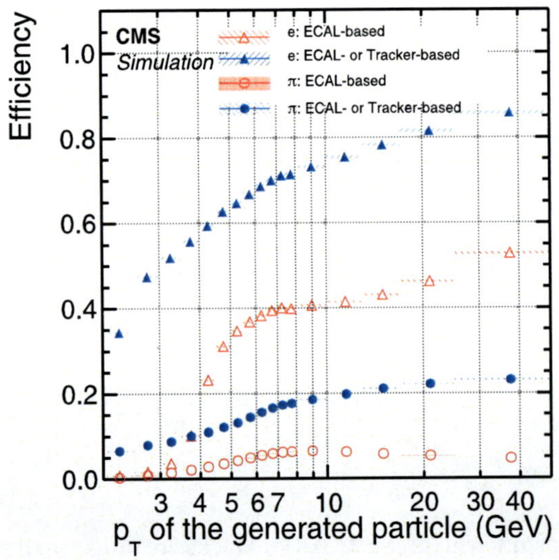

Fig. 2.23 Comparison between ECAL- (hollow symbols) and tracker-driven seeding (solid symbols) efficiencies for electrons (triangles) and pions (circles), as a function of p_T. Figure taken from [42]

Reconstruction

Electrons and photons have similar behaviour in the ECAL. Thus, the underlying principle used for their reconstruction is similar, with the additional handle of the tracker information for the case of electrons. When dealing with these objects, the main challenge is the presence of the bremsstrahlung effect: due to a large amount of material present in the tracker, electrons probably radiate photons traversing the layers, while photons can emit e^+e^- pairs, with the subsequent emission of bremsstrahlung photons. In turn, it becomes more complex to link electromagnetic signatures in the tracker and the ECAL.

An electron candidate is seeded from a GSF track only if the corresponding ECAL cluster is not linked with more than three tracks. The response of the HCAL is used to discriminate between electromagnetic and hadronic showers, as the candidate is neglected if the HCAL response exceeds 10% of the ECAL supercluster. Nevertheless, in some cases, the ECAL clusters might be too displaced from the corresponding GSF track, mainly because of the bremsstrahlung effect mentioned above. Hence, ECAL clusters are associated with the electron candidate if they are linked to the tangent to the GSF track at a given tracker layer, as schematically illustrated in Fig. 2.24. ECAL superclusters are supplemented with compatible bremsstrahlung clusters in order to define a *refined supercluster*, which is used to build the final electron candidate and its GSF track.

As shown in Fig. 2.24, the main GSF track and the main ECAL supercluster can have a large separation, which may lead to an incorrect estimate of the particles' charge. The sign of the GSF track's curvature is used to extract the charge, neglecting early bremsstrahlung emissions: for this reason two additional measurements are used as well. If a KF track (cf. Sect. 2.2.2) was built together with the GSF one, the charge

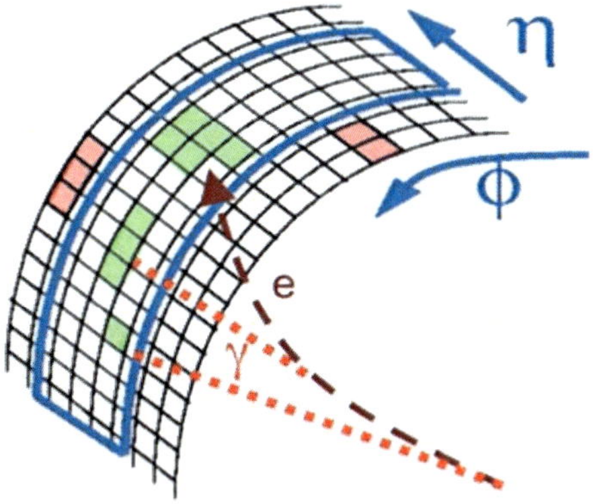

Fig. 2.24 Schematic representation of the PF reconstruction for electrons. Three ECAL superclusters (green boxes) are identified by the blue area in the (η, ϕ) plane. The electron GSF track is depicted in brown, along with two sample tangent lines (orange) corresponding to the radiated bremsstrahlung photons. Figure taken from [42]

associated to it is taken as a second estimator. A complementary way of determining the particle's charge consists in looking at the angular distance $\Delta\phi$ between the vector originating from the IP and pointing towards the ECAL supercluster, and the one pointing towards the first hit associated to the GSF track. The most probable value among the three is used as an estimator of the electron's charge.

The electron's momentum measurement presents the same issues as the charge. Therefore it is carried out in subsequent steps. First of all, the electrons are divided into mutually exclusive categories according to the goodness-of-fit of their tracks. The supercluster energy is directly accessible by summing the response recorded by the ECAL crystals. Nonetheless, this method would naturally miss possible energy losses in the tracker, leakages into the HCAL, and pileup effects, which can be recovered by means of multivariate techniques. After a final calibration of the electron's energy and an estimation of its uncertainty, it is combined with the GSF track momentum to be given as inputs to a second BDT to build the final electron's candidate.

In order to enhance the electrons identification and to discriminate between isolated and non-isolated leptons, the analyses make use of a dedicated BDT, trained in six different (η, ϕ) bins. The additional isolation and identification criteria applied on electrons and specific of the $H \to ZZ \to 4\ell$ analysis are presented in Chap. 4.

2.3.3 *Muons*

Along with electrons, muons are the most important objects for analysing the $H \to ZZ \to 4\ell$ decay, as they are used to resolve completely the final state of this channel. Hence, precise and accurate reconstruction and identification of these particles are crucial for the ultimate precision of the analysis.

Unlike electrons, muons traverse the detector essentially as *minimum ionizing particles* (MIP) when interacting with the material on their path. Along with neutrinos, they are the only particle capable of traversing the entire CMS apparatus before being detected by the outermost muon chambers. Having the magnet placed before the muon stations (cf. Sect. 2.2.2) implies that the particle's trajectory will present two different curvatures: a helpful handle when it comes to momentum estimation.

Tracking

The CMS detector was originally designed to detect muons with extremely large efficiency. The DTs and CSCs hits are combined to form a muon seed, which is linked together with the RPCs information to build a *standalone muon track*. Two additional ways of constructing a muon track are considered in the PF approach: *global muon tracks* are obtained by refitting the hist associated to a standalone muon track and the corresponding tracker hits, if the two can be geometrically matched; conversely, *tracker muons* are built exploiting only the tracker's information, if $p_T > 0.5$ GeV and $p > 2$ GeV.

Global or tracker muons make up to 99% of the muons produced and in most of the cases the candidate fits both the categories. If the two also share inner tracker hits, they are merged into a single candidate. The combination of these two methods provides good efficiency up to low-p_T values. On the other hand, standalone muons present slightly worse track parameters, mainly because of the missing tracker information.

Reconstruction

The muons reconstruction is less challenging than for the electrons. For all those particles with $p_T < 200$ GeV, the charge and momentum are estimated directly from the track's parameters, after a fit on the tracker hits. In all the remaining cases, the parameters are estimated from the track with the lowest χ^2, selected among those that can be built combining the tracker and muon chambers' information.

Additional analysis-specific selection criteria are applied on PF muons in order to enhance the reconstruction and isolation efficiencies, as well as the signal purity. Besides specific cuts on the muon variables, centrally provided *muon identification* (ID) *working-points* (WP) are defined to help analysers select a specific topology of muons. There are three WPs - loose, medium and tight - which increase the true positive rate, while reducing the overall efficiency. The additional isolation and identification criteria applied on muons and specific of the $\mathrm{H} \rightarrow \mathrm{ZZ} \rightarrow 4\ell$ analysis are presented in Chap. 4.

2.3.4 Jets

The number of jets in a given event is an excellent handle to discriminate between different production mechanisms of the Higgs boson. An example is given in Fig. 2.25, where the number of jets reconstructed in a ggH and VBF samples are compared.

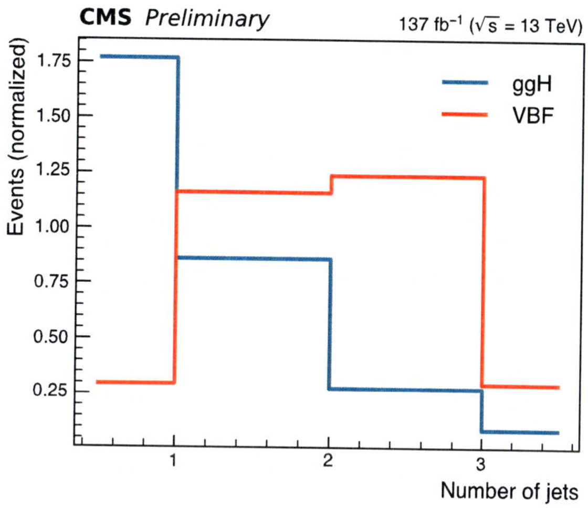

Fig. 2.25 Comparison between the number of jets in H $\rightarrow$ ZZ $\rightarrow 4\ell$ simulated events originating from gluon fusion (ggH) and vector boson fusion (VBF) processes at $\sqrt{s} = 13$ TeV, corresponding to 137 fb^{-1}

This information is used extensively in the analysis for the computation of matrix element discriminants and the definition of the reconstructed categories, as further detailed in Chap. 5. Hence, the proper identification and reconstruction of jets is a fundamental ingredient for studying the Higgs boson properties in a decay channel with a final state as clean as the four-lepton one.

Reconstructed jets are defined as narrow cones of particles produced by the hadronization of a gluon or quark [48]. The main challenge in the jets' reconstruction is their large occupancy in the detector, which makes it complicated to disentangle hits corresponding to isolated particles from those belonging to the jet cone. This analysis makes use of the so-called anti-k_T reconstruction algorithm [49]: a sequential recombination algorithm in which particles are clustered according to a distance parameter proportional to their transverse momentum k_T. More precisely, a particle i is merged to the entity j, being it another particle or a jet, if:

$$\min\left(k_{Ti}^{2p}, k_{Tj}^{2p}\right) \frac{\Delta R_{ij}}{R^2} < k_{Ti}^{2p}, \tag{2.8}$$

where R is a free parameter of the anti-k_T algorithm and defines the radius of the jet cone area. The spatial distance between the i and j objects is ΔR_{ij}. The power parameter p is characteristic of all the sequential recombination algorithms used for jets clustering. Its value is set to $p = -1$, from which comes the "anti-" suffix in the algorithm name.

PF objects are given as inputs to the FASTJET package [50] to perform an anti-k_T clustering with $R = 0.4$ and to produce the jets used in the H $\rightarrow$ ZZ $\rightarrow 4\ell$ analysis. In addition, jets are cleaned from the presence of charged hadrons not belonging to the primary interaction vertex, in order to reduce the pileup dependence.

Once the jet has been reconstructed, its momentum and charge are computed as the vectorial sum of all the particles clustered into the jet. The use of the PF approach

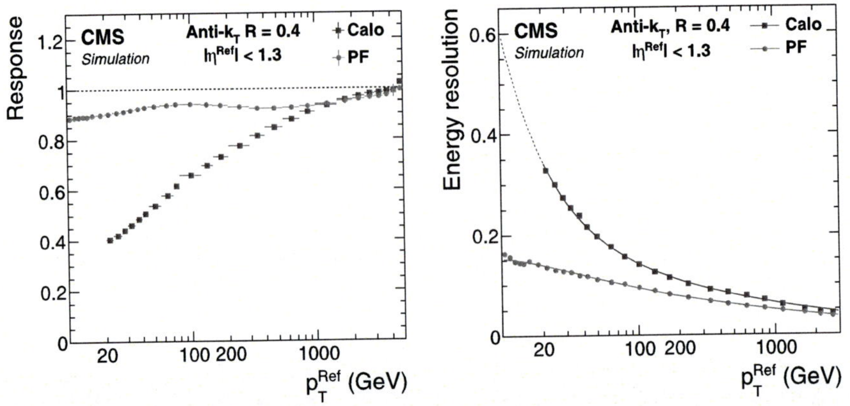

Fig. 2.26 Response (left) and relative resolution (right) for the jet transverse momentum in the barrel. Jets are defined as anti-k_T jets with $R = 0.4$. Figures taken from [42]

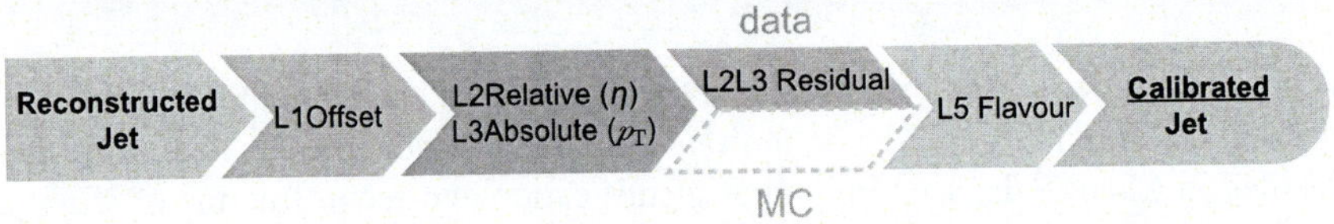

Fig. 2.27 Different factors that contribute to the jet energy corrections in CMS. Each level of correction takes care of a specific effect and is applied sequentially after the preceding ones

allows to reconstruct up to 90% of the jet components, substantially improving the performance of a calorimeter-based reconstruction, as shown in Fig. 2.26.

The reconstructed jet energy does not correspond to the one of the original parton, mainly because of detector effects[5] and theoretical uncertainties on the underlying hadronization model. These effects are compensated with jet energy corrections (JEC), applied in dedicated (p_T, η) bins [51, 52]. Figure 2.27 shows the different factors that contribute to the JECs.

Each factor, except for the residual corrections, is applied to both data and simulation and takes care of compensating a specific contribution that degrades the true jet energy. The PF approach, combined with the application of JEC, leads to an overall energy resolution lower than 20% for jets with transverse momentum larger than 30 GeV, improving by almost a factor two the result one would get with a pure calorimeter-based reconstruction.

According to the topology of the events studied in a particular analysis, additional identification criteria can be introduced on jets, similarly to what is done for the

[5] For example pileup corrections, electronic noise, and non-uniformities in the HCAL and ECAL responses.

leptons. The analysis presented in this thesis makes use of tagging algorithms to identify jets originating from b-quarks [53], as well as dedicated pileup corrections to cure for discrepancies between the jet observables in data and in MC.

References

1. Baconnier Y (1993) LHC: the large hadron collider accelerator project
2. The large hadron collider: conceptual design. 10 1995
3. ATLAS Collaboration (2012) Observation of a new particle in the search for the standard model Higgs boson with the ATLAS detector at the LHC. Phys Lett B 716:1
4. CMS Collaboration (2012) Observation of a new boson at a mass of 125 GeV with the CMS experiment at the LHC. Phys Lett B 716:30
5. CMS Collaboration (2013) Observation of a new boson with mass near 125 GeV in pp collisions at $\sqrt{s} = 7$ and 8 TeV. JHEP 06:081
6. Evans L, Bryant P (2008) LHC machine. JINST 3(08):S08001–S08001
7. Mobs E The cern accelerator complex
8. CMS Collaboration (2018) Measurement of the inelastic proton-proton cross section at $\sqrt{s} =$ 13 TeV. JHEP 07:161
9. Contardo D, Klute M, Mans J, Silvestris L, Butler J (2015) Technical proposal for the phase-II upgrade of the CMS detector. Technical report CERN-LHCC-2015-010. LHCC-P-008. CMS-TDR-15-02, Geneva
10. CMS Collaboration (2017) The phase-2 upgrade of the CMS endcap calorimeter. Technical report CERN-LHCC-2017-023. CMS-TDR-019, CERN, Geneva
11. CMS Collaboration. Public luminosity results
12. HiLumi Collaboration. The HL-LHC project. https://hilumilhc.web.cern.ch/content/hl-lhc-project
13. ATLAS Collaboration (2008) The ATLAS experiment at the CERN large hadron collider. JINST 3:S08003
14. LHCf Collaboration (2008) The LHCf detector at the CERN large hadron collider. JINST 3:S08006
15. ALICE Collaboration (2008) The ALICE experiment at the CERN LHC. JINST 3:S08002
16. CMS Collaboration (2008) The CMS experiment at the CERN LHC. JINST 3:S08004
17. TOTEM Collaboration (2008) The TOTEM experiment at the CERN large hadron collider. JINST 3(08):S08007–S08007
18. LHCb Collaboration (2008) The LHCb Detector at the LHC. JINST 3:S08005
19. MoEDAL Collaboration (2009) Technical design report of the MoEDAL experiment. CERN CDS, 6
20. CMS Collaboration. Cutaway diagrams of CMS detector
21. Neutelings I. CMS wiki pages, how to draw diagrams in latex with tikz
22. Acquistapace G et al (1997) CMS, the magnet project: technical design report. CERN CDS, 5
23. Collaboration CMS (2010) Precise mapping of the magnetic field in the CMS barrel yoke using cosmic rays. JINST 5:T03021
24. CMS Collaboration (1997) The CMS tracker system project: technical design report. CERN CDS
25. CMS Collaboration (2000) The CMS tracker: addendum to the technical design report. CERN CDS
26. CMS Collaboration. Images for US LHC CMS detector upgrade project. https://cms-docdb.cern.ch/cgi-bin/PublicDocDB/
27. CMS Collaboration (2014) Description and performance of track and primary-vertex reconstruction with the CMS tracker. JINST 9(10):P10009–P10009

28. CMS Collaboration (2012) CMS technical design report for the pixel detector upgrade. CERN CDS, 9
29. CMS Collaboration (2015) Technical proposal for the phase-II upgrade of the CMS detector. 6
30. CMS Collaboration (1997) The CMS electromagnetic calorimeter project: technical design report. CERN CDS
31. Bayatian GL et al (2006) CMS physics: technical design report volume 1: detector performance and software. CERN CDS
32. Adzic P et al (2007) Energy resolution of the barrel of the CMS electromagnetic calorimeter. JINST 2:P04004
33. CMS Collaboration (1997) The CMS hadron calorimeter project: technical design report. CERN CDS
34. Lutz B (2012) Upgrade of the CMS hadron outer calorimeter with SiPM sensors. J Phys: Conf Ser 404:012018
35. Collaboration CMS (2018) Performance of the CMS muon detector and muon reconstruction with proton-proton collisions at $\sqrt{s} = 13$ TeV. JINST 13(06):P06015
36. Bell AJ et al (2010) Fast beam conditions monitor BCM1F for the CMS experiment. Nucl Instrum Meth A 614:433–438
37. CMS Collaboration (2017) CMS luminosity measurements for the 2016 data taking period. Technical report CMS-PAS-LUM-17-001, CERN, Geneva
38. CMS Collaboration (2018) CMS luminosity measurement for the 2017 data-taking period at $\sqrt{s} = 13$ TeV. Technical report CMS-PAS-LUM-17-004, CERN, Geneva
39. CMS Collaboration (2019) CMS luminosity measurement for the 2018 data-taking period at $\sqrt{s} = 13$ TeV. Technical report CMS-PAS-LUM-18-002, CERN, Geneva
40. CMS Collaboration (2013) CMS technical design report for the level-1 trigger upgrade. CERN CDS, 6
41. TOTEM Collaboration (2011) First measurement of the total proton-proton cross section at the LHC energy of $\sqrt{s}$ =7 TeV. EPL 96(2):21002
42. Collaboration CMS (2017) Particle-flow reconstruction and global event description with the CMS detector. JINST 12(10):P10003
43. Billoir P (1989) Progressive track recognition with a Kalman like fitting procedure. Comput Phys Commun 57:390–394
44. Azzurri P (2009) Track reconstruction performance in CMS. Nucl Phys B, Proc Suppl 197:275–278
45. Collaboration CMS (2014) Description and performance of track and primary-vertex reconstruction with the CMS tracker. JINST 9(10):P10009
46. Adam W, Frühwirth R, Strandlie A, Todor T (2005) Reconstruction of electrons with the Gaussian-Sum filter in the CMS tracker at the LHC. J Phys G: Nucl Part Phys 31(9):N9–N20
47. Baffioni S, Charlot C, Ferri F, Futyan D, Meridiani P, Puljak I, Rovelli C, Salerno R, Sirois Y (2007) Electron reconstruction in CMS. Eur Phys J C 49:1099–1116
48. Strologas J (2016) Performance of Jet reconstruction in CMS at 13 TeV. PoS ICHEP2016:736
49. Cacciari M, Salam GP, Soyez G (2008) The anti-k_t jet clustering algorithm. JHEP 04:063
50. Cacciari M, Salam GP, Soyez G (2012) FastJet user manual. Eur Phys J C 72:1896
51. CMS Collaboration (2017) Jet energy scale and resolution in the CMS experiment in pp collisions at 8 TeV. JINST 12(02):P02014
52. Dordevic M (201+) Measurements of the CMS jet energy scale and resolution at 13 TeV. PoS ICHEP2018:688
53. CMS Collaboration (2013) Performance of the DeepJet b tagging algorithm using 41.9/fb of data from proton-proton collisions at 13TeV with phase 1 CMS detector. CERN CDS

Chapter 3
The CMS Endcap Calorimeters Upgrade for the HL-LHC

The High-Luminosity LHC (HL-LHC) is expected to start its operations by the end of 2027. It is designed to deliver a peak instantaneous luminosity of 5×10^{34} cm^{-2}s^{-1}, thus giving access to a total integrated luminosity of 3000 fb^{-1} and therefore increasing the discovery potential of the LHC. The HL-LHC will allow more precise measurements of the SM properties and enhance its sensitivity to rare processes, possibly unveiling the presence of previously unknown particles and BSM scenarios.

The higher luminosities of the HL-LHC will also result in exceedingly high pile-up rates, with $\mathcal{O}(200)$ events per bunch crossing and unprecedented radiation levels, with fluences of up to 10^{16} n_{eq}/cm^2 and doses of around 2 MGy, thus posing several technical challenges for the operation of the detectors and the entire infrastructure. For example, the LHC tunnel will have to feature the most advanced superconducting magnets, vacuum pipes, cryogenic systems, and superconducting radiofrequency caverns.

The CMS Collaboration, as well as the other LHC experiments, are planning a series of major upgrades of the sub-detectors, expected to be commissioned during the second and the third long shutdowns (cf. Fig. 2.3), to maintain the current physics performance in the harsh environment of the HL-LHC.

A brief overview of the CMS upgrade plans is given in Sect. 3.1, while Sect. 3.2 focuses on the High Granularity Calorimeter (HGCAL) that will replace the current endcap calorimeters.

The HGCAL will feature silicon-based active layers for a total active area of about 600 m^2, and it will be the first large-scale silicon-based imaging calorimeter employed in a high-energy physics experiment. Hence, validating the detector design and assessing its physics performance are cornerstone aspects for the successful realization of this project. For this purpose, a long series of tests of HGCAL prototypes have been carried out in particle beams since the end of 2016, leading to the test of the first large-scale prototype of the HGCAL in October 2018. On this occasion, the prototype was exposed to positron and hadron beams with momenta ranging from 20 to 300 GeV at the CERN-SPS beamline, thus allowing a comprehensive

M. Bonanomi, *Response of the High Granularity Calorimeter HGCAL and Characterisation of the Higgs Boson*, Springer Theses,
https://doi.org/10.1007/978-3-031-26833-5_3

characterization of the properties of this innovative detector. The findings of the test beam data analysis are presented in Sect. 3.3, with particular regard to the measurement of the response of the electromagnetic compartment of the HGCAL prototype.

3.1 Upgrades of CMS

The ultimate goal of the CMS upgrades programme [1, 2] for the HL-LHC phase is twofold. On one side, many of the current detector components cannot cope with the harsh environment expected at the HL-LHC, thus requiring either a complete replacement or a substantial upgrade of the existing modules. On the other hand, the current physics performance must not deteriorate during the high luminosity research programme, which will target a completely new set of precision measurements, direct searches for rare processes, and possible hints of any BSM physics. The perfect synthesis of these two aspects requires to face three major challenges, such as:

- **Unprecedented radiation doses** impose a complete replacement of the tracker and the endcap calorimeter systems, the need for new technologies for the EB, and a substantial improvement of the electronics systems in the barrel calorimeters and the muon detectors.
- An **increase in the pileup rate** for which are mandatory highly granular readouts, wherever the detector allows it, the introduction of precision timing detectors, and novel approaches to pileup mitigation.
- A **high luminosity** that will cause in an increased data stream, for which substantial improvements of the L1 Trigger primitives and of the overall TriDAS system will be fundamental.

Figure 3.1 depicts a cross section of the CMS detector, showing the different upgrades foreseen for the HL-LHC. The subdetectors in the green boxes will be completely replaced, while the purple boxes indicate new detector systems, not present in the current design of CMS, that will be installed.

The tracking system will be entirely replaced to increase the granularity of the detector, thus enhancing the reconstruction performance, while at the same time reducing the overall material budget: pixel detectors of smaller size with respect to the current ones will constitute the inner tracking system, while the outer tracking stations will feature strips and macro pixel sensors, extending the coverage up to $\mid \eta \mid = 3.8$ [3]. This design choice will improve longitudinal and transverse resolutions, and it will bring lower fake rates, thus allowing the reconstruction of L1 trigger tracks up to $\mid \eta \mid = 2.4$.

The endcap calorimeters will be replaced with the HGCAL, a high granularity sampling calorimeter that will provide, thanks to its features, enhanced showers separation and particle identification, as well as additional precise timing information [4].

With a similar goal of increasing the overall granularity and providing additional timing measurements, an upgrade of the ECAL and HCAL barrel electronic readout is foreseen [5]. The redundancy of the current muon detection system that features DTs,

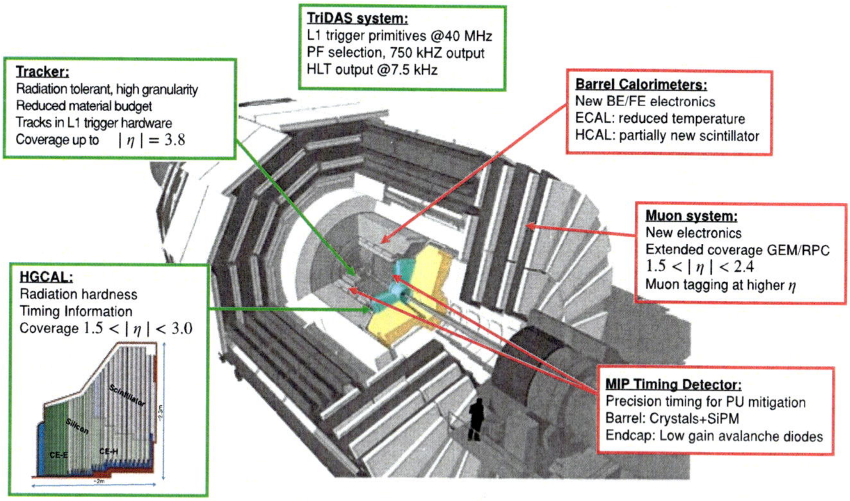

Fig. 3.1 Cross section of the CMS detector with the indication of the upgrades foreseen for the HL-LHC [1, 2]. The green boxes represent subdetectors that will be completely replaced, while the purple boxes indicate either new modules or extensions of the current ones

RPCs, and CSCs, will be increased with the installation of Gas Electron Multiplier chambers and a new generation of RPCs [6], which will also extend the coverage up to $|\eta| = 2.8$ and $|\eta| = 2.4$, respectively.

Multiple MIP timing detectors (MTD) [7] will be placed in front of the barrel and endcap calorimeters to increase the available timing information on charged particles.

The TriDAS systems will undergo a complete replacement [8], with increased throughput. The HLT will continue to have access to the total detector information, even at the maximum input rate of 750 kHz expected at the top of the HL-LHC operations, to which it will provide a $\times 100$ reduction rate for a bandwidth output of 7.5 kHz.

3.2 The High Granularity Calorimeter

The current Endcap Calorimeters (CE) of the CMS detector were designed to sustain integrated luminosities up to 500 fb^{-1}, expected to be exceeded by far during the Phase-II operations. The transparency of the lead-tungstate crystals currently employed in the endcaps would not survive the unprecedented radiation flux typical for the HL-LHC environment in this detector region, where doses of the order of 2 MGy and fluences up to 10^{16} n_{eq}/cm^2 are expected. Hence, the CMS Collaboration intends to replace the current endcap calorimeters with a High Granularity CALorimeter (HGCAL) [4] in the context of the Phase-II upgrades programme [1, 2].

The HGCAL will comprise an electromagnetic (CE-E) and a hadronic (CE-H) compartment. The former will feature 26 active layers, interspersed with CuW, Cu, and Pb absorbers, while the latter is composed of 21 layers that exploit stainless steel as passive material. The electromagnetic compartment will extend for a total of 27.7 radiation-lengths (X_0), equivalent to 1.3 nuclear radiation lengths (λ_n), resulting in a total detector depth of 10 λ_n when combined with the 21 layers of the CE-H.

To meet the requirements of a radiation hard detector, the CE-E and the front part of the CE-H will employ silicon as active material, for a total area of about 600 m^2 to be covered. Hexagonal silicon sensors will be used to optimize the coverage of such a large surface and reduce the detector's total cost. The design, production, and validation of these hexagonal sensors, hereafter referred to as *modules*, is one of the most challenging aspects of the HGCAL project. A more detailed discussion about silicon as active material and the structure of the HGCAL modules is given in Sect. 3.2.1. The HGCAL will feature 8" sensors with an active thickness of 120, 200, or 300 μm, depending on the detector region. Each module will comprise several single readout diodes, hereafter referred to as *cells* or *pads*, with a 0.5 or 1.0 cm^2 active area, for a total of about six million cells read out individually in the ultimate detector operation. A schematic representation of the subdivision of a silicon module into its readout units is shown in the left part of Fig. 3.2.

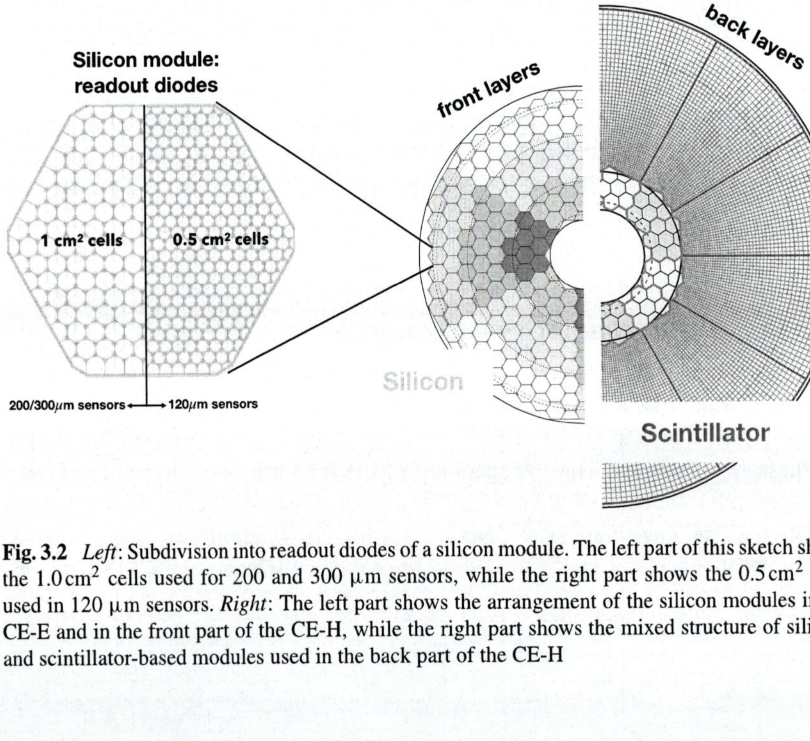

Fig. 3.2 *Left*: Subdivision into readout diodes of a silicon module. The left part of this sketch shows the 1.0 cm^2 cells used for 200 and 300 μm sensors, while the right part shows the 0.5 cm^2 cells used in 120 μm sensors. *Right*: The left part shows the arrangement of the silicon modules in the CE-E and in the front part of the CE-H, while the right part shows the mixed structure of silicon- and scintillator-based modules used in the back part of the CE-H

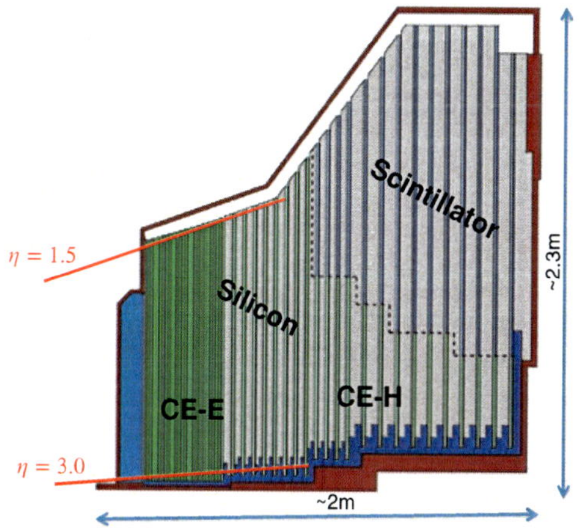

Fig. 3.3 Cross section view of the HGCAL. The CE-E and front CE-E compartments comprise silicon-based components, shown in green. The SiPM scintillator tiles are employed in the back of the hadronic compartment and are depicted in blue in the figure. The expected coverage of the HGCAL, ranging from $\eta = 1.5$ up to $\eta = 3.0$ is shown by the red lines

The substantially lower doses expected in the endcaps at about 4 m from the interaction point allow segmented plastic scintillators, with silicon photomultipliers (SiPM) for the readout, as the active material for the back part of the CE-H. Squared scintillator tiles with sizes ranging from 4 cm^2 up to 30 cm^2 will be used, for a total of about 400 m^2 of active area covered in scintillators.

The arrangement of the silicon modules in the CE-E and in front part of the CE-H, as well as the mixed technology of silicon sensors and scintillator tiles employed in the back part of the CE-H, are shown in the right part of Fig. 3.2.

A longitudinal cross section of the HGCAL is shown in Fig. 3.3, along with the difference between the CE-E and CE-H active materials and the lines defining the expected detector coverage of $1.5 <| \eta |< 3.0$. The calibration of the detector will be performed *in situ* with minimum ionizing particles (MIP). The effects of radiation damage must be limited in order to ensure a reasonable detection of MIPs throughout the detector lifetime and to ensure a signal-to-noise ratio above one. To achieve this, the HGCAL will have to be operated at a constant temperature of -30 °C or less, obtained with a dedicated CO_2 cooling system.

Because of its highly granular readout design, the HGCAL is often referred to as an *imaging*, or 5D, calorimeter: the high granularity in the longitudinal and transverse planes can be exploited to completely resolve both electromagnetic and hadronic shower profiles, starting from the Cartesian coordinates reconstructed in the active layers of the detector. This information is complemented by the measurement of the energy deposited within the calorimeter, thus enhancing the pileup mitigation and the particles identification and discrimination. The "fifth dimension" corresponds to the timing information provided by the readout ASICs of the HGCAL, which will feature an expected resolution of $\mathcal{O}(10\text{ ps})$ and it will mark a real innovation in calorimetry. This information will be used for pileup mitigation, rejecting hits recorded outside a certain time interval Δt, for vertex location of triggered hard interactions within the busy environment of the HL-LHC, as well as to enhance the inputs available for the Particle Flow reconstruction. In addition, the readout ASICs are also expected to provide information directly to the L1 trigger system, making the development of the frontend electronics even more challenging.

3.2.1 Silicon Modules

The choice of silicon as the active material for the modules is driven by the tight constraints dictated by the expected physics performance and the environment in which the detector will operate. More precisely, silicon sensors ensure the detection of both minimum ionizing particles (MIPs) and the precision measurement of high energetic showers. The former is a key requirement for the operational performance of the HGCAL, as MIPs are intended to be used for in situ calibrations of the detector. The latter can be ensured only with full containment of the showers, for which the compactness of the detector, achieved with thin silicon modules and a proper mixture of absorbers, is fundamental to keep under control the Molière radius of the calorimeter. Silicon sensors meet these requirements and have the advantage of producing fast signals ($\mathcal{O}(10\ \text{ns})$), fundamental to keep up with the 40 MHz rates expected in the busy environment of the HL-LHC. In addition, silicon modules can be produced in a relatively short time scale, profiting from the longstanding industrial efforts in the field.

The validation of the silicon modules structure and of their expected performance has been one of the main pillars of the project since after the approval of the HGCAL Technical Proposal (TP) in 2015 [1]. Several proof-of-concept modules have been produced by the Japanese company Hamamatsu Photonics K.K. (HPK), and an increasing number of modules have been exposed to test beams between 2016 and 2017. The findings of these test beams have been incorporated in the 2018 HGCAL Technical Design Report (TDR) [4], and they have set the basis for understanding the properties of the silicon modules involved in the design of such a complex detector.

From the early tests of single modules in 2016 to the beam test of the first large scale prototype of the HGCAL in 2018, 6" silicon modules have been employed to reflect the original design presented in the TP [1]. During this period, the size of the sensors changed from 6" to 8" to reduce the overall number of modules. The production and testing of the first 8" prototypes are in progress at the moment of writing this thesis.

Each module is made of as a stacked structure comprising the printed circuit board (PCB), where the front-end electronics are installed, the silicon sensor, a gold-plated Kapton sheet providing HV connection to the sensor back-plane and electrical insulation from the baseplate, made of materials such as CuW or Cu in order to ensure enough rigidity to support the module, whilst minimizing the total radiation length and dissipating the heat via a dedicated cooling system. A schematic view of the structure of a typical HGCAL module is given on the left side of Fig. 3.4 and the picture of a 6" prototype used for the beam tests is shown on the right side.

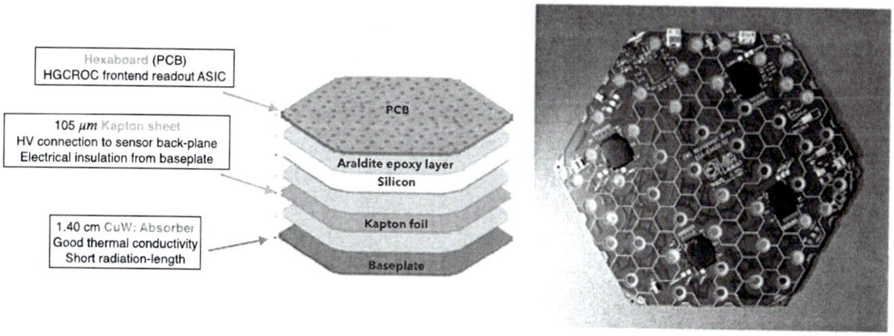

Fig. 3.4 *Left*: Schematic view of the stacked layers of a CE-E silicon module. *Right*: Complete 6” prototype module used in the 2018 beam tests

One of the most challenging aspects in the design of this module is the front-end electronics [9]. The readout ASICs foreseen for the final design have to operate over a wide dynamic range (0.2 fC–10 pC), featuring low noise levels ($<2000\ e$) and reduced power consumption, and they should be capable of transmitting the high granularity information of the HGCAL to the L1 trigger. The silicon sensors will require a bias voltage of O(200 V) that is expected to steadily grow up to 800 V to cope with their ageing in the HL-LHC operations. In addition, in order to fully exploit the HGCAL as an *imaging calorimeter*, the ASICs need to have an intrinsic single-channel resolution of the order of 40 ps for signals above 50 fC, thus achieving an ultimate timing resolution of $\mathcal{O}(10\,\text{ps})$. These requirements have to be met all together with the additional constraint of the reduced space available for the integration of the ASIC into the active layers, resulting in a front-end readout electronics not available in market nowadays. Hence, the CMS Collaboration, in collaboration with the Omega Microelectronic Center in École Polytechnique, is working on the design of the *HGCROC*, a readout chip specific for the HGCAL applications. In the right part of Fig. 3.4, where a 6” prototype used for beam tests is shown, one can identify the ASICs as the four black chips installed onto the PCB. The ultimate design of the *HGCROC* is not finalized yet for production, therefore prototypes with preliminary functionalities have been developed for the beam test campaigns. As detailed in the next section, the *Skiroc2* [10] chip, originally developed by the CALICE Collaboration [11], was used for the early tests, while for the 2018 test beam the *Skiroc2-CMS* ASIC [12] was used.

3.3 Test Beams of HGCAL Prototypes

A long series of test beams started in late 2016 to validate the detector design and the physics performance of the HGCAL. After the first test of a single module in 2016, stacks of up to 16 modules featuring the *Skiroc2* ASIC were tested [13].

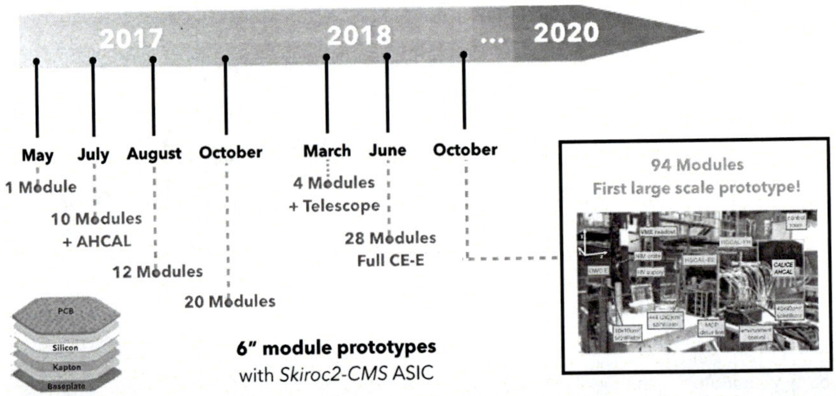

Fig. 3.5 Schematic depiction of the HGCAL test beam series. All the test beams featured 6" silicon modules, equipped with PCB and ASICs of evolving complexity. The *Skiroc2-CMS* ASIC, developed by the OMEGA laboratory specifically for HGCAL test beam applications, was employed in all test beams after the initial one in 2016. A more detailed description of the test beam setup is given in the text

The first module supporting the *Skiroc2-CMS* chip was available by the beginning of 2017, and the first beam test of such a module took place in May 2017, after which the complexity of the system and the number of modules involved in the test beam campaigns have been increased continuously. A schematic illustration of the HGCAL test beam series is given in Fig. 3.5: all the beam tests carried out after 2016 featured 6" silicon module prototypes with the structure described in Sect. 3.2.1 and equipped with the *Skiroc2-CMS* ASIC developed by the CALICE Collaboration.

The experience gained throughout these years and the findings of the several tests lead to major improvements in the design of the PCB and the development of the *Skiroc2-CMS* ASIC, designed specifically for the HGCAL operations in beam tests: both these aspects naturally converged in the prototype assembled for the October 2018 beam test, where the first large scale prototype of the HGCAL was tested at the SPS beamline. The detector under test comprised fully equipped electromagnetic and hadronic compartments, as well as an analog hadronic calorimeter (AHCAL), which was installed to mimic the scintillator+SiPM-based compartment expected in the final HGCAL design. The large plethora of results obtained from this test beam lead to the preparation of different papers, three of which already published in the Journal of Instrumentation (JINST) [14, 22, 30].

Section 3.3.1 presents the experimental area and the HGCAL prototype tested in October 2018, while Sect. 3.3.2 describes the improvements introduced in the context of this thesis to the reconstruction of the *Skiroc2-CMS* ASIC pulse shape signal.

3.3.1 The October 2018 Beam Test Setup

The following section presents the experimental setup of the October 2018 beam test at the SPS beamline, with particular emphasis on the detectors placed upstream of the beam test area and relevant for the analysis. In the description of the HGCAL prototype, the focus is put on the description of the electromagnetic compartment, of which the design and performance have been validated in the context of this thesis.

The CERN H2A Beamline

The test beam data were collected at the EHN1 Extension located at the CERN North Area [15] with positron and pion beams extracted from the PPE172 (H2A) beamline. The SPS delivers 400 GeV/*c* protons to the whole North Area: here the beam is directed towards three stripping targets, placed along the different beamlines, and used to initiate the production of test beams. In the case of H2A a beryllium target (*T2*) is employed. Figure 3.6 depicts a schematic view of the PPE172 test area, showing the different steps of the beam generation after the extraction of the primary 400 GeV/*c* proton beam from the SPS.

The primary proton beam produces secondary beams of hadrons, electrons, and positrons by impinging on a 500 mm beryllium *T2* target, where the thickness is chosen specifically to obtain a comparable rate in the production of the two secondary beam species: electromagnetic particles (electrons, photons), hadronic particles (kaons, protons, and charged pions), and muons. As sketched in Fig. 3.6, a sweeping magnet and a converter are placed immediately after the target to select specific particles, according to their charge and momenta. The selection of the final momenta is performed exploiting a combination of dipoles and collimators, which result in a nominal combined acceptance of $\Delta p/p = 0.2$–2% [15]. The major drawback of using bending magnets for the momenta selection is the energy loss due to synchrotron radiation (SR), which induces an additional spread. More precisely, the power radiated by a particle of mass m, with a given momentum p, when subject to a perpendicular magnetic field B is proportional to:

$$P \propto \frac{B^2 p^2}{m^4}, \tag{3.1}$$

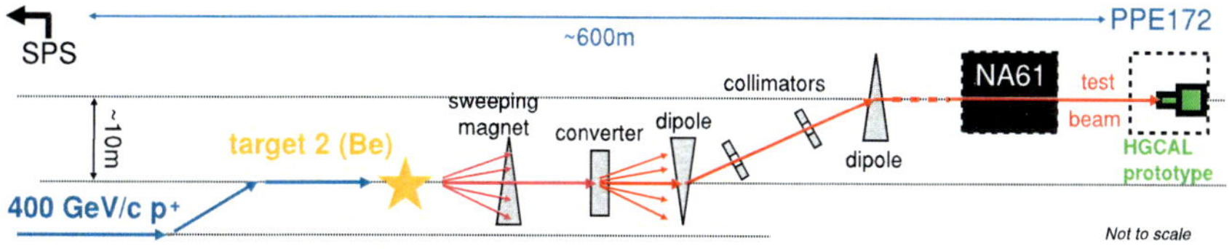

Fig. 3.6 Schematic view of the PPE172 (H2A) of EHN1. The different elements used to convert the primary 400 GeV/*c* proton beam, extracted from the SPS, into the final test beam are also indicated with the corresponding labeling. Figure taken and edited from Ref. [16]

which scales with the fourth power of the mass of the particle. This correction is negligible for hadrons, which do not suffer SR losses thanks to their large masses. However, the same does not hold for positrons with momenta larger than 100 GeV, which undergo significant SR losses, as their small mass cannot keep under control the power radiated due SR effect. Such an effect cannot be neglected in the analysis of positron data, where an accurate measurement of the beam momentum is mandatory for the precise estimation of the electromagnetic energy resolution and the characterization of the showers' profiles presented in Sect. 3.4. The effect of the SR losses is responsible for an additional spread of the beam momentum and for beam energy values lower than the nominal ones. The final momenta of positron beams at the entrance of the experimental area, along with the corresponding spread, are given in Table 3.1.

The length of the H2A beamline from the SPS injection to the beginning of the PPE172 experimental area (cf. Fig. 3.6) is approximately 600 m and comprises several passive material components that can cause additional energy losses after the interaction with the beam. Hence, in order to have an accurate benchmark for the results of the data analysis presented in Sect. 3.4, a detailed `GEANT4` [17, 18] MC simulation of the full H2A beamline is used. The beam composition, including the effect of residual impurities and the energy losses due to SR, and its propagation through the entire beamline are modelled with the `G4Beamline` tool [19]. The findings of the Data to MC comparison are further detailed in the Sect. 3.4.

A zoomed view of the PPE172 beamline in the proximity of the experimental area is given in Fig. 3.7. Before reaching the HGCAL prototype, the beam passes through four delay wire chambers (DWC) [20] installed between 32 m and 1.6 m before the first layer. This long lever arm can be used to reconstruct the particles' track and to extrapolate the impact points at each layer of the detector, thus providing an external reference to measure the pointing resolution of each layer, as discussed in Sect. 3.4.5.

Table 3.1 Nominal momenta for positron beams provided by the H2 beamline and momenta measured by the spectrometer placed upstream the HGCAL CE-E prototype, after the last dipole

Nominal momentum [GeV/c]	Final momentum [GeV/c]	Final momentum spread [GeV/c]	Final momentum resolution %
20	20.00	0.06	0.3
30	30.00	0.08	0.3
50	49.97	0.12	0.3
80	79.91	0.19	0.2
100	99.81	0.22	0.2
120	119.64	0.28	0.2
150	149.16	0.35	0.2
200	197.40	0.47	0.2
250	243.84	0.60	0.2
300	287.65	0.79	0.3

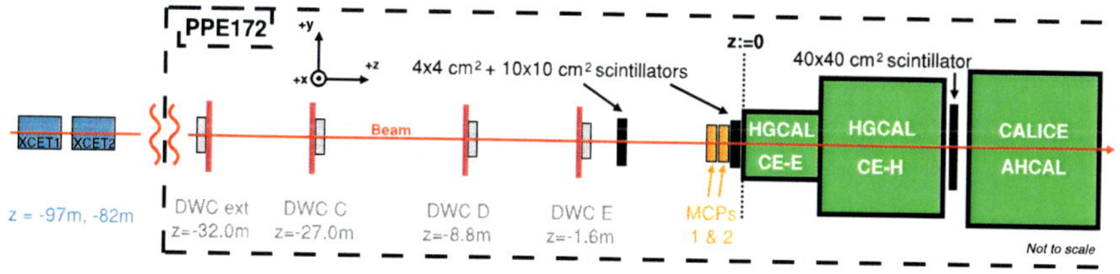

Fig. 3.7 Zoomed view of the PPE172 beamline in the proximity of the experimental area. The upstream detectors installed along the beamline are shown together with their location with respect to the first layer of the HGCAL prototype under test. Figure taken and edited from Ref. [16]

The signals of two scintillators of 4×4 cm^2 and 10×10 cm^2, located after the last DWC, are used in coincidence to trigger the readout of the silicon sensors. The third scintillator of 40×40 cm^2, placed after the HGCAL prototype, is used as a veto to identify possible hadronic contamination present in the electromagnetic showers. Two Microchannel Plates (MCP) [21] are placed before the face of the prototype and they are used as an external reference for the measurement of the timing performance of the HGCAL prototype. More precisely, their fast signal waveform is readout with a fast digitizer and the MCP timestamp used as an external reference is taken from the analysis of such a waveform.

After traversing the entire PPE172 beamline, the beam arrives at the face of the HGCAL prototype and it interacts with the first layer of the electromagnetic compartment.

The HGCAL Prototype

The HGCAL prototype assembled for the 2018 test beam comprises a fully instrumented electromagnetic compartment (CE-E) of 28 layers, a hadronic one (CE-H) with 66 modules,[1] and the CALICE AHCAL [11] to mimic the scintillator tiles with SiPM readout. The CE-E prototype has had a total length of $\sim$27.2 X_0 and $\sim$1.3 λ_0 to which the CE-H prototype and the CALICE AHCAL added 3.5 λ_0 and 4 λ_0, respectively, thus reflecting the ultimate design of the HGCAL. A picture of the experimental setup is shown in Fig. 3.8, with labels for most of the elements described above.

The electromagnetic and the hadronic compartments are equipped with 6" silicon modules, reflecting the design outlined in Sect. 3.2.1. All the modules were equipped with the *Skiroc2-CMS* ASIC [12], developed specifically for the HGCAL applications: in addition to a dual-gain amplification, it provides a supplementary measurement of the time over threshold (ToT) used for the final pulse shape reconstruction in cases where the low-gain channel saturates. The pulse shape, after a pre-amplification stage, is sampled with a frequency of 40 MHz and stored to a 13 channels Switch Capacitor Array (SCA). In addition, the *Skiroc2-CMS* ASIC pro-

---

[1] Disposed into groups of seven, with a central module surrounded by six others in a "daisy" structure, for the first nine layers of the CE-H and followed by three single module layers.

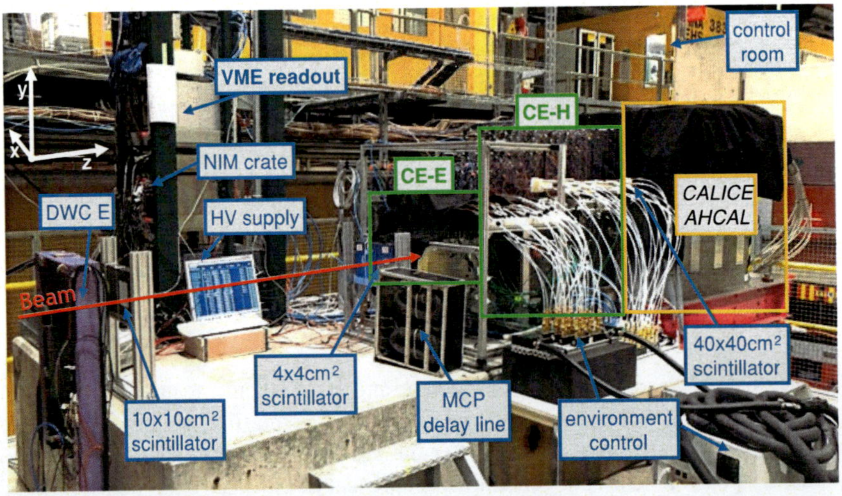

Fig. 3.8 The experimental setup of the 2018 test beam. The green and orange boxes represent the HGCAL prototype, while the blue boxes depict the additional detectors and the readout components. Figure taken and edited from Ref. [16]

vides a measurement of the pulse time-of-arrival (TOA), used for the assessment of the timing performance of the HGCAL prototype.

Figure 3.9 provides a more detailed representation of the CE-E structure, being it the object of the analysis presented in the following. The 28 modules are disposed into 14 "cassettes", each comprising a module on the two sides of an absorber made of copper (Cu) and copper-tungsten (CuW). The stacked structure of each module reflects the one described in Sect. 3.2.1, where the silicon sensor is sandwiched between a CuW absorber plate and the PCB. A cassette is defined by a structure starting with a lead plus stainless steel absorber and ending with the PCB of the second module within it, as shown in the zoomed region of Fig. 3.9. Each cassette comprises two active modules, mounted on a 6.0 mm Cu support. The zoomed region of Fig. 3.9 highlights an important feature of such a structure, where different amounts of passive materials precede odd and even layers. This observation plays a fundamental role in reconstructing longitudinal shower shapes, as discussed in detail in Sect. 3.4.3.

Monte Carlo Simulation

Test beams are the cornerstone for validating the design and assessing the physics performance of any new detector, such as the HGCAL. In these scenarios, the purity of the beam and the well-defined environment ensure complete control of all the aspects from the data taking to the subsequent analysis. In test beams and in the real use case applications, one is also interested in having an external benchmark against which to test the data analysis findings. For this reason, it is common to compare the experimental data to the expectation from detailed Monte Carlo simulations of the experimental setup.

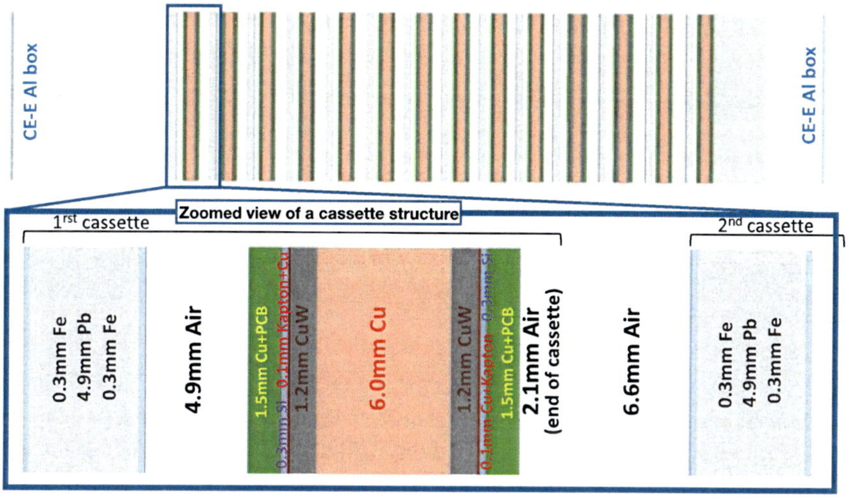

Fig. 3.9 Schematic representation of the electromagnetic compartment (CE-E) of the HGCAL prototype under test. The zoomed view depicts the structure of a cassette, which comprises two active sensors, the Cu and CuW support and absorber, and the PCB

A precise knowledge of the dimensions and the material composition of all the elements that define the experimental setup is fundamental for their accurate modelling in the simulations. The `GEANT4` [17, 18] toolkit is used to model the HGCAL prototype studied in the context of this thesis. The relative position of the simulated beam gun with respect to the face of the detector is estimated from data measuring both the offset and the inclination of the beams with respect to the center of the detector. The interaction of particles with matter is simulated using the `FTFP_BERT_EMN` physics list[2] of `GEANT4` and the recording of an event is performed only when the same trigger condition applied to data is met.

The HGCAL prototype sensors are modelled as a "sensitive detector" of `GEANT4`, meaning that the software has direct access to the hits of all the particles in the active layers. The electronic noise in the Si pads is included in the simulation, whilst possible non-uniformities in the response and cross-talk effects are not. Conversely to the detector under test, the DWCs placed upstream the experimental setup are not modelled as a sensitive detector. Therefore, the DWCs hits are taken from an extrapolation of the beam gun position and its direction and they are smeared according to the nominal DWCs resolution of 500 μm in both x and y planes.

The simulation of the detector is complemented with detailed modelling of the H2A full beamline and of all the detectors located upstream the beam test area, of which a schematic representation is given in Fig. 3.10. The `G4Beamline` tool [19] is used to simulate the H2A beamline from the T2 stripping target up to the

---

[2] The choice of this specific physics list is motivated by the increased precision on the simulation. More precisely, this list includes a description of the e^+e^- multiple scattering below 100 MeV, as well as an improved description of the bremsstrahlung and of the Compton scattering.

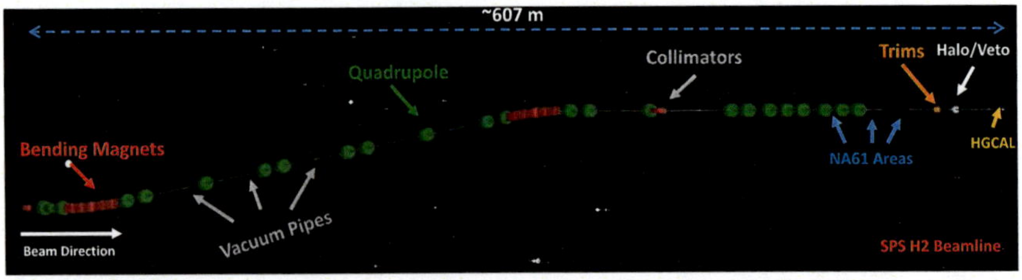

Fig. 3.10 Simulation of the H2A experimental area. All the elements included in the H2A beamline are simulated and are depicted with different colors, as indicated by the corresponding labels

experimental area, as well as to simulate the propagation of the beam and to model in a realistic manner the beam population at the entrance of the HGCAL.

Several discussions have been carried out with the H2A beamline experts in the context of this thesis to fine-tune the dedicated simulation and to achieve the most accurate description possible of the experimental setup, both in terms of the modelling of the detector and of the composition of the beam population at the entrance of the HGCAL prototype. Eventually, a beam composition of 90% of positrons and 10% of hadrons is found to model the expected beam population properly and is propagated throughout the beamline, up to the experimental area. For the first time all the details of the detector geometry, the elements surrounding the experimental setup, as well as the accurate description of the beams population have been included in what turns out to be the best MC simulation employed in the context of the HGCAL test beam campaigns. The ultimate sensitivity to upstream materials of the final modelling is estimated to be of the order of 0.2 X_0 and it is treated as a systematic uncertainty in the context of the analysis.

3.3.2 *Skiroc2-CMS Pulse Shape Reconstruction*

The *Skiroc2-CMS* ASIC features 64 channels connected to the silicon sensors' hexagonal cells and to non-hexagonal cells at the boundaries of the sensors. Each channel of the chip features low- and high-gain shapers and TOT and TOA circuits. The shapers have analog-to-digital converters (ADC) that sample the pulse shape every 25 ns, saving the information to a 13 SCA rolling array memory. In the context of the October 2018 test beam, each module was equipped with four ASICs (cf. Fig. 3.4 right) connected to a total of 128 channels, leaving half of them unconnected to facilitate the routing of the channels of such a complex board. A more detailed description of the data format and the DAQ system can be found in Ref. [22].

The reconstruction of the HGCAL data consists of the conversion of the raw data extracted from the ASIC into maps, hereafter referred to as calorimeter *hits*, comprising the energy (E_i) and time (T_i) deposit in a given cell, as well as its cartesian location (x_i, y_i, z_i). Pedestals are determined for each SCA and subtracted.

A subsequent common mode noise reduction is performed for each ASIC. Further details about these two steps of the reconstruction workflow are provided in Ref. [22]. After pedestal and common-mode noise subtractions, the time-ordered SCA raw values are used to reconstruct pulse shapes and extract the amplitudes in the low and high-gain regimes. The pulse shape data are fitted as a function of time starting from the analytical expression suggested in Ref. [23] for a bipolar waveform, with some additional modifications:

$$S(t) = \begin{cases} A_0 \left[\left(\frac{t-t_0}{\tau} \right)^2 - \frac{1}{n+1} \left(\frac{t-t_0}{\tau} \right)^{n+1} \right] \exp^{-\alpha(t-t_0)/\tau} & \text{if } t > t_0 \\ 0 & \text{if } t < t_0 \end{cases} \tag{3.2}$$

where A_0 is the pulse shape amplitude in either the low or high-gain regimes, and it is expected to be directly proportional to the charge deposited in the silicon sensors, t_0 is the starting time of the pulse shape, and τ is the time constant of the shapers. Conversely to the expression proposed in Ref. [23], the parameter n is left freely floating in the optimization of the fitting function, and an additional parameter α is introduced to properly take into account the undershoot of the *Skiroc2-CMS* pulse shape. The optimization of the analytical expression of Eq. (3.2) was carried out in the context of this thesis, where dedicated studies were performed to support the introduction of the α parameter and find the most suitable values of all the parameters that enter Eq. (3.2). For this purpose, data collected with a test stand were used to obtain a continuous description of the signal shape from the injection of a predefined charge, increased at constant time intervals. The analytical expression of Ref. [23] and the optimized version of Eq. (3.2) are compared in Fig. 3.11, where it is clear that the latter results in a better description of both the maximum amplitude (A_0) and of the undershoot of the pulse shape.

The additional parameters α, n, and τ introduced were further optimized using test beam data and they were fixed in the reconstruction framework to the following values [14]:

$$\alpha = 0.97 \pm 0.01, \quad \tau = 25.82 \pm 0.09 \text{ ns}, \quad n = 2.92 \pm 0.02$$

and only the t_0 and A_0 parameters of Eq. (3.2) are left floating in the fitting of pulse shapes.

This choice of the values for α, n, and τ allows for the reconstruction of an approximately uniform distribution of t_0 in test beam data, expected from the asynchronous time of arrival of the particles with respect to the clock cycle. As mentioned above, 13 points randomly sampled over the entire pulse shape are stored in the SCA memory and used to fit the pulse shape after an initial sorting in time. Figure 3.12 shows the reconstructed distribution of t_0 for the low-gain shaper of a sample run at 300 GeV. With a clock cycle of 40 MHz, one would expect a ~25 ns time interval in which t_0 can be distributed. The t_0 distribution of Fig. 3.12 can be considered approximately uniform within 45 and 70 ns, thus corroborating the expectation and supporting the introduction of the α parameter in Eq. (3.2).

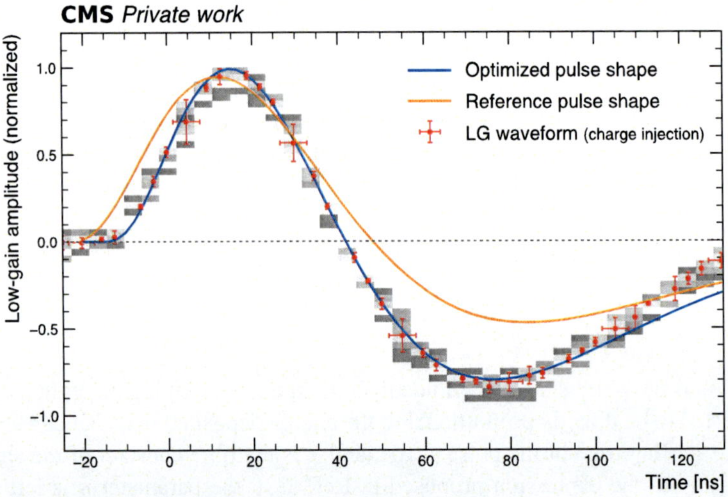

Fig. 3.11 Normalized pulse shape from low-gain shaper extracted from injection data as described in the text. The orange curve represents the analytical expression suggested in [23], while the blue curve shows a fit with Eq. (3.2) where the α, n, and τ parameters are allowed to float in the fit

Fig. 3.12 Distribution of the pulse shape maximum extracted from Eq. (3.2). Shifting the pulse shape in such a way that the rising edge starts at 0 ns, the $t - t_0$ distribution shown here corresponds to the expected position in time of the pulse shape maximum (A_0)

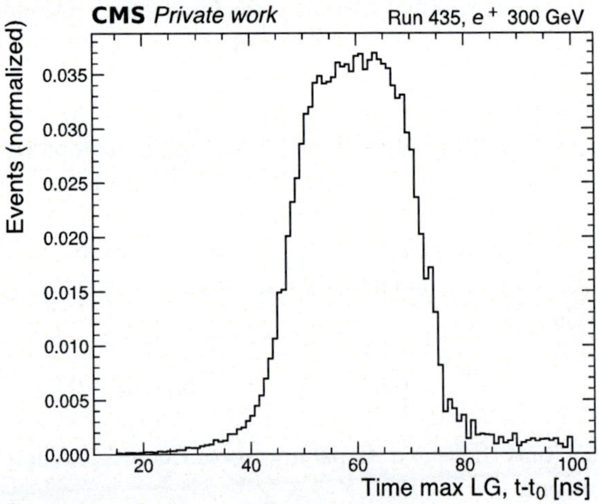

The ASICs operate in the linear region of the gain shapers or of the TOT when both the gains are saturated to ensure a direct proportionality between the reconstructed amplitude A_0 and the collected charge. Figure 3.13 shows an example of high- vs low-gain and low-gain vs TOT curves for a specific *Skiroc2-CMS* ASIC channel.

Two different slopes are derived for the linear regions for the high- vs low-gain and low-gain vs TOT regimes, and they are used to retrieve the gain-intercalibrated signal of each cell. More precisely, the reconstruction procedure selects either one of

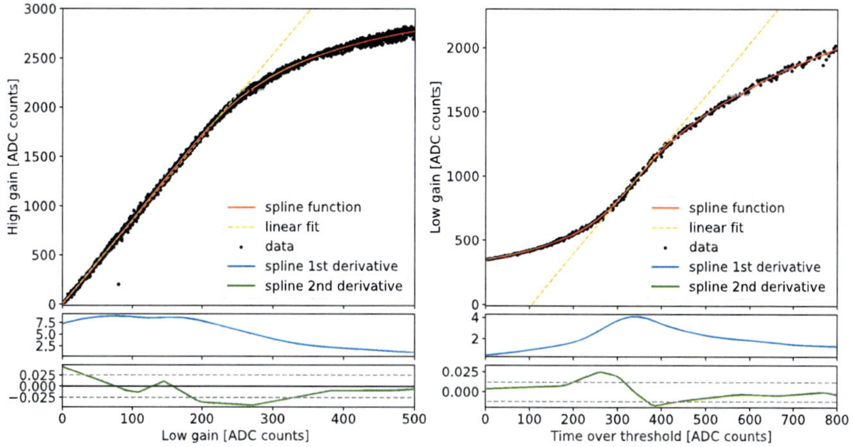

Fig. 3.13 Shapers distributions for a 280 GeV electromagnetic shower in the CE-E prototype. *Left*: High-gain amplitude as a function of the low-gain. In this specific example, the high gain value of ~1700 ADC counts corresponds to the HG_{sat} quantity in Eq. (3.3). *Right*: Low-gain amplitude as a function of the TOT. The orange lines represent the linear fit used to identify the operational regime of each shaper. In this specific example, the low gain value of ~1300 ADC counts corresponds to the LG_{sat} quantity in Eq. (3.3). Figure taken from Ref. [14]

the non-saturated gains or the TOT and converts it to the high gain equivalent A'_{HG} according to the following relation:

$$A'_{\mathrm{HG}} = \begin{cases} A^{\mathrm{HG}} & \text{, if } A^{\mathrm{HG}} < \mathrm{HG}_{sat} \\ A^{\mathrm{LG}} \cdot m_{\mathrm{HG/LG}} & \text{, if } A^{\mathrm{HG}} > \mathrm{HG}_{sat} \text{ and } A^{\mathrm{LG}} < \mathrm{LG}_{sat} \\ A^{\mathrm{TOT}} \cdot m_{\mathrm{LG/TOT}} \cdot m_{\mathrm{HG/LG}} & \text{, otherwise} \end{cases} \quad (3.3)$$

where the $m_{\mathrm{LG/TOT}}$, and $m_{\mathrm{HG/LG}}$ coefficients are calibration constants determined in the linear region of the LG vs TOT and HG vs LG curves as described above. The quantities HG_{sat} and LG_{sat} determine the transition points between the linear and saturated regimes for the HG and LG, respectively. Some of these constants could not be derived for few channels, as the response recorded in the detector was not large enough to construct the calibration curves of Fig. 3.13, especially for the TOT-to-low-gain curves. In these cases the average in a given chip or module was used for the calibration.

The final step of the calorimeter hits reconstruction consists in converting the pulse shape amplitude into the corresponding energy deposit in the active sensors in MIP units. The conversion factor involved in this step ($M^{-1}_{\mathrm{HG/MIP}}$) corresponds to the MPV of the Landau convoluted to a Gaussian function used to fit the high-gain signal distribution of 200 GeV muons, as described in Ref. [14, 16]. Further details about the calibration of the HGCAL prototype using test beam data can be found in Ref. [14, 16]. A total of ~85% of the individual channels involved in the beam test were calibrated. In the remaining cases the calibration constants are taken from the average of the neighbouring cells or from the corresponding readout chip. Eventually, the reconstructed energy deposit associated to each hit can be written as:

$$E^{\text{Si}}\ [\text{MIP}] = A'_{\text{HG}} \cdot M^{-1}_{\text{HG/MIP}}. \tag{3.4}$$

where A'_{HG} is obtained from Eq. (3.3) and $M^{-1}_{\text{HG/MIP}}$ is the conversion factor mentioned above. Most of the sensors involved in the 2018 test beam had an active thickness of 300 µm, but few 200 µm modules were installed in some layers. The MIPs energy deposit is expected to be 86 keV in the former and 57 keV in the latter [14, 16].

3.4 Performance of CE-E Prototype in Positron Beams

As a result of the reconstruction workflow described in the previous section, collections of *calorimeter hits*, hereafter referred to as *rechits*, can be created starting from the raw data extracted from the silicon sensors. More precisely, the *rechits* consist of a map comprising all the relevant "high-level" information for each event:

$$\text{Raw data} \rightarrow \textit{rechits} = \left\{x_i, y_i, z_i, E_i^{\text{Si}}, \text{TOT}_i, \text{TOA}_i\right\} \forall \text{ event}.$$

Figure 3.14 shows the distributions of the reconstructed cartesian coordinates ($\{x_i, y_i, z_i\}$) as a function of the reconstructed energy ($\{E_i^{\text{Si}}\}$) in the HGCAL prototype for an electromagnetic (left) and a hadronic (right) shower, demonstrating the *imaging* power arising from the high granularity of this calorimeter. These event displays demonstrate not only the capability of reconstructing hits down to the single-cell level, but also the very different signatures between electromagnetic and hadronic showers. It would be reductive to say that the right figure reflects the signature of *any* hadronic shower, as the propagation of this kind of cascades is strongly stochastic and at most unpredictable. However, the substantially different signatures in the calorimeter are a profitable handle when it comes to identify and reject possible contamination in the beam population under study.

Hence, before presenting the results of the assessment of the CE-E prototype properties, it is worth highlighting the different preselection cuts introduced in the analysis to ensure the highest positron-purity of the dataset considered. These consist in the removal of known detector imperfections as well as in the application of suitable selections, based on the properties of electromagnetic showers, defined to reject possible hadronic contamination present in the positron beams.

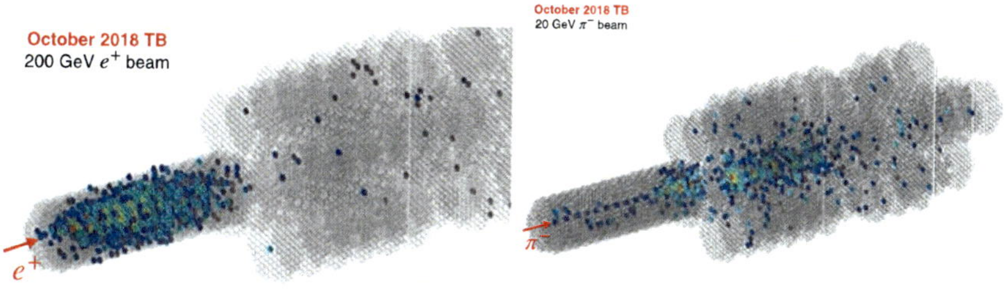

Fig. 3.14 Event displays for a 200 GeV electromagnetic shower and a 20 GeV hadronic cascade. The reconstructed cartesian coordinates ($\{x_i, y_i, z_i\}$) are used to identify the positions of the showers in the prototype and the color scale is proportional to the reconstructed energy of each hit ($\{E_i^{\mathrm{Si}}\}$)

3.4.1 Analysis Preselection Cuts

The prompt identification of defects that may alter the detector performance is crucial in the early stages of the analysis, especially when these issues were not detected during the data acquisition campaigns. Hence, data quality monitoring (DQM) plays a fundamental role both online, i.e., during data-taking operations, and offline, i.e., in the initial steps of the analysis and refinement of the reconstruction and analysis procedures.

All the *rechits* with reconstructed energy (cf. Eq. (3.4)) lower than 0.5 MIP are not selected. More precisely, a threshold of 0.5 MIP ensures the rejection of noisy hits, as the intrinsic noise in the high-gain shaper of the *Skiroc2-CMS* ASIC was found to be around 1/7 MIP, as reported in Ref. [12]. During the data acquisition phase a specific chip of the first layer[3] was found to have a malfunctioning connector, thus causing fake energy deposits in the reconstructed response of the corresponding module. Therefore, the hits of this particular ASIC are masked both in data and simulation samples, for a total of ~1.6 % channels not being considered. However, neither electromagnetic nor hadronic showers are expected to have substantial energy deposits in the first layer of the CE-E prototype, and this cut has the net effect of resolving fake rechits caused by the malfunctioning chip. Figure 3.15 shows the effect of the two preselection cuts aforementioned on the data samples collected with 20 GeV positron beams.

The H2A full beamline simulation (cf. Sect. 3.3.1) study led to the ultimate estimate of a ~10% hadronic contamination to the positron beams, especially at low energies. Therefore, a dedicated cut is applied to reduce as much as possible any residual hadronic contamination, selecting only events with less than 50 hits in the hadronic compartment of the prototype under test. The reason for this specific value is motivated by the findings reported in Ref. [16]. To further enhance the purity of the dataset considered for the analysis, events with less than 95 % of the total

[3] Chip 0 in layer 1 of the setup shown in Fig. 3.14.

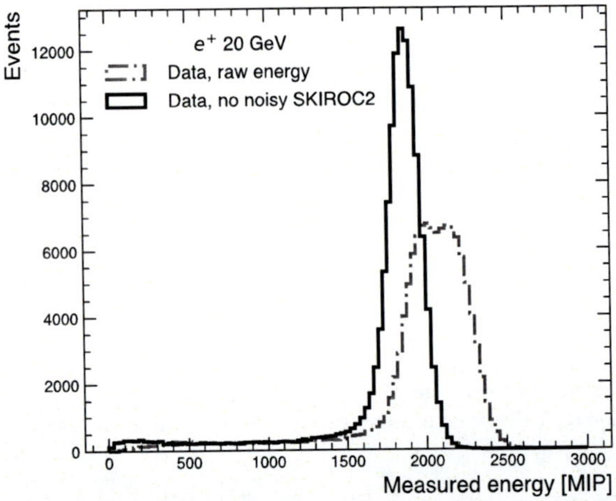

Fig. 3.15 The reconstructed energy distributions in the CE-E prototype are shown, for 20 GeV positron beam data, before (dashed histogram) and after (solid histogram) the application of the 0.5 MIP threshold and the masking of the malfunctioning ASIC in the first layer

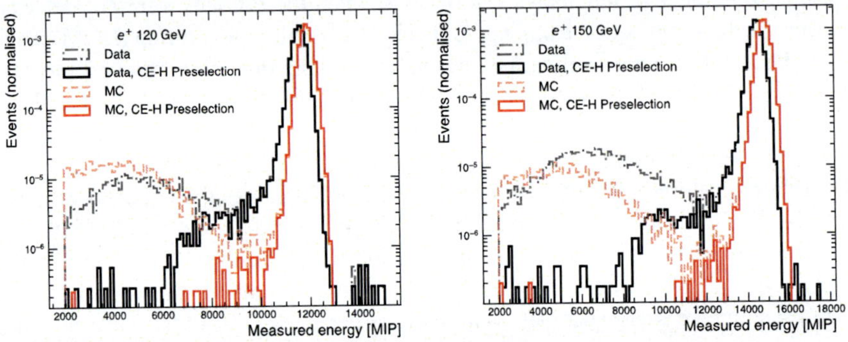

Fig. 3.16 Reconstructed energy distributions in the CE-E prototype for 120 (left) and 150 GeV (right) positron beams. The distributions are shown for data (black histograms) and simulation (red histograms) before (dashed histograms) and after (solid histograms) the application of the cuts requiring less than 50 hits in CE-H and 95% of the energy deposit in the CE-E. The residual discrepancy observed in the reconstructed energy scales is of the order of 3.5% and it is corrected for in the analysis with the application of an energy-independent scale factor, as further detailed in the text

energy deposit contained in the CE-E are rejected. The combined effect of these two cuts is illustrated in Fig. 3.16 for 120 and 150 GeV positron beams, where data and simulation are compared. The reconstructed energy scale in data is observed to be a few percent lower than the one in the simulation: this effect is further discussed in the next section and is taken into account in the assessment of the CE-E prototype's performance by introducing a global scale factor to match the two energy scales.

Table 3.2 Summary of the cuts applied to enhance the purity of the dataset considered for the analysis

Definition	Effect
$E_i^{\mathrm{Si}} > 0.5$ MIP	Removal of noisy rechits
Reject chip 0 in layer 1	Removal of a malfunctioning chip detected during the DAQ
Nr. of CE-H hits < 50	Rejection of hadronic contamination from the electromagnetic showers
95% energy deposit in CE-E	Enhance purity of electromagnetic showers
2×2 cm^2 acceptance window	Limit lateral losses

In events passing these cuts, only rechits in the CE-E are considered for the analysis

An additional cut considering only events falling within a $2{\times}2\,\mathrm{cm}^2$ acceptance window is applied to restrict the analysis to high-efficiency cells and limit the effect of lateral losses. This region is identified by extrapolating the DWCs reconstructed tracks up to the first layer of the CE-E prototype, and it is applied consistently on both data and simulation. The information on the number of tracks reconstructed in the DWCs is also exploited to reject events with more than one incoming track, thus reducing the possibility of selecting early showers and multiple tracks.

In addition, given that the focus of this study is the assessment of the physics performance of the electromagnetic compartment of the detector, only the *rechits* in the first 28 layers of the HGCAL prototype, after the application of these preselection cuts, are considered for the analysis.

Table 3.2 summarizes all the cuts applied in the analysis and discussed above.

3.4.2 Energy Resolution

The most straightforward application of a calorimeter is the reconstruction of shower energy, and its performance is often quantified by measuring the energy resolution and linearity. In this regard, the HGCAL is no different from any other calorimeter, and therefore it is logical to start exploring its characteristics by assessing these two quantities.

Three complementary methods are used in this analysis to reconstruct the energy deposit in the electromagnetic compartment: the "visible energy," the Sampling Fraction, and the dEdx method. Each of them is extensively discussed in the following.

Visible Energy

After the calibration of the detector and the mapping of the raw signals into *rechits* as described in Sect. 3.3.2, the energy deposit in the active sensors can be expressed as of Eq. (3.4). The preselection cuts presented in Sect. 3.4.1 ensure that the analysis is limited to high-purity positron beams and that a single track initiates the electromagnetic cascade propagating through the CE-E. Hence, the most straightforward way to retrieve the total shower energy is to perform an unclustered sum of all the *rechits* energy in the 28 active-layers of the CE-E, hereafter referred to as "visible energy":

$$E_{\text{CE-E}} = \sum_{i=1}^{28} \sum_{j=1}^{N_{\text{hits}} \in \text{layer}} E_{i,j} = \sum_{i=1}^{28} E_i^{Si} \tag{3.5}$$

This expression does not consider either the energy deposit in the passive layers or the additional upstream and lateral losses. While the former can be recovered by using appropriate layer weights (w_i), as discussed in the two following sections, the latter is an unknown quantity in real-life applications. Nevertheless, the estimation of these contributions is possible with the dedicated Monte Carlo simulation of the experimental setup, where the energy deposit in the absorbers is known on an event-by-event basis. Hence, the total energy deposit in the calorimeter can be expressed as:

$$E_e = \sum_{i}^{28} (w_i \times E_i^{Si}) + E_{\text{Up}} + E_{\text{Lat}} + E_{\text{Leak}} \,, \tag{3.6}$$

where the E_{Up} and E_{Lat} terms represent the upstream and lateral energy losses, respectively, while the E_{Leak} term accounts for possible energy leakage at the end of the calorimeter, even though this contribution is expected to be negligible when dealing with electromagnetic showers. These contributions are not considered in the later stages of the analysis, as they are expected to have a sub-percent effect on the overall energy reconstructed in the calorimeter and they are covered by the systematic uncertainty assigned to the knowledge of the beamline composition. The layer-dependent weights w_i are necessary to recover the energy deposit in the absorbers and reconstruct the total shower energy in GeV units. However, the "visible energy" method Eq. (3.5) still allows a complete assessment of the performance of the calorimeter. This method corresponds to the application of unitary weights in Eq. (3.6), as discussed in the following.

Gaussian functions describing the core regions of the reconstructed energy distributions are used to measure the energy resolution and linearity of the electromagnetic compartment of the HGCAL prototype. More precisely, an iterative fit procedure is applied to the interval $[\mu - 1\sigma; \mu + 2.5\sigma]$, where the asymmetric range is chosen to avoid the introduction of biases due to the presence of left tails in the distributions caused by upstream losses and bremsstrahlung of the showers. The estimators of the mean ($\hat{\mu}$) and standard deviation ($\hat{\sigma}$) are interpreted as the mean reconstructed energy ($\langle E \rangle$) and the resolution (σ_E) of the shower, respectively. Figure 3.17 depicts the normalized energy sums reconstructed in data and the corresponding Gaussian functions resulting from the fit procedure described.

The reconstructed energy scale in the simulation is found to be on average 3.5% larger than in data, as shown in the bottom panel of Fig. 3.18. Hence, a global scale factor of 3.5% is applied to data to correct for this discrepancy. It is worth to mention

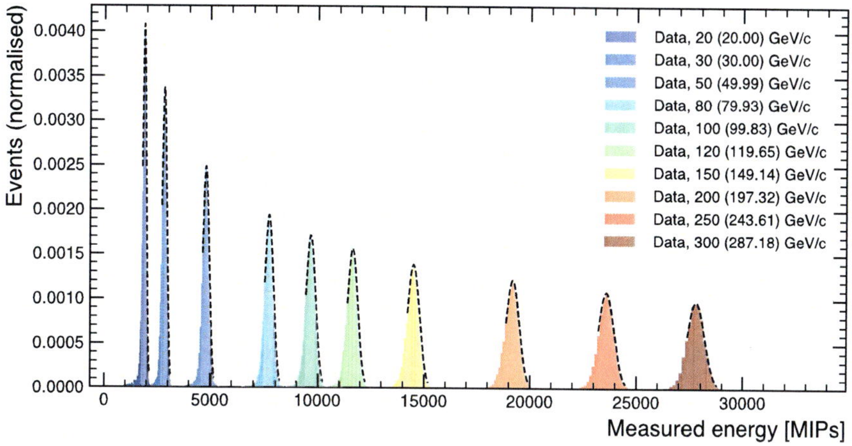

Fig. 3.17 Distributions of the reconstructed visible energy in the electromagnetic compartment of the HGCAL prototype used in the October 2018 test beam. The Data distributions are shown for positron beams with energies ranging from 20 to 300 GeV and are normalised to unity. The dashed lines represent the Gaussian fit in the core region $[\mu - 1\sigma; \mu + 2.5\sigma]$. The numbers in brackets represent the actual beam energy as measured in the H2 beamline and used in the following steps of the analysis

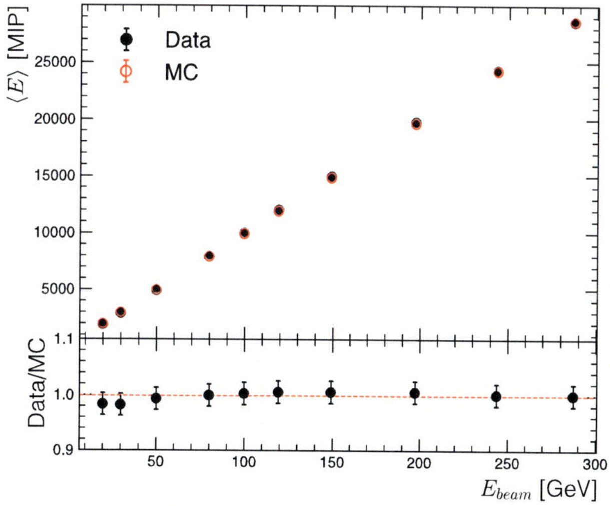

Fig. 3.18 Average energy deposit reconstructed in the active layers of the CE-E with the visible energy method as a function of the beam energy. The black points represent the data and the red open circles represent the MC simulation. The points in the lower panel correspond to the ratio between the data and MC reconstructed energies, while the red dashed line defines the 3.5% average discrepancy between the two energy scales

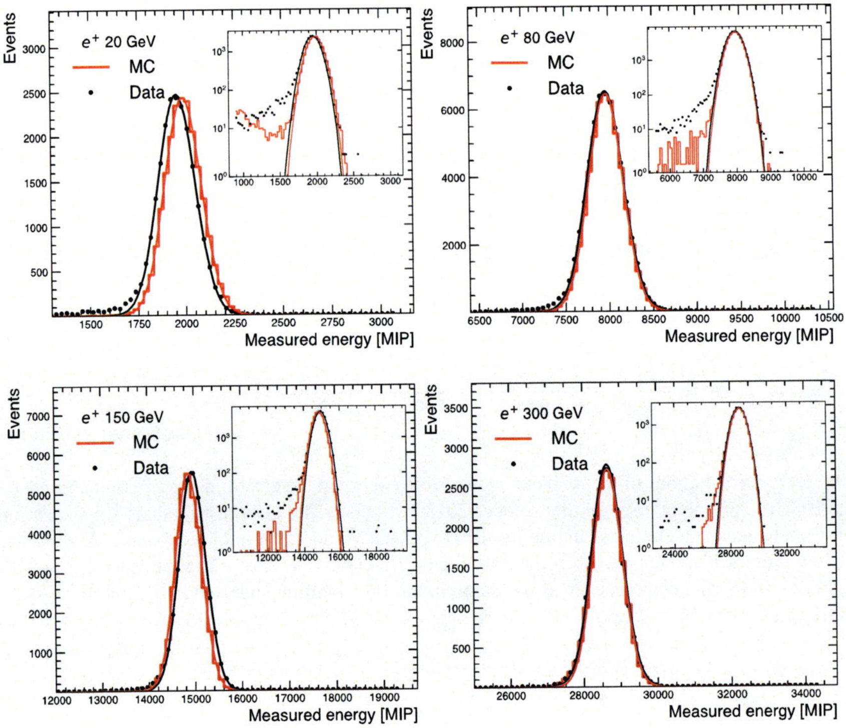

Fig. 3.19 Measured energy distributions for data (points) and simulation (histograms) for nominal positron beam energies ranging from 20 to 300 GeV. The measured energy is the total unclustered energy reconstructed in the Si pads. A factor of 1.035 is applied to the energy measured in data

here that the ultimate goal of this scaling is to observe a closure in the reconstructed energy distributions, but it does not impact directly on the assessment of the detector performance, as the observables studied in the following involve ratios where this scale factor factorizes out.

The resulting energy distributions, expressed in units of MIP (cf. Eq. (3.4)), after the application of this 3.5% scale factor, are shown in Fig. 3.19 for 20, 80, 250, and 300 GeV positron beams.

An overall qualitatively good agreement is observed throughout the entire energy range, even though a residual shift of a few percent between data and simulation energy scales is observed, especially at low energies. This discrepancy is of the order of 2.5% at low energies and it becomes negligible for higher beam energies, as shown in more detail in Fig. 3.20, and it can be realistically imputed to the difficulty of reproducing completely the beam population observed in data, especially in its hadronic and neutral components.

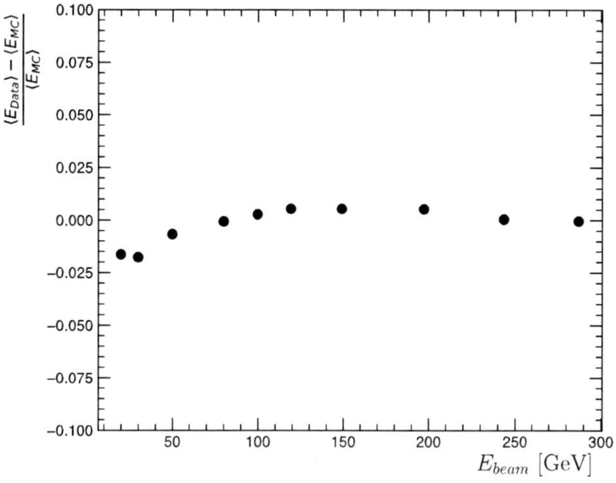

Fig. 3.20 Residual difference between the energy scales reconstructed in data and in simulation after the application of the 3.5% scale factor discussed in the text. A remarkable agreement is observed for energies above 50 GeV, while a residual 2% discrepancy is present at low energies

The reconstructed energy deposit in the CE-E (cf. Eq. (3.5)) is expected to be proportional to the true beam energy (E_{beam}) of the impinging shower:

$$E_{beam} = m^{-1} \cdot \langle E \rangle + b \tag{3.7}$$

The above equation can be used to assess the calorimeter's *linearity*, defined as the energy range for which the relation in Eq. (3.7) is satisfied. For completeness an additional offset (b) is introduced in the fit and its value is expected to be compatible with 0. The results are presented in Fig. 3.21a, where the average energy deposit $\langle E \rangle$ extracted from the gaussian fit is plotted against the true beam energy. It is worth to remark that for the validation of the CE-E performance all the synchrotron radiation effects and energy losses have to be taken into account to avoid the introduction of unwanted biases or overestimates in the analysis. Therefore all the observed quantities are tested against the energies reported in Table 3.1, resulting from the precise measurement provided by the H2A beamline spectrometer.

Equation (3.7) is found to properly describe the trend of $\langle E \rangle$ as a function of E_{beam} for both data and simulation. The goodness of fit is corroborated by the χ^2/ndf values obtained. The values of the offset b are found to be compatible with 0, consistent with the expectation.

A more precise assessment of the CE-E prototype's linearity is performed by means of a linear regression on Eq. (3.7) comparing the $\langle E \rangle / m$ values extracted from the fit with the true beam energies. Figure 3.21b shows the $(\langle E \rangle / m - E_{beam}) / E_{beam}$ residuals for both data and simulation. An optimal linearity is found of the points extracted from the simulation, with flat residuals across the entire energy range, thus confirming the expected linearity of the CE-E prototype under test. On the other hand,

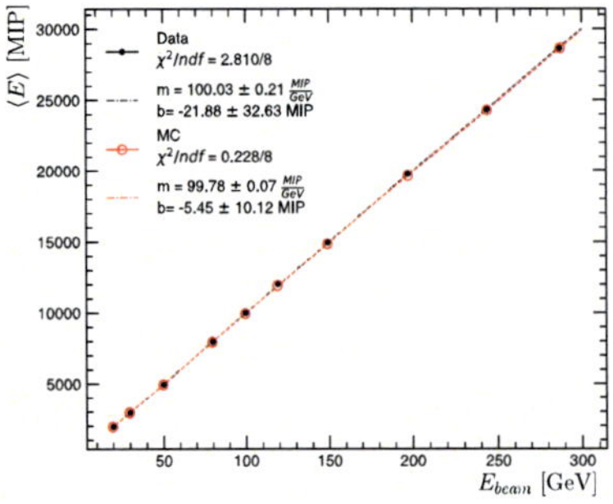

(a) Energy linearity measured for data (black points) and simulation (red circles) using the "visible energy" method discussed in the text. An energy independent scale factor of 3.5 % is applied data to match the energy scale observed in MC.

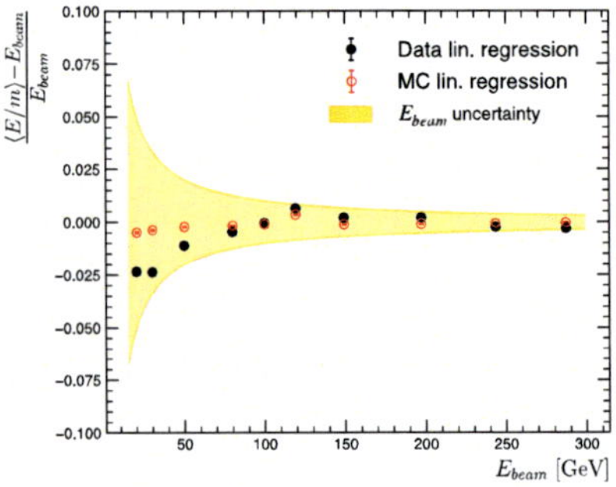

(b) Residuals of the linear regression used to measure the energy linearity of the CE-E prototype using the "visible energy" method. The yellow band corresponds to the relative error on the beam energy due to the uncertainty of the dipole currents.

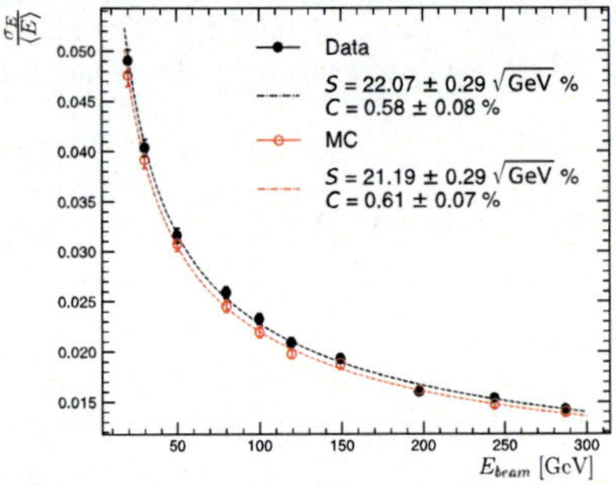

(c) Energy resolution measured for data (black points) and simulation (red circles) using the "visible energy" method. The dashed lines represent the fit with the analytical expression of eq. (3.8). The values of the stochastic and constant terms extracted from the fit are reported in the legend.

Fig. 3.21 Measurement of the energy resolution and linearity of the CE-E prototype using the "visible energy" method

a residual non-linearity of the order of $\mathcal{O}(2.5\%)$ is observed in data at low energies. While this is effect is small and not necessarily dramatic for the final performance of the HGCAL, it is a cornerstone in assessing the ultimate performance expected for this detector. Possible causes of this non-linearity can be undetectable defects in the data acquisition or early bremsstrahlung effects that are not properly considered in the simulation. In any case, all the values measured are found to be in agreement with the expectation from the simulation within the systematic uncertainty on beam energy, dominated by the knowledge on the current of the dipoles used for the measurement.

Notwithstanding this non-linearity and the discrepancy between data and simulation's energy scales, it is still possible to provide a complete characterization of the physics performance of the CE-E prototype through observables unaffected by these peculiarities. For instance, the results presented in Sect. 3.4.3 on the assessment of the longitudinal shower shapes are valid independently on the absolute energy scale. The same holds for another important property of calorimeters: the *energy resolution*, defined as the ratio of the energy distribution's width (σ_E) to the mean ($\langle E \rangle$), with the two parameters extracted from the fitting procedure discussed above. The CE-E prototype energy resolution is shown in Fig. 3.21c as a function of the beam energy and it is found to range from 5% at 20 GeV to less than 1.5% at 300 GeV. A remarkably good agreement between data and simulation is observed throughout the entire energy range, especially at the highest energies. The results are interpreted with the theoretical, analytical expression used to parametrize resolution in calorimeters [24]:

$$\left(\frac{\sigma_E}{\langle E \rangle}\right)^2 = \frac{S^2}{E_{beam}} + C^2 \tag{3.8}$$

where S is also referred to as *stochastic* term and describes the fluctuations intrinsic to the shower's development. It is particularly relevant in sampling calorimeters as the one studied in this analysis, as the energy deposit in the active media can have large fluctuations due to the presence of absorber layers. These are often referred to as *sampling fluctuations*, arise from the number of charged particles crossing the active layers, and are a possible limiting factor for the energy resolution of this kind of calorimeters. A uniform sampling can be adopted to reduce them as much as possible and to avoid drastic limitations in the physics performance. Alternatively, one can use *sampling fractions* (SF) to determine the energy deposit in the absorbers and possibly recover, or even improve, part of the calorimeter's energy resolution. The use of the SF method in the context of this analysis and the corresponding results are discussed in the next section.

The C term in Eq. (3.8) is also referred to as *constant* term and describes energy independent contributions to the resolution, such as nonuniformities arising from the detector geometry, its mechanical structure, and aging. Solid-state calorimeters, such as the one studied here, are often designed targeting C values of the order of 1 % or lower.

Equation (3.8) is usually supplemented with a term (N) proportional to E_{beam}^{-1} to take into account effects coming from the electronic noise. This term is not included in the parametrization used for the fits for two reasons: first of all, such a term would introduce non-negligible correlations with the S term, being both energy-dependent; in addition, the electronic noise is not included[4] in the simulation of the sensors. Hence, one would have to use different parametrizations in the description of the results extracted from data and simulation. In the early stages of the analysis, a fit on data was performed including the noise term in Eq. (3.8): while a non-negligible correlation with the stochastic term was observed, its value was found

[4] However it is modelled by introducing an additional gaussian smearing.

to be compatible with 0, thus supporting the argrment for its neglection in the final results.

The results of the fits with Eq. (3.8) are shown by the dashed lines in Fig. 3.21c. Both data and simulation are observed to have a good agreement with the theoretical model. The stochastic and constant terms for data are measured to be $S = 22.1 \pm 0.3\ \sqrt{\text{GeV}}\ \%$ and $C = 0.58 \pm 0.08\ \%$, both in agreement with the results extracted from the simulation.

All these results are found to be compatible with the expectations from the theory and from the corresponding modelling in the MC simulation, confirming the good performance of the detector under test. The analysis strategy detailed in this section allows a clear quantification of the CE-E prototype's properties, leading to measurements of its energy resolution and linearity and validating the application of such a detector over a wide energy range. As mentioned above, the reconstruction of the shower's energy via the unclustered sums of the active layers' deposits neglects the energy losses in the absorbers, while their recovery is a fundamental input for the ultimate reconstruction performance of the HGCAL. In the next two sections the sampling fraction and dEdx methods are outlined: both consist of appropriate weights (w_i in Eq. (3.6)) to recover the energy deposit in the passive layers to reconstruct the total energy of electromagnetic cascades accurately.

The Sampling Fraction Method

Sampling fraction constants (SFs) are commonly used in sampling calorimeters for the calibration of the energy response. This method consists in the application of layer-dependent fractions:

$$\text{SF}_i = \frac{E_i^{Si}}{E_i^{Si} + E_i^{\text{Abs}}}. \tag{3.9}$$

where E_i^{Abs} corresponds to the amount of energy deposit in the passive layers placed before the active sensors. Hence, imposing $w_i = \text{SF}^{-1}$ in Eq. (3.6), one can reconstruct the total energy of the electromagnetic cascade. This information is usually extracted from simulation and it relies on the precise modelling of the experimental setup. The SFs of the CE-E prototype layers are computed starting from the dedicated simulation introduced in Sect. 3.3.1 that is used to retrieve the true energy deposit in each passive layer of the CE-E prototype. In addition, given the regular sampling of the detector, one can average the expression of Eq. (3.9) over all the CE-E layers and thus obtain a global scale factor:

$$SF = \frac{\sum_{i=1}^{28} \left(E_i^{Si}\right)}{\sum_{i=1}^{28} \left(E_i^{Si} + E_i^{\text{Abs}}\right)}. \tag{3.10}$$

These weights are used to reconstruct the total energy deposit in the calorimeter as:

$$E^{\text{SF}} = \text{SF}^{-1} \times E_{\text{tot}}^{\text{Si}}, \tag{3.11}$$

Table 3.3 Inverted sampling fraction constants used for the estimation of the energy loss in the absorbers. The weights are derived from the MC simulation of the CE-E prototype

Beam energy [GeV]	SF^{-1}
20	108.6
30	108.5
50	108.5
80	108.6
100	108.6
120	108.7
150	108.8
200	108.9
250	109.0
300	109.0

In a test beam scenario one has direct access to the true beam energy and this information was used to study the possible energy-dependence of the sampling fraction of Eq. 3.10. Hence, global SF are derived from the simulation of positron-induced cascades with energies between 20 and 300 GeV and are used to assess the energy-independence of this method. The inverted SF values used in Eq. (3.11) are reported in Table 3.3 for all the beam energies. As a result of the uniform sampling of the CE-E prototype, the inverted SF values are essentially stable across the entire energy range studied. In a real life application one would derive a single set of SF and apply it to all the shower energies. In the following the performance of the CE-E prototype has been assessed with energy-dependent weights as no big fluctuation in their value is observed.

The use of SF weights is one of the most common methods employed in sampling calorimeters [24] for the recovery of the energy loss in the absorbers, especially because it often gets to a reconstructed energy that "closes" perfectly to the true shower energy and it results in a stable linearity [25].

The energy resolution and linearity of the CE-E prototype are studied with the same strategy detailed in the previous section and the results are presented in Fig. 3.22.

A good linearity is observed for both data and simulation, as shown in the χ^2/ndf values of Fig. 3.22a. The $\langle E \rangle$ trend as a function of E_{beam} follows Eq. (3.7), with the offset b found to be compatible with 0. For both data and MC the $\langle E \rangle$ is found to be 3% lower than the true beam energy ($m = 0.97$ in the linear regression): while the SF method is expected to "close" perfectly to the true beam energy, possible upstream and lateral losses can be responsible of this marginal discrepancy. The residuals of the linearity are shown in Fig. 3.22c and corroborate the results found with the "visible energy" method: the $(\langle E \rangle / m - E_{beam}) / E_{beam}$ residuals are flat throughout the entire energy range for the simulation, whereas a non-linearity of about 2.5% is observed in data at low energies. Similar to what observed for the visible energy method presented above, also in this case the CE-E is found to be linear within the systematic uncertainty on the knowledge of the beam energy, as shown in Fig. 3.22b. The measurement of the CE-E resolution is reported in Fig. 3.22c and the

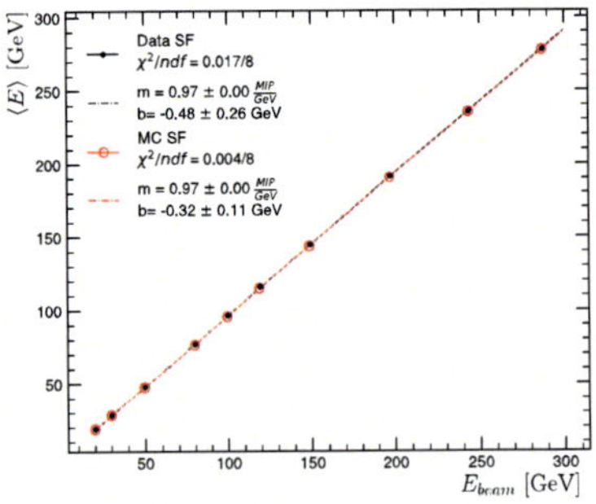

(a) Energy linearity measured for data (black points) and simulation (red circles) using the SF method discussed in the text. An energy independent scale factor of 3.5 % is applied data to match the energy scale observed in MC.

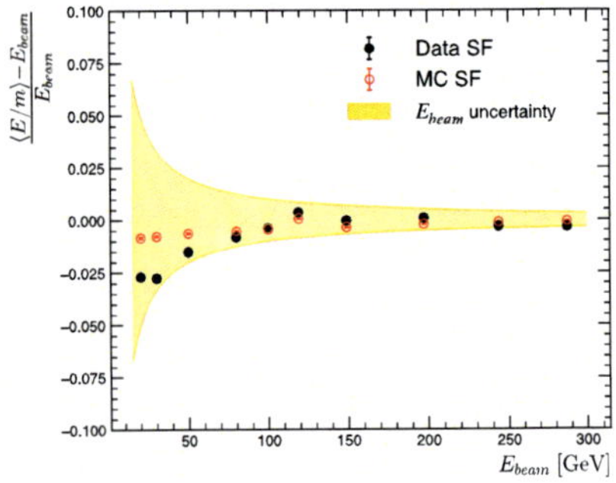

(b) Residuals of the linear regression used to measure the energy linearity of the CE-E prototype. The yellow band corresponds to the relative error on the beam energy due to the uncertainty of the dipole currents.

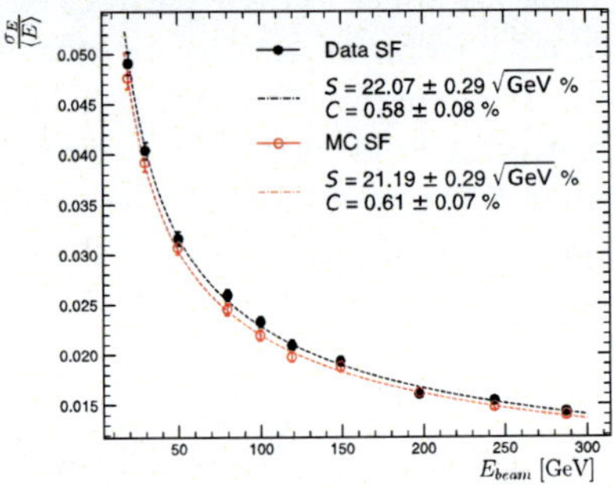

(c) Energy resolution measured for data (black points) and simulation (red circles) using the SF method. The dashed lines represent the fit with the analytical expression of eq. (3.8). The values of the stochastic and constant terms extracted from the fit are reported in the legend.

Fig. 3.22 Measurement of the energy resolution and linearity of the CE-E prototype using the SF method

results are found to be in agreement with the ones presented in the previous section, where only the information of the active layers is used. More precisely, when SF weights are applied, the stochastic and constant terms for data are measured to be $S = 22.1 \pm 0.3\ \sqrt{\text{GeV}}\,\%$ and $C = 0.58 \pm 0.08\,\%$, in agreement with the expectation from simulation.

The results presented in this section validate the use of the SF method to recover the energy deposit in the passive layers and reconstruct the total shower energy. On the other hand, it is not surprising that these results are identical to those obtained with the "visible energy" method, as in practice the SF approach consists of applying a constant weight (SF^{-1}) to the unclustered energy sums used in the previous section.

The dEdx Method

The conventional method used to calibrate the energy response in sampling calorimeters is based on dE/dx weights. More precisely, this method aims to recover of the energy deposit in passive layers by extrapolating the information from the mean of the dE/dx distribution for a MIP in the absorbers. Hence, the dEdx method, conversely to the sampling fraction one outlined in the previous section, is sensitive to the detector's geometry and the active layers' material.

Given a sampling calorimeter, the mean dE/dx deposit for a MIP in the ith absorber layer of thickness Δx_i^{Abs} can be written as $\Delta E_i^{\text{Abs}} = \frac{dE^{\text{Abs}}}{dx} \cdot \Delta x_i^{\text{Abs}}$. Being n_i^{Abs} the total number of MIPs crossing the layer, the total energy deposited can be expressed as:

$$E_i^{\text{Abs}} = n_i^{\text{Abs}} \times \Delta E_i^{\text{Abs}} \tag{3.12}$$

The number of particles (interpreted as MIPs) n_i is estimated from the average of the energy deposits in the two adjacent active layers $i-1$ and i. Hence:

$$\begin{aligned} n_1^{\text{Abs}} &= \frac{1 + E_1^{Si}[\text{MIP}]}{2}, \\ n_i^{\text{Abs}} &= \frac{E_{i-1}^{Si}[\text{MIP}] + E_i^{Si}[\text{MIP}]}{2} \quad \text{for } i = 2, ..., 28. \end{aligned} \tag{3.13}$$

where n_1^{Abs} is the MIP average energy deposit in the first layer and is calculated assuming a single MIP traversing it. The number of MIPs in the active sensors, labelled as $E_i^{Si}[\text{MIP}]$ in Eq. (3.13) is obtained from the ratio of the high-gain ADC readout extracted from the ith layer to the most probable value (MPV) of the MIP distribution in ADC.

Hence, the reconstructed energy of electromagnetic cascades in the CE-E prototype can be expressed from the combination of Eqs. (3.12) and (3.13):

$$\begin{aligned} E^{\text{dEdx}} &= \frac{\Delta E_1^{\text{Abs}}}{2} + \sum_{i=1}^{27} \left(\frac{\Delta E_i^{\text{Abs}} + \Delta E_{i+1}^{\text{Abs}}}{2} + \Delta E_i^{Si} \right) \times E_i^{Si}[\text{MIP}] + \\ &+ \left(\frac{\Delta E_{28}^{\text{Abs}}}{2} + \Delta E_{28}^{Si} \right) \times E_{28}^{Si}[\text{MIP}] = \\ &= \frac{\Delta E_1^{\text{Abs}}}{2} + \sum_{i=1}^{28} \left(W_i E_i^{Si}[\text{MIP}] \right), \end{aligned} \tag{3.14}$$

The quantity ΔE_i^{Si} is defined in an analogous way to ΔE_i^{Abs} and corresponds to the MPV extracted from the reconstructed energy deposit of a MIP in the ith active layer. The values of the quantities that enter Eq. (3.14) are computed from inputs[5] extracted from the PDG [26] and are reported in Table 3.4.

[5] I.e. the information on the materials and on the mean energy deposit of a MIP.

Table 3.4 Input parameters for the application of the dEdx method. The mean energy loss for a MIP in the ith absorber layer (ΔE_i^{Abs}) is reported in the second column and is extracted from Ref. [26]. The average of the mean energy deposit in the absorbers preceding and following the ith active layer are given in the third column ($(\Delta E_i^{Abs} + \Delta E_{i+1}^{Abs})/2$) and enter directly Eq. (3.14) for the reconstruction of the total shower energy. The sole exception is for the last active layer of CE-E, where $(\Delta E_{28}^{Abs})/2$ is used directly. The difference observed for the layers from 21 to 23 arises from the use of copper-tungsten baseplates instead of the standard copper ones used in the other layers. The fourth column represents the peak value of the energy deposit by a MIP in the active sensors. This value is extracted from a simulated sample of 150 GeV muons and is the same used for the reconstruction procedure detailed in Sect. 3.3.2. The sum of the values in the third and fourth columns corresponds to the weight that enters in Eqs. (3.14) and (3.6). The last column specifies the position of each layer in units of radiation length

Layer number i	Energy loss ΔE_i^{Abs} [MeV]	Average $(\Delta E_i^{Abs} + \Delta E_{i+1}^{Abs})/2$ [MeV]	Energy loss ΔE_i^{Si} [MeV]	Depth z [X_0]
1	10.2	11.29	$88.92\ 10^{-3}$	0.9
2	12.3	9.85	$88.92\ 10^{-3}$	1.9
3	7.4	9.85	$88.92\ 10^{-3}$	2.9
4	12.3	9.85	$88.92\ 10^{-3}$	3.9
5	7.4	9.85	$88.92\ 10^{-3}$	4.8
6	12.3	9.85	$88.92\ 10^{-3}$	5.8
7	7.4	9.85	$88.92\ 10^{-3}$	6.7
8	12.3	9.85	$88.92\ 10^{-3}$	7.7
9	7.4	9.85	$88.92\ 10^{-3}$	8.6
10	12.3	9.85	$88.92\ 10^{-3}$	9.6
11	7.4	9.85	$88.92\ 10^{-3}$	10.5
12	12.3	9.85	$88.92\ 10^{-3}$	11.5
13	7.4	9.85	$88.92\ 10^{-3}$	12.4
14	12.3	9.85	$88.92\ 10^{-3}$	13.4
15	7.4	9.85	$88.92\ 10^{-3}$	14.3
16	12.3	9.85	$88.92\ 10^{-3}$	15.3
17	7.4	9.85	$88.92\ 10^{-3}$	16.3
18	12.3	9.85	$88.92\ 10^{-3}$	17.2
19	7.4	9.85	$88.92\ 10^{-3}$	18.2
20	12.3	9.85	$88.92\ 10^{-3}$	19.2
21	7.4	11.36	$88.92\ 10^{-3}$	20.1
22	15.4	11.36	$88.92\ 10^{-3}$	21.3
23	7.4	11.36	$88.92\ 10^{-3}$	22.2
24	15.4	11.36	$88.92\ 10^{-3}$	23.4
25	7.4	9.85	$88.92\ 10^{-3}$	24.3
26	12.3	9.85	$88.92\ 10^{-3}$	25.3
27	7.4	9.85	$57.\ 10^{-3}$	26.2
28	12.3	6.17	$57.\ 10^{-3}$	27.2

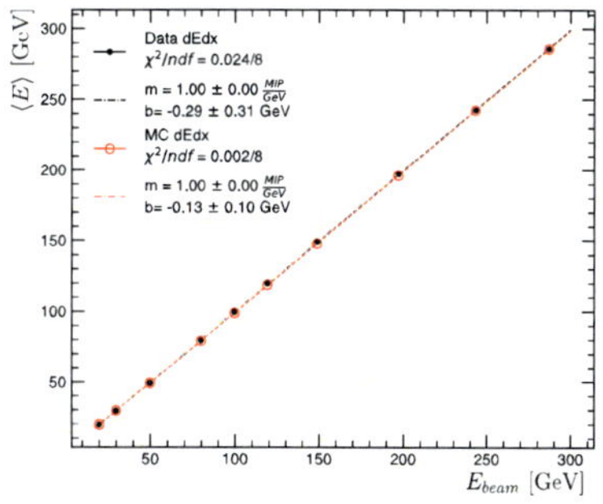

(a) Energy linearity measured for data (black points) and simulation (red circles) using the dEdx method discussed in the text. An energy independent scale factor of 3.5 % is applied data to match the energy scale observed in MC.

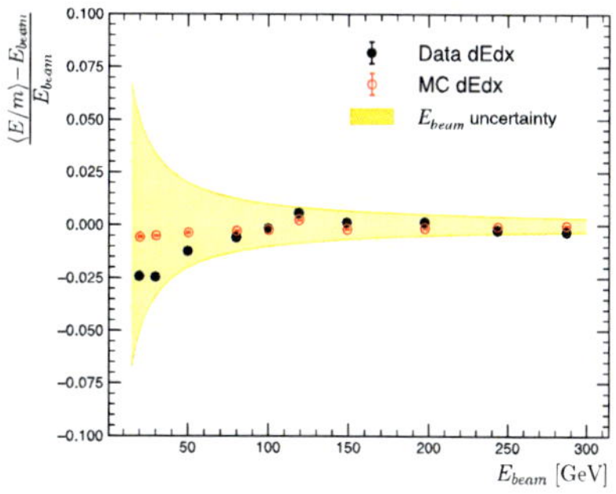

(b) Residuals of the linear regression used to measure the energy linearity of the CE-E prototype with the dEdx method. The yellow band corresponds to the relative error on the beam energy due to the uncertainty of the dipole currents.

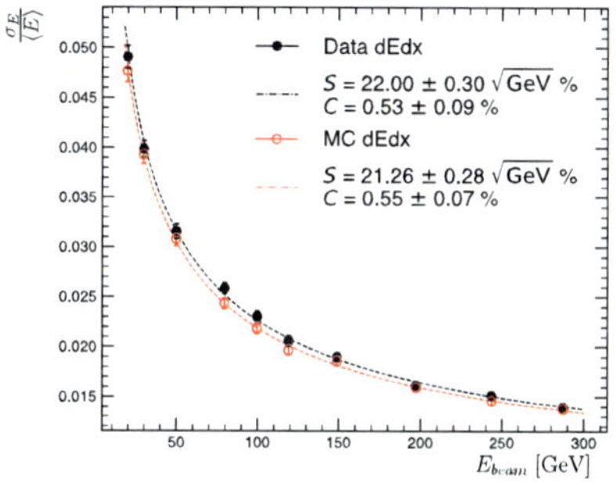

(c) Energy resolution measured for data (black points) and simulation (red circles) using the dEdx method. The dashed lines represent the fit with the analytical expression of eq. (3.8). The values of the stochastic and constant terms extracted from the fit are reported in the legend.

Fig. 3.23 Measurement of the energy resolution and linearity of the CE-E prototype using the dEdx method

The energy resolution and linearity measured with the dEdx method are reported in Fig. 3.23. This method yields slightly better linearity than the SF one, as shown in Fig. 3.23a: the fitted value of the m coefficient in Eq. (3.7) is found to be $m = 1.0$ and the additional offset (b) introduced in the fit is measured to be compatible with 0, consistently with the findings highlighted in two previous sections for the "visible energy" and SF methods. The residuals of the linear fit on $\langle E^{\mathrm{dEdx}} \rangle$ are shown in Fig. 3.23b and are corroborated by the results obtained using SF weights for the passive layers' compensation. Also when the dEdx method is used to calibrate the response of the detector a residual non-linearity of the order of $\mathcal{O}(2.5\%)$ is observed in data at low energies, whereas above 50 GeV a flat trend characterizes the linearity, thus matching the expectations from the simulation. The constant term of the energy resolution improves with respect to the measurements performed on visible and SF weighted energy sums. The stochastic and constant terms for data are measured to be $S = 22.0 \pm 0.3\ \sqrt{\text{GeV}}\%$ and $C = 0.53 \pm 0.09\%$ and are found to be in agreement with the expectation from simulation, as shown in Fig. 3.23c.

Summary of the energy resolution and linearity measurements

The application of the dEdx method leads to slightly more precise measurements of the energy resolution and linearity if compared to the results presented above for the "visible energy" and the sampling fractions methods, reflecting the larger sensitivity of the dEdx weights to the detector's geometry and composition.

The measurements of the CE-E prototype energy resolution with the three methods outlined in this section are summarised in Table 3.5. These findings are in agreement with the results reported in Ref. [25], where the ultimate performance of the HGCAL detector and a comparison between the SF and dEdx methods are presented.

The resolution measured with the dEdx calibration yields better stochastic and constant terms when compared to the other methods presented. Overall, all the results are found to be in statistical agreement with each other. The stochastic and constant terms measured on test beam data are found to be in agreement with the expectations from the MC simulation for all the reconstruction methods studied.

The CE-E is found to be linear within a 2.5% level, with the measurement on data well represented by the expectations from the simulation. The residual non-linearity of the order of 2.5% observed for all the three methods studied can be imputed to the difficulty of modelling precisely the beam population and its interactions with the passive materials encountered upstream the CE-E prototype, but it is compatible with the relative error on the beam energy due to the systematic uncertainty on the dipole currents.

These results mark a significant step forward in understanding the HGCAL, both in the assessment of its physical properties and in the validation of the detector's design. The measurements of the energy resolution and linearity presented here

Table 3.5 Positron energy resolution stochastic and constant terms for measured energies in Si (Vis), dEdx calibrated energies (dEdx) and SF calibrated energies (SF) for data and simulation. A good agreement between data and MC in the stochastic and constant term is observed in both cases. In data, the constant term in the case of dEdx calibration is slightly higher than the one for the measured energies

Positron energy resolution

	Vis S [$\sqrt{\text{GeV}}$] %	Vis C %	SF S [$\sqrt{\text{GeV}}$] %	SF C %	dEdx S [$\sqrt{\text{GeV}}$] %	dEdx C %
Data	22.0 ± 0.3	0.6 ± 0.1	22.0 ± 0.3	0.6 ± 0.1	22.0 ± 0.3	0.5 ± 0.1
MC	21.0 ± 0.3	0.7 ± 0.1	21.0 ± 0.3	0.7 ± 0.1	21.1 ± 0.2	0.6 ± 0.1

are obtained from the analysis of the data collected with the very first large scale prototype of the HGCAL and their remarkable agreement with the expectations from simulation makes confident that the HGCAL will meet the expected physics requirements foreseen in the TDR [4].

The measurement of the energy resolution and linearity of the CE-E prototype is fundamental in validating the detector design and in the assessment of its physics performance. Nevertheless, these results do not make use of the most remarkable features of the HGCAL: its high granularity and the fine segmentation that allow a complete reconstruction of the longitudinal and transverse shower shapes. The large scale of the detector tested in October 2018 and the variety of beam energies to which it was exposed permit an extensive characterization of its properties: from the reconstruction of the shower profiles to the measurement of the pointing resolution.

3.4.3 Longitudinal Shower Shapes

The HGCAL is often referred to as *imaging*, or 5D, calorimeter as it is designed to provide a complete characterization of the shower's propagation, as qualitatively illustrated by the event displays in Fig. 3.14. As mentioned in the introduction, the "five dimensions" provided by the HGCAL are the three spatial coordinates that can be used to reconstruct the shower profiles as presented in this section, the precision measurement of the shower energy, presented in the previous sections, and the timing information.

The energy weighted longitudinal position can be used to reconstruct the showers' center of gravity (COG_z), defined as:

$$COG_z = \frac{\sum_{i=1}^{28} E_i^{\mathrm{Si}} \cdot z_i \ [\mathrm{X}_0]}{\sum_{i=1}^{28} E_i^{\mathrm{Si}}}. \tag{3.15}$$

This observable can be thought of as the longitudinal barycenter of the electromagnetic shower, and it can be used to identify the region of the detector closest to the shower maximum, and to assess the complete containment of the cascade. The distributions of the COG_z, also referred to as *shower depth*, are shown in Fig. 3.24 for 20, 30, 150, and 300 GeV positron beams, where the simulation is normalized to the number of events collected in data to ease the comparison between the two. Especially in the core of the distributions, the simulation is found to model fairly well the data.

The average shower depth, $\langle COG_z \rangle$ is expected to have a $\log(E_{beam})$ dependence (cf. Ref. [26]). This comes from the fact that the maximum number of shower particles is proportional to the beam energy. More precisely, the propagation of the shower implies exponential growth of the number of particles as a function of the shower depth and therefore $\log(N_{\mathrm{particles}}) \propto \log(E_{beam})$. Figure 3.25 shows the comparison

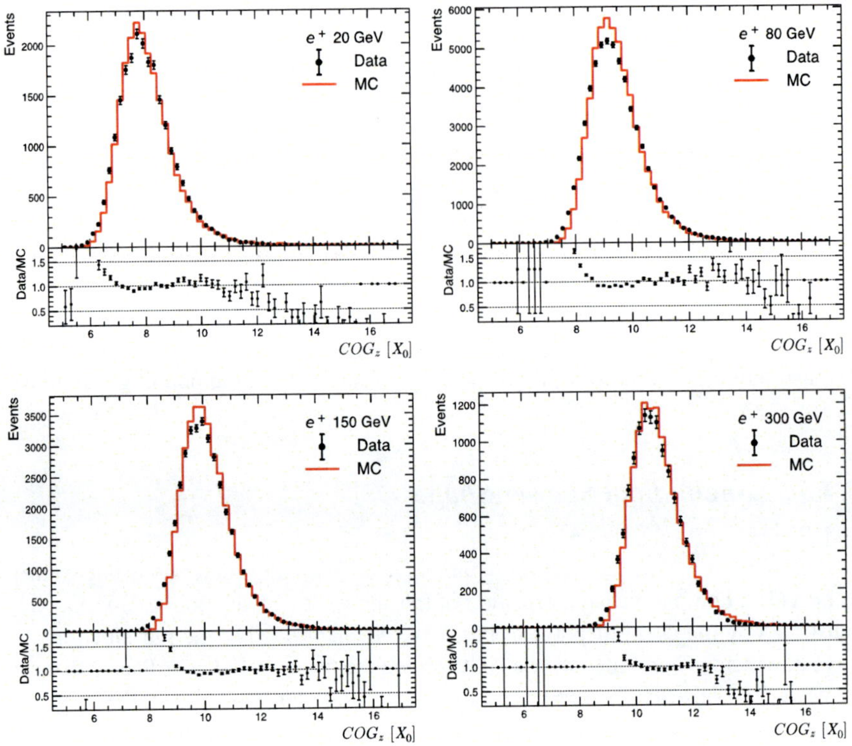

Fig. 3.24 Shower depth distributions, comparison between data (black points) and simulation (red histogram) for positron nominal beam energies ranging from 20 to 300 GeV

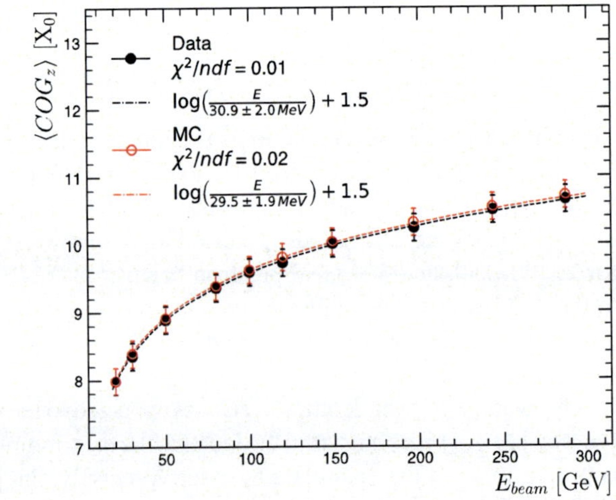

Fig. 3.25 Average shower depth, extracted from the shower depth distributions, as a function of the beam energy. The points in black represent data, while the red open circles are for the simulation. The fit with the expected $\log(E_{beam})$ dependence is shown with the dashed lines

of the $\langle COG_z \rangle$ values between data and simulation, as well as the fit with the expected relation:

$$\langle COG_z \rangle = \log\left(\frac{E_{beam}}{E_c}\right) + 1.5 \tag{3.16}$$

where E_c represents the only free parameter of the fit and can be thought, in first approximation, as the calorimeter's critical energy.[6]

Notwithstanding the shower depth distributions provide an insight into the average position of the shower in the calorimeter, this information still does not exploit the full power of the HGCAL arising from the fine granularity. A more accurate description of the propagation of the showers can be achieved from the study of the longitudinal profiles. Starting from the energy deposit in each layer:

$$E_{\text{layer}} = \sum_{i=1}^{N_{\text{hits}} \in \text{layer}} E_i$$

and taking the average energy per layer, $\langle E_{\text{layer}} \rangle$, one can reconstruct the longitudinal development of electromagnetic cascades throughout the 28 layers of the CE-E.

Figure 3.26 shows the comparison between data and simulation for the longitudinal profiles at 20, 100, and 300 GeV.

Two remarkable features arise from observing these profiles: first of all, one can note the disagreement in the average energy deposits in the 10th layer, placed at a depth of ~9.6 X_0. The positron beam hitting a specific cell with a reconstruction efficiency lower than the neighbouring pads was responsible for this issue. Hence, dedicated calibration constants were re-derived and applied to the dataset used for the analysis results presented in this thesis. Nonetheless, a residual local inefficiency remained, resulting in the disagreement observed here. The second striking feature of the longitudinal profiles presented in Fig. 3.26 is the oscillations of the entire profile, with a more pronounced effect around the shower maximum. To study this phenomenon, one has to decouple two different effects that may contribute to it: the first is the phenomenology of the showers in the calorimeter, while the latter is the intrinsic sampling structure of the detector.

The photon emission mainly drives the propagation of an electromagnetic shower in a material due to bremsstrahlung and photon decay into electron-positron pairs (cf. Ref. [26]). When the average losses due to bremsstrahlung equal those from ionisation, the multiplicative behavior of the shower ceases, and the cascade begins to fade away. This (simplistic) description is complicated by the presence of additional particles emitted throughout the propagation of the shower, such as δ-rays, i.e., secondary electrons produced along the primary shower axis that have sufficient energy to produce further ionization. Hence, the presence of these δ-rays may alter the response of the calorimeter, as discussed in the following.

[6] As explained better in the following, E_c strictly represents the critical energy of the calorimeter only for homogeneous media.

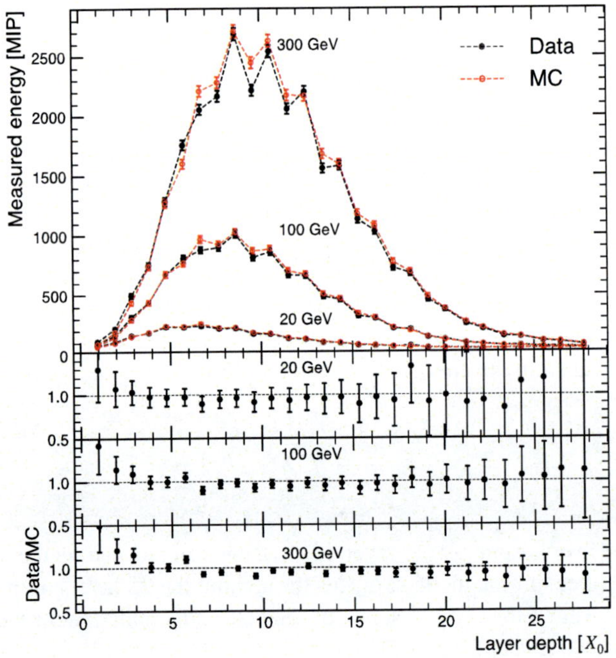

Fig. 3.26 Longitudinal profiles in the electromagnetic compartment of the HGCAL prototype for 20, 100, and 300 GeV positron beams

The second aspect that may impact the reconstructed longitudinal profiles is the sampling structure of the CE-E prototype. Conversely to homogeneous calorimeters, sampling ones enable a characterization of the longitudinal profiles of the showers. However, only part of the total shower energy is deposited in the active material, as part of the cascade loses its energy in the passive layers and is not recorded by the detector. The ratio between the energy deposited in the active medium and the total energy deposited in the calorimeter is referred to as sampling fraction. This quantity depends on the materials that compose the calorimeter and the kind of particle that originated the shower.

For this reason, it is typical to characterize electromagnetic sampling calorimeters through their e/mip, defined as the ratio between the mean of the reconstructed Si energy distributions of electromagnetic showers and minimum ionizing particles which deposit the same energy in the calorimeter. The complex structure of the CE-E prototype studied results in odd and even layers having a different e/mip factor due to the intrinsic difference in the amount of passive material that precedes them (cf. Fig. 3.9)). Consequently, particles produced by the shower near the exit boundary of the passive layers may have enough energy to escape from them and therefore alter the response of the sensitive sensors.

Particular configurations of the MC simulation have been produced to investigate the origin of the different energy deposits observed in odd and even layers and possibly isolate the two aforementioned effects. More precisely, the longitudinal profiles reconstructed from the nominal MC configuration (cf. Sect. 3.3.1) are compared to the ones obtained from a simulation where the effect of backscattering is suppressed and to those reconstructed in a MC without secondary electrons and positrons, corresponding to δ-rays produced in the absorber material in front of the active layers. The former is obtained from a simulation where only tracks produced with an angle $\theta > 90°$, i.e., travelling in the same direction of the electromagnetic shower, are retained, while the latter results from removing all the δ-rays from the shower. The three longitudinal profiles are compared in the left column of Fig. 3.27. More precisely, the longitudinal profile obtained from the simulation without backscattering (Fig. 3.27c) is observed to be more regular with respect to the one reconstructed in the nominal MC (Fig. 3.27a), and an additional improvement is achieved removing all the secondaries from the simulation MC (Fig. 3.27e).

These observations lead to the conclusion that a mixture of two effects indeed causes the oscillations in the longitudinal profiles of Fig. 3.26: the presence of backscattered particles from the absorbers after the active layers, as well as the direct production of δ-rays, within the passive materials in front of the silicon sensors, that are consequently propagated with the shower. Even though the removal of these two effects results in a more regular profile, a residual wiggling is observed in the longitudinal profiles reconstructed with the two alternative MC configurations, as shown in Fig. 3.27c, 3.27e. These oscillations are removed completely when the longitudinal profiles are reconstructed using only the cells in a ring (E7) around the pad with the maximum energy deposit (Fig. 3.27d, 3.27f), but are still present if the same information is used to reconstruct the profile in the nominal MC (Fig. 3.27b), as shown in the right column of Fig. 3.27.

This effect can be attributed to the peculiar structure of the calorimeter under study. As a result of this, the interaction of the electromagnetic cascade with the passive layers tends to produce charged particles near the exit boundary of the latter, resulting in a larger response in the thinner sensitive layers of the detector. In addition, since the odd and even active layers are preceded by different amounts of passive material[7] (cf. Fig. 3.9) they feature a slightly different e/mip factor, thus resulting in a distinct absorption of the soft components, i.e. subject to more multiple scattering, of the shower. This effect tends to be more pronounced in the periphery of the shower, where the material difference between odd and even layers becomes more relevant, as highlighted in Fig. 3.28, where a longitudinal profile is reconstructed using only the energy deposited outside the E7 ring for the simulation without $e^{\pm}$ secondaries.

The fact that the wiggling is still observed in this longitudinal profile but not in the one reconstructed using only the E7 ring (cf. Fig. 3.27f), corroborates the hypothesis that the intrinsic layout composition of the calorimeter is responsible *per se* of an oscillation in the longitudinal profiles. More precisely, this residual wiggling can be identified as the soft component of the electromagnetic shower interacting with the

[7] Fe+Pb support of the cassette, Air, and Cu+PCB for the former, Kapton, Cu, CuW for the latter.

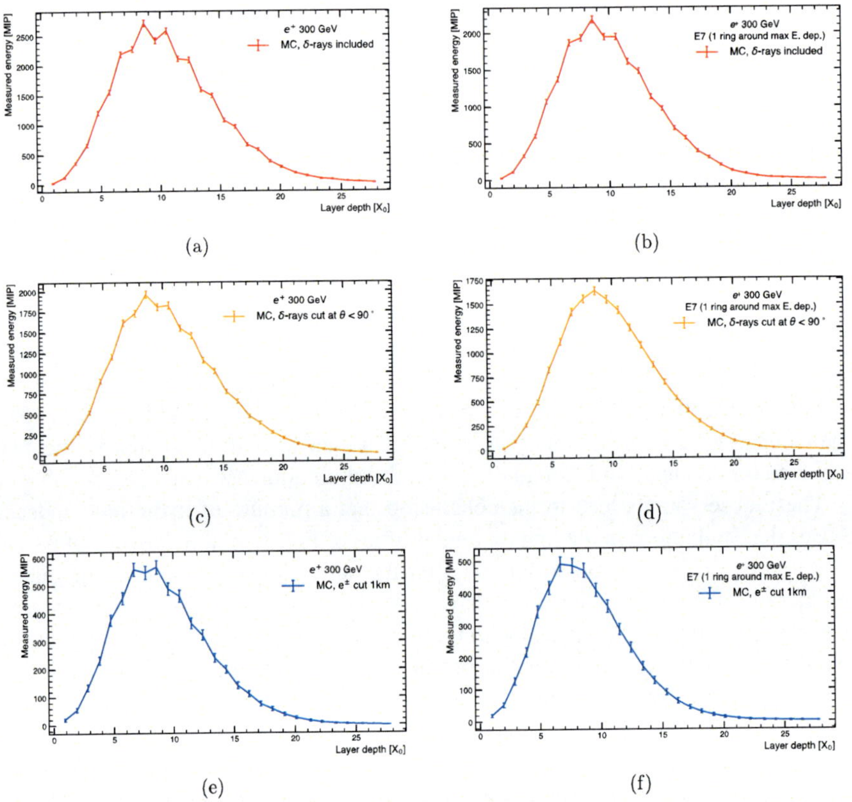

Fig. 3.27 Longitudinal profiles for simulated samples of 300 GeV positron beams. The profiles are shown for a full simulation (red), a configuration in which the backscattering of δ-rays is turned off (orange), and a simulation without secondary electrons and positrons (blue). For all the three cases the profiles are reconstructed using the full silicon sensors (left column) and only the first ring around the maximum energy deposit in each layer (right column)

odd and even passive layers, which feature a different e/mip factor and therefore result in a different absorption of the particles produced within the absorbers.

Hence, this thorough study allows identifying the three effects responsible for the oscillations observed in the longitudinal profiles: the backscattering, the forward propagation of δ-rays, and the intrinsic difference between the e/MIP factors of odd and even layers. This result is a cornerstone in assessing the performance of the detector, as the complete understanding of this phenomenon can be used in the ultimate operation of the HGCAL to compensate for the different responses of the layers and to reconstruct a smooth longitudinal profile.

In spite of this wiggling, the longitudinal profiles reconstructed with test beam data (cf. Fig. 3.26) can still be used to validate the complete lateral containment of electromagnetic showers qualitatively. In fact, significant deposits are observed

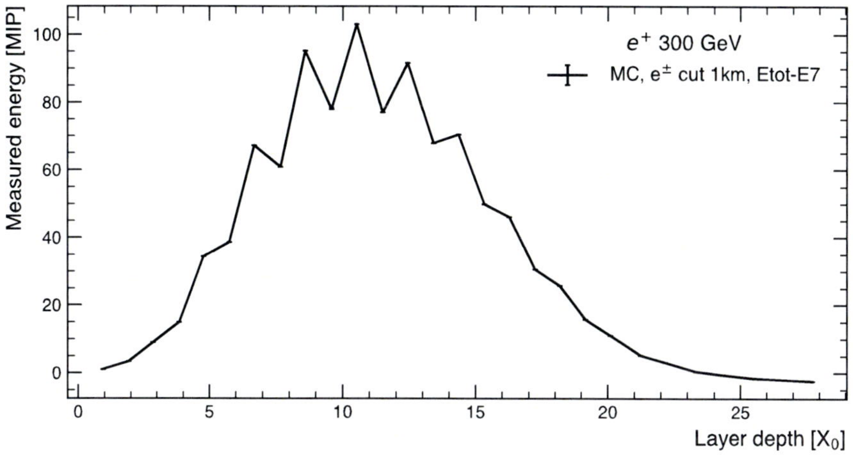

Fig. 3.28 Longitudinal profile reconstructed using the energy deposit in the cells outside the ring around the pad with the maximum energy deposit. The simulation without secondary electrons and positrons described in the text is used. The residual wiggling observed can be attributed to the soft component of the electromagnetic shower interacting with the odd and even layers, which feature a different e/mip factor

between the 5th and 13th layers, while after the 23th layer practically no energy deposit is observed in the calorimeter, even at the large beam energies.

For a more quantitative description of the longitudinal propagation of electromagnetic showers one can exploit the following relation, also known as *Longo's equation* [27]:

$$\left\langle \frac{dE(t)}{dt} \right\rangle = E_0 \frac{(\beta t)^{\alpha-1} \beta \exp(-\beta t)}{\Gamma(\alpha)}, \tag{3.17}$$

where t represents the shower depth, expressed in radiation length units, E_0 is the total energy deposit in the calorimeter, and α and β are two additional free parameters introduced in the fit to describe the shape and the scaling of the longitudinal profile. Notwithstanding Eq. (3.17) is primarily designed to describe the propagation of electromagnetic showers in homogeneous media, it can also be used as a first order approximation for sampling calorimeters, as highlighted in Ref. [27]. More precisely, the α and β parameters can be used to infer the shower's average center of gravity, $\langle COG_z \rangle$, and the depth of the maximum energy deposit, T:

$$\langle COG_z \rangle = \frac{\alpha}{\beta} \tag{3.18}$$

and

$$T = \frac{\alpha - 1}{\beta} = \langle COG_z \rangle - \frac{1}{\beta}. \tag{3.19}$$

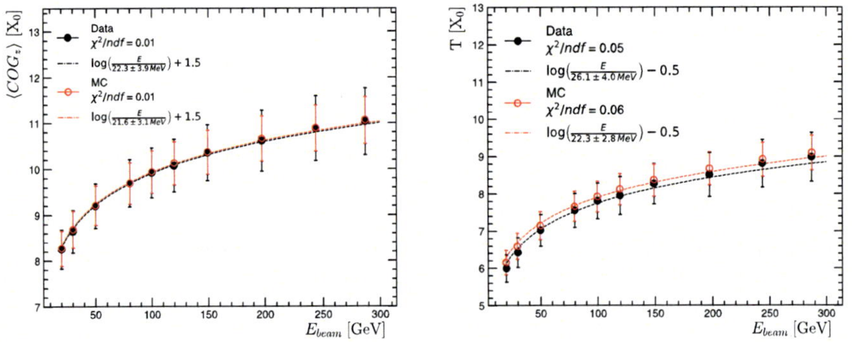

Fig. 3.29 Average shower depth (left) and shower maximum (right) as a function of the beam energy. Black solid points represent data, while red circles correspond to the simulation. The dashed lines correspond to the fit with the expected $\log(E/E_c)$ relations defined in the text

the $\langle COG_z\rangle$ values extracted from the fit of the longitudinal profiles as of Eq. (3.18) are displayed in the left plot of Fig. 3.29. Their trend reproduces the one shown in Fig. 3.25, where the $\langle COG_z\rangle$ values are extracted directly from the definition (cf. Eq. (3.15)). Since Eq. (3.15) provides a mathematical definition of $\langle COG_z\rangle$, hence valid rigorously for both homogeneous and sampling calorimeters, this result corroborates the use of Eq. (3.17) for the description of the longitudinal profiles in the sampling calorimeter under study. The values of T obtained from the fit of the longitudinal profiles as of Eq. (3.19) are displayed in the right plot of Fig. 3.29.

The average center of gravity and the shower maximum are expected to manifest a $\log(E_{beam}/E_c)$ dependence, where E_c is the only free parameter of the fit. In homogeneous media, this parameter represents the critical energy of the detector. This is not exactly true in sampling calorimeters, as the complex mixture of materials in the active layers and in the absorbers makes it more difficult to compare the fitted value with a theoretical expectation. For this reason, this parameter is rather used in the analysis as a reference observable to compare data and simulation and the results obtained with different approaches. More precisely, the expected relation between the shower maximum (T) and the beam energy (E_{beam}) is [26, 27]:

$$T = \log\left(\frac{E_{beam}}{E_c}\right) + C_j,\ j = e, \gamma \tag{3.20}$$

where $C_e = -0.5$ for electron-induced showers and $C_\gamma = +0.5$ for photon-induced cascades. In the following the value of this parameter is fixed to $C_e = -0.5$. This relation can be used together with Eq. (3.19) to extract the expected $\log(E_{beam})$ dependence of $\langle COG_z\rangle$. This is obtained fixing $\beta = 0.5$ in Eq. (3.19) and combining it with Eq. (3.20):

$$\langle COG_z\rangle = \log\left(\frac{E_{beam}}{E_c}\right) + 1.5 \tag{3.21}$$

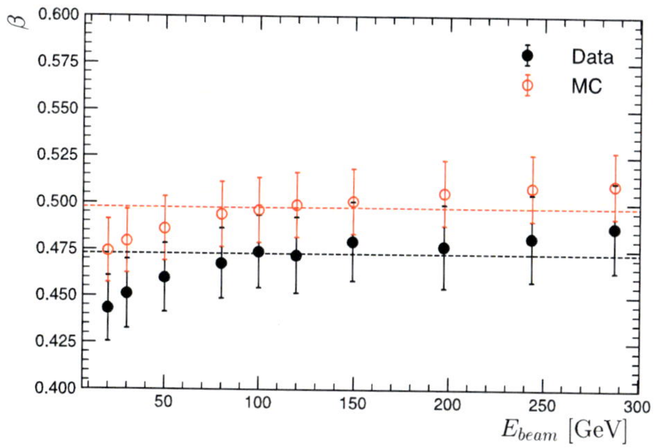

Fig. 3.30 Distribution of the β parameter, extracted from the fit of longitudinal profiles using Eq. (3.17), as a function of the beam energy (E_{beam}). The black (red) dashed line represent the average value of the β distribution for data (simulation)

where the value of β is fixed to 0.5 following the findings of the studies reported in Ref. [26, 27]. In order to corroborate this assumption one can derive the values of β as a function of E_{beam} from the fit of the longitudinal profiles with Eq. (3.17). Figure 3.30 shows the distributions of β as a function of E_{beam}: while for simulation all the values are in agreement with the suggested one of 0.5, a slightly more pronounced energy dependence is observed for data.

Nonetheless, the average values are found to be $\langle \beta_{\text{MC}} \rangle = 0.49 \pm 0.32$ and $\langle \beta_{\text{Data}} \rangle = 0.47 \pm 0.41$, supporting the $\beta = 0.5$ hypothesis in the derivation of Eq. (3.21).

The results of the fit of the $\langle COG_z \rangle$ and T distributions, as a function of E_{beam}, using Eqs. (3.20)–(3.21) are represented by the dashed lines in Fig. 3.29.

The values of the effective critical energy extracted from the fits are reported in Table 3.6. The measurements of E_c are found in statistical agreement between data and simulation in all three cases. As expected, the results obtained from the fit of the nominal $\langle COG_z \rangle$ are more precise than the ones that involve the fit of the longitudinal profiles. The results from the three different methods show a good agreement within 2 σ, therefore validating the use of Eq. (3.17) for the description of the longitudinal profiles in sampling calorimeters.

As mentioned above, this result must be taken with grain of salt: the critical energy is a well-defined quantity for a single material, but it is more complicated to measure it for a sampling calorimeter, where one should take into account not only the material of the absorbers, but also air gaps and any other passive material encountered by the developing shower. Hence, the most direct consequence of the good agreement among the E_c values reported in Tab. 3.6 is the validation of Eq. (3.17) for the description of the longitudinal profiles in the CE-E prototype.

It is worth closing this section demonstrating a possible application of the longitudinal profiles parametrization with Eq. (3.17), showing how it can be used to

Table 3.6 Effective critical energy extracted from the $\log(E_{beam})$ fit of the mean shower depth and of the shower maximum

	CE-E prototype critical energy [MeV]		
Method	Mean shower depth direct calculation	Mean shower depth	Shower maximum
		From longitudinal fit	
Data	30.9 ± 2.0	22.3 ± 3.9	26.1 ± 4.0
Monte Carlo	29.5 ± 1.9	21.6 ± 3.1	22.3 ± 2.8

mitigate the effects arising from the presence of low-efficiency cells and misreconstructed energies. The 10th layer of the electromagnetic compartment is one of such cases, as observed in Fig. 3.26. Hence, one can try to recover the response of this layer exploiting Eq. (3.17) to fit the longitudinal profiles on an event-by-event basis and to extrapolate the expected response of this layer. The comparison between the energy in the 10th layer reconstructed from test beam data and the one inferred from the fit of the longitudinal profiles is shown in the left plot of Fig. 3.31, where the former is observed to be $\mathcal{O}(13\%)$ larger than the latter. Correcting for this factor the average response of this layer in the longitudinal profile results in an optimal agreement with the expectation from the simulation, as shown in the right plot of Fig. 3.31, thus corroborating the analysis strategy and the application of Eq. (3.17) for real use case scenarios.

The results presented in this section serve as an additional validation of the HGCAL design, highlighting the power of the fine longitudinal sampling in the context of a complete characterization of electromagnetic cascades. In fact, the

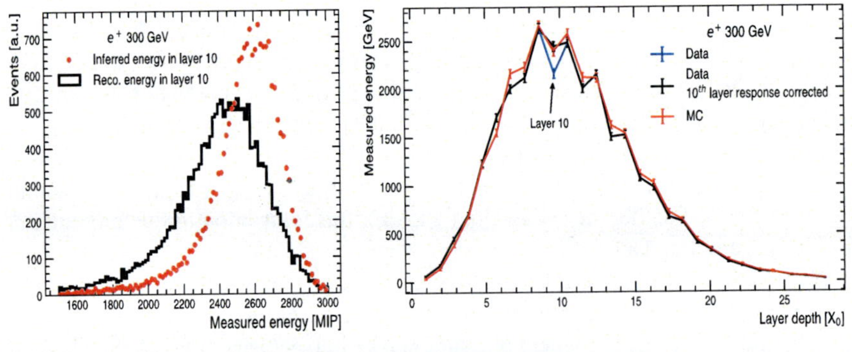

Fig. 3.31 *Left*: The distribution of the reconstructed average energy deposit in the 10th layer for a 300 GeV positron beam is represented by the black histogram, while red points show the same distribution extrapolated from a fit of the longitudinal profiles on an event by event basis. *Right*: Longitudinal profiles for a 300 GeV positron beam. Data are shown in blue and simulation is shown in red. The black points correspond to the longitudinal profile with the response of the 10th layer extrapolated from the analytical expression of Eq. (3.17) as detailed in the text

reconstruction of the longitudinal profiles demonstrates the total containment of electromagnetic showers, it allows to identify the position of the shower maximum, confirming the expected dependence as a function of the beam energy, and validates the analytical description used for their parametrization (cf. Eq. (3.17)).

3.4.4 Transverse Shower Shapes

The results presented in the two previous sections exhibit two important features of the CE-E prototype tested in October 2018:

- Its capability of reconstructing precisely the total shower energy, including a good recovery of the deposit in the absorbers with different techniques.
- The possibility of exploiting the fine granularity of the detector to resolve completely the longitudinal profile of the shower, thus confirming the total containment of the cascade and identifying the position of the shower maximum.

These two complementary aspects already provide a good characterization of electromagnetic cascades, which becomes optimal when supplemented with the information on the transverse shower shape.

To study the transverse containment of the showers, the cell with the maximum energy deposit in each layer is taken as a reference against which to compare the deposits in the neighbouring cells. Two different observables are introduced in the analysis to quantify the lateral spread of the electromagnetic cascades:

- $E(r_0)/E(r_1)$: defined as the ratio between the energy in the cell with the maximum deposit ($E(r_0)$) and that contained in the surrounding ring ($E(r_1)$);
- $E(r_1)/E(r_2)$: defined as the ratio between the energy deposited within one ($E(r_1)$) and two ($E(r_2)$) rings around the central pad.

The comparison of their distributions between data and simulation is shown in Fig. 3.32 for the 9th and 10th layers for a positron beam of 300 GeV. All the distributions are well-bounded between 0 and 1, indicating a complete lateral containment of the showers. A mild disagreement between data and simulation is observed, especially for $E(r_0)/E(r_1)$, and it might be caused by a possible residual mismodelling in MC of the hadronic contamination and by the missing information on the cross-talk effects between the cells. In particular, during the test beam data analysis it was observed that cross-talk effects of a few percent could significantly affect the lateral shower shapes, especially in the early layers, where the energy deposit is not large enough to completely dominate on electronic artifacts. Hence, the residual disagreement observed here between data and simulation should rather be imputed to the mismodelling of the physics or the materials, but can be safely considered negligible in terms of the validation of the design of the detector under test.

An additional observable sensitive to the transverse containment of electromagnetic cascades is the Molière radius, defined as the radius of a cylinder containing 90 % of the shower's energy. The material chosen for the active sensors in the design

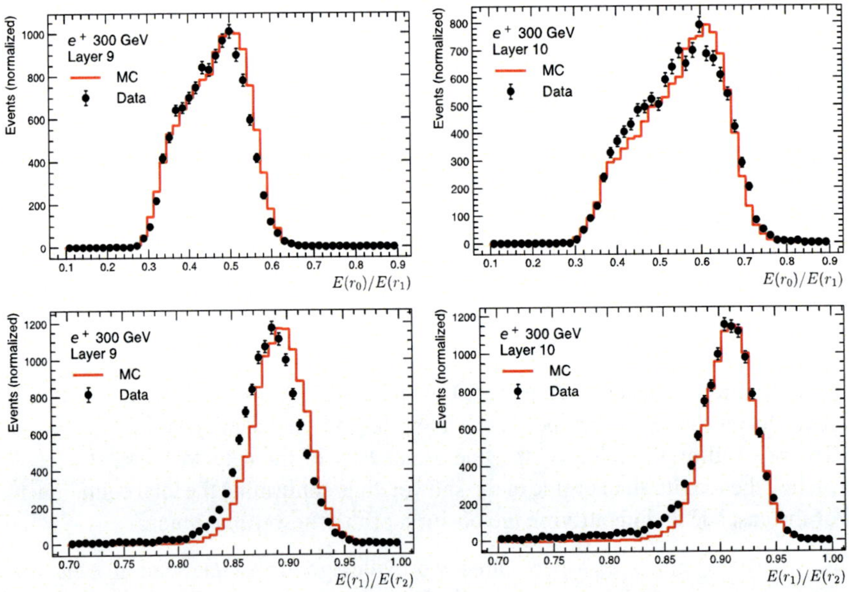

Fig. 3.32 Distributions of the $E(r_0)/E(r_1)$ (top row) and $E(r_1)/E(r_2)$ (bottom row) observables for data (black points) and simulation (red histogram) in the 9th (left column) and 10th (right column) layer of the prototype

of a calorimeter has to be thin enough to guarantee small Molière radii when combined with the total thickness of the absorbers. Hence, having proven that the CE-E prototype ensures a complete lateral containment of the showers, it is also interesting to measure its characteristic Molière radius to assess the complete transverse containment of the showers. The following empirical relation defines this quantity:

$$R_M = 0.0265 \cdot \mathrm{X}_0(Z + 1.2)$$

where Z is the atomic number and X_0 the radiation length of the active layers' material. While the X_0 value is well defined for homogeneous media, the same does not hold in sampling calorimeters. Similarly to what observed for the critical energy, the measurement of the Molière radius presented in the following has to be taken with grain of salt: the goal of this study is to provide a complementary proof of the complete lateral containment of the showers, rather than extracting the ultimate Molière radius for such a complex mixture of materials.

In order to measure this observable, an approach similar to the one used for the reconstruction of transverse shower shapes is adopted. More precisely, after having determined the main shower axis from the extrapolation of the DWCs tracks to the HGCAL prototype, one measures for all the layers the energy deposited within cylinders of increasing radii:

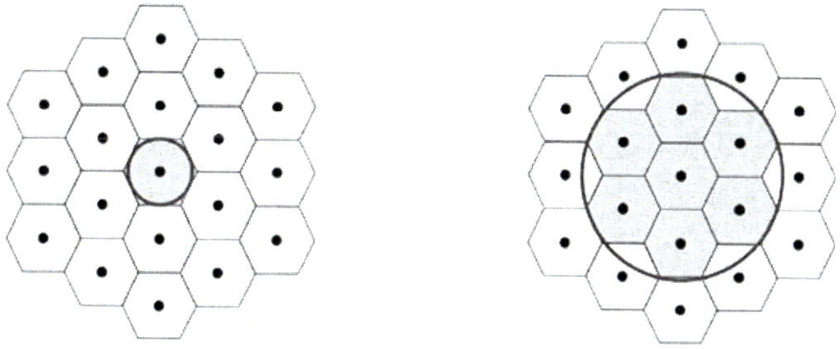

Fig. 3.33 Schematic representation of the geometry of a module, along with the definition of the r_0 and r_1 rings used for the assessment of the transverse containment in the CE-E prototype

$$E(r) = \sum_{i=1}^{28} E_i^{\mathrm{Si}}(r) \tag{3.22}$$

where $E_i^{\mathrm{Si}}(r)$ is the energy deposit in the ith layer within a circle of radius r centered around the cell with the maximal energy deposit. This situation corresponds to what illustrated in Fig. 3.33, under the assumption of a beam perfectly aligned to the z-axis of the HGCAL prototype.

The quantity in Eq. (3.22) is computed at five different values of r and its ratio with respect to the total energy deposit in the CE-E prototype ($E(r)/E_{\mathrm{tot}}$) is used to derive the Molière radius R_M. More precisely, at each value of r the average energy deposit $\langle E(r)/E_{\mathrm{tot}}\rangle$ is computed and parametrized as a function of r with the following relation:

$$\langle E(r)/E\rangle = 1 - A \cdot e^{-B \cdot r}. \tag{3.23}$$

Eventually, R_M is measured by applying the definition and solving:

$$\langle E(R_M)/E\rangle = 0.9$$

The trends of $\langle E(r)/E\rangle$, along with the curves obtained from a fit with Eq. (3.23), are shown in Fig. 3.34 for 50 and 200 GeV positron beams. The first point is not included in the fit, as it corresponds to small values of r, comparable with the cell size, at which the deposited energy strongly depends on the impact point of the cascades. In both cases the values of $\langle E(r)/E\rangle$ tend to 1.0 at large values of r, validating the CE-E design for what concerns the transverse containment of electromagnetic showers.

The analysis is repeated separately for each beam energy, and the measurements of R_M are reported in Table 3.7. An overall good agreement between data and simulation is observed for all the beam energies. The measured R_M values get more stable at higher energies, reflecting the more significant contribution of the multiple scattering

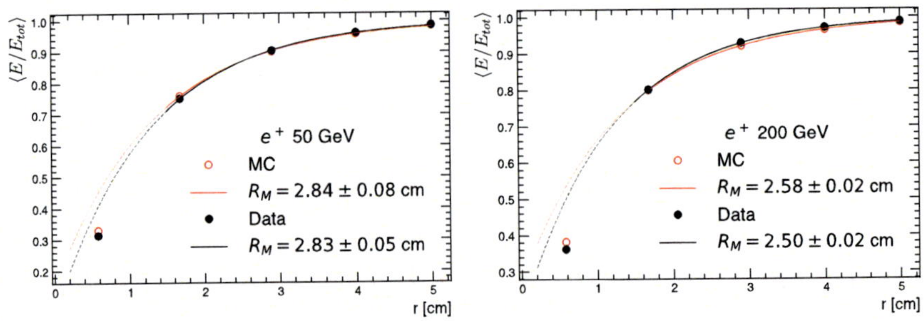

Fig. 3.34 Measurement of the transverse containment for 50 (left) and 200 (right) GeV positron beams. The points correspond to the $\langle E(r)/E\rangle$ quantity defined in the text. Black points represent data, while red circles are for the simulation

Table 3.7 Values of the radial containment observable R_M measured at different beam energies

Nominal beam energy [GeV]	R_M data [cm]	R_M MC [cm]
20	3.1 ± 0.1	3.1 ± 0.1
30	3.0 ± 0.1	2.97 ± 0.08
50	2.84 ± 0.08	2.83 ± 0.05
80	2.73 ± 0.05	2.70 ± 0.02
100	2.70 ± 0.04	2.65 ± 0.01
120	2.67 ± 0.04	2.61 ± 0.01
150	2.63 ± 0.03	2.54 ± 0.02
200	2.58 ± 0.02	2.50 ± 0.02
250	2.55 ± 0.02	2.47 ± 0.03
300	2.52 ± 0.02	2.45 ± 0.04

in the air for low-energy electrons and the wider profiles of the corresponding beams. The different values observed in Table 3.7 confirm the difficulty of measuring this observable precisely in such a complex calorimeter, and the R_M observable should be interpreted as a quantifier of the radial shower containment rather than the actual Molière radius of the CE-E prototype. It is worth observing that the R_M values measured stabilize at higher energies, corroborating the hypothesis of the different beam profiles and scattering effects being responsible for the larger Molière radii measured at low energies.

A small Molière radius is a fundamental property for the ultimate performance of the HGCAL in the harsh environment of the HL-LHC. In fact, this detector is designed for particle identification and reconstruction over an extremely busy background, for which a great separation power is fundamental.

To conclude this comprehensive characterization of the CE-E prototype, the next section presents a study of the pointing resolution of the active layers of the detector. Needless to say, an good position resolution is fundamental for the application of such a detector in the context of the Particle Flow algorithm [28] for an optimal linking between tracks and clusters (cf. Sect. 2.3.1) as well as for the offline particle identification and reconstruction.

3.4.5 Pointing Resolution

The studies on the longitudinal and transverse shower profiles highlight the capability of the HGCAL to completely resolve and reconstruct the evolution of electromagnetic cascades and identify the position of their maximum within the calorimeter. The fine sampling and high granularity of the HGCAL allow complementing these purely calorimetric observables with positioning and angular resolutions, which in turn can be used to enhance the PF reconstruction improving the linking between tracks and energy clusters in the calorimeters.

The analysis strategy adopted here closely follows the one performed on the first test beam of the HGCAL prototype in 2016 [13]. The DWCs placed upstream of the CE-E prototype (cf. Sect. 3.3) are used as an external reference to infer the expected impact position in each layer. These values are compared with the position reconstructed in each CE-E layer using an energy-weighted approach, and the difference between these two quantities is interpreted as the residuals. The width of these residuals distributions is taken as an estimator of the layer's pointing resolution.

As mentioned in Sect. 3.3.1, the DWCs are not introduced in the simulation as "sensitive detectors."[8] Therefore the DWC reconstructed track is obtained from an extrapolation of the beam gun position and angle, to which a smearing proportional to the DWC expected resolution is applied.

The impact position on each layer is extracted by an energy-weighted average of the reconstructed Cartesian coordinates:

$$\vec{x}_{\text{reco}} = \frac{\sum_i \omega(E_i)\vec{x}_i}{\sum_i \omega(E_i)} \tag{3.24}$$

where the sum runs over all the cells within two rings around the pad with the maximum energy deposit. The optimization of this selection region was performed in the context of the 2016 test beam studies [13], where it was found that considering such a region, instead of the full sensor area, optimizes the final position resolution measurement. The $\omega_i(E_i)$ are defined as log-energy dependent weights with the following expression:

$$\omega(E_i) = \max\left(0, a + \ln\left(\frac{E_i}{E_{\text{layer}}}\right)\right), \quad a = 3.5 \tag{3.25}$$

[8] In GEANT4 a *sensitive detector* is defined as such if it constructs one or more hit objects and is capable of recording them, possibly by simulating the expected signal extraction by the electronics.

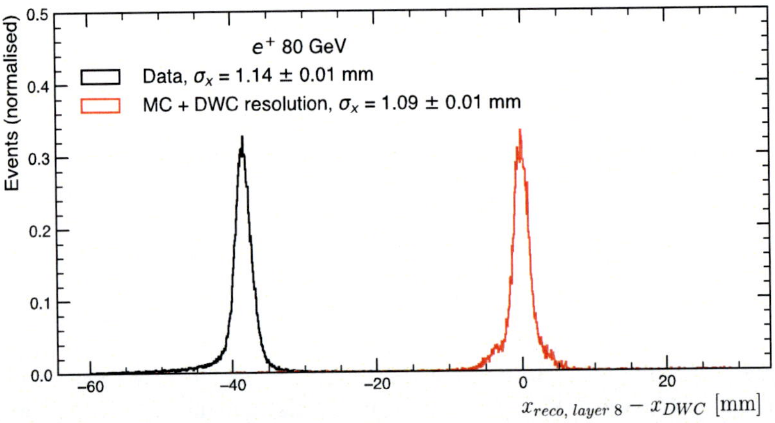

Fig. 3.35 Residuals of the DWC track extrapolation at layer 8, placed at a depth of about 7.7 X_0, for a positron beam energy of 80 GeV

Since the structure of the modules adopted for the October 2018 beam test is unchanged with respect to the 2016 one, the same analysis strategy is followed.

The residuals are computed as described above for each layer, for both data and simulation samples. Their distributions at the 8th layer, placed at a depth of about 7.7 X_0, are shown in Fig. 3.35 for a positron beam energy of 80 GeV. Similar distributions are derived for all the layers and beam energies and are fitted with a Gaussian function from -2.0σ to $+2.0\sigma$ around the peak.

The fitted values of the standard deviation σ and mean μ are interpreted as reliable estimators of the layer's pointing resolution and of the misalignment between the DWC and HGCAL z axes, respectively. The trend of μ as a function of the layer position is shown in the left plot of Fig. 3.36, along with the straight line fit used to infer the angle between the DWC and HGCAL z axes, which is found to be $\alpha_x = 12$ mrad. This result corroborates the findings obtained from the studies on the tracking of MIPs using the HGCAL prototype presented in Ref. [16], where a similar value for the angle between the DWC and the HGCAL z axes is reported. The measurement of the pointing resolution at the 8th layer of the electromagnetic compartment of the detector is shown in the right plot of Fig. 3.36. The DWC+HGCAL pointing resolution of each layer i can be parametrized with the function [29]:

$$\sigma_{x,i}^2 = \frac{a^2}{E_{beam}} + b^2. \tag{3.26}$$

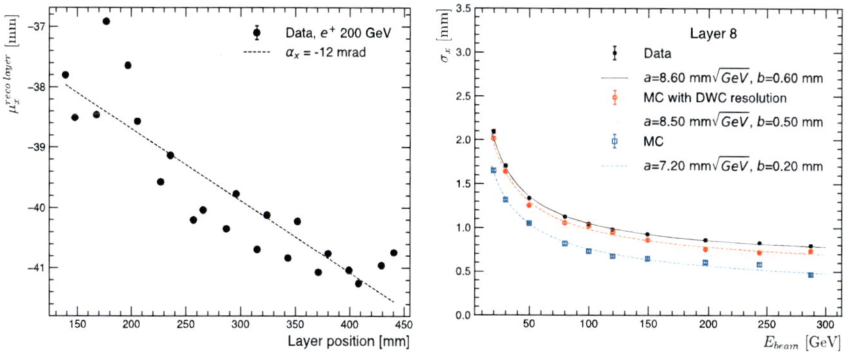

Fig. 3.36 *Left*: Position of the mean value of the residuals distribution (μ_x) as a function of the layer depth. The slope of $\alpha_x = 12$ mm resulting from a linear fit of the points can be interpreted as the angle between the DWC and the HGCAL z axes. *Right*: Pointing resolution measured at layer 8, placed at a depth of about 7.7 X_0, for data (black points) and simulation of the experimental setup (red open circles). The ultimate performance of the CE-E prototype is assessed by using a MC simulation without the inclusion of the DWC pointing resolution (blue open squares)

The single layers are found to have a sub-mm pointing resolution, measured on data. It is worth to observe that the measured pointing resolution is better than the nominal one expected for the silicon modules[9] of $\sigma \sim 3$ mm for all the beam energies.

In order to estimate the ultimate pointing resolution of the CE-E layers, one has to subtract the effect of the DWC intrinsic resolution from the results discussed above. The results of the pointing resolution at 7.7 X_0 obtained without the DWC resolution applied to the simulation are shown by the blue open squares in the right plot of Fig. 3.36. At high energies, where the DWC resolution is lower than 0.7 mm, these results can be interpreted as the intrinsic resolution of the electromagnetic compartment at 7.7 X_0, a depth, close to the expected position of the shower maximum derived in the previous section (cf. Sect. 3.4.3). An overall good agreement between data and simulation is observed across the entire energy range, thus validating also the modelling of the DWCs in the simulation.

3.5 Conclusion

The results presented in this chapter enabled a comprehensive characterization of the electromagnetic compartment of the HGCAL prototype tested in October 2018 at the CERN-SPS H2A beamline. The data analyzed were collected with a 28 layers sampling calorimeter comprising 6" Si sensors interspersed with CuW absorber planes, mimicking the ultimate design structure of the HGCAL CE-E. The detector

---

[9] A rough and conservative estimate of the nominal spatial resolution of silicon sensors with pitch p is $\sigma \sim p/\sqrt{12}$. The Si pad sensors of the modules adopted for the beam test are 1.1 cm, therefore their expected spatial resolution is of $\sigma \sim 3.2$ mm.

was exposed to positron beams with energies ranging from 20 to 300 GeV, exploited to characterize its physics performance and validate its design.

All the results have been benchmarked against a dedicated Monte Carlo simulation that models both the experimental apparatus, including all the upstream detectors and beamline components, and the beam population at the entrance of the first layer of the CE-E prototype. Part of the work of this thesis was dedicated to the fine-tuning of the simulation, with several discussions carried out with beamline experts that lead to the most precise MC simulation ever used in an HGCAL test beam analysis. In addition, dedicated studies aiming to improve the pulse shape reconstruction were conducted, leading to the definition of the analytical expression currently used in all the HGCAL test beam analyses.

Particular effort has been put in the optimization of specific preselection cuts (cf. Sect. 3.4.1) designed with the ultimate goal of enhancing as much as possible the purity of the data analyzed and of limiting any bias arising from detector malfunctionings or residual hadronic contamination present in the beams.

The energy resolution and linearity of the CE-E prototype have been measured at first on the unclustered energy sums (cf. Sect. 3.4.2) obtained from the sum of all the reconstructed hits in the active layers of the detector, after the calibration of all the cells using MIPs. More accurate measurement of these properties has been performed to reconstruct the total energy deposit in the calorimeter through Sampling Fractions (cf. Sect. 3.4.2) and dEdx (cf. Sect. 3.4.2) weights, introduced in the analysis to recover the energy deposit in the passive layers. A residual $\sim$3.5% discrepancy between the energy scales in data and simulation has been observed and corrected for in the analysis. The stochastic and constant terms of the energy resolution are measured to be $S = 22\sqrt{\text{GeV}}\%$ and $C = 0.6\%$, respectively, in agreement with the expectations from the simulation. The application of SFs or dEdx weights leads to values in statistical agreement with those obtained with the "visible energy" method. The linearity of the CE-E prototype studied is measured to be of $\mathcal{O}(2\,\%)$ throughout the entire energy range.

The assessment of longitudinal and transverse shower shapes enabled a complete validation of the imaging properties of the HGCAL prototype. The Longo parametrization has been used to fit the reconstructed longitudinal profiles (cf. Sect. 3.4.3) to extract measurements of the average shower depth and the position of the maximum, as well as to test the $\log(E_{beam})$ dependence expected for these observables. The analytical expression used for the fit of the longitudinal profiles has been proven to give a realistic description of the electromagnetic showers, and it has been shown how it can be used to retrieve the response of missing or malfunctioning layers, even around the shower maximum.

The transverse shower shapes are studied by means of observables sensitive to the lateral containment (cf. Sect. 3.4.4), finding an excellent agreement between data and simulation. Additional measurements of the energy deposit within cylinders of increasing radius around the most energetic cells confirm a complete transverse containment of the showers and can be used to extract a qualitative measurement of the Molière radius.

The pointing resolution has been measured for layers close to the shower maximum to complement the characterization of the imaging properties of the CE-E prototype under test. For each layer, the reconstructed impact position within the HGCAL is compared to the estimated position extracted from an interpolation of the DWC tracks and the difference between these two distributions is taken as the residuals, from which the pointing resolution is extracted by means of a Gaussian fit. A sub-mm pointing resolution is measured for all the layers close to the shower maximum, and it is found to be in agreement with the expectations from the simulation once the finite DWC resolution is included therein.

The results presented in Sects. 3.4.1, 3.4.2, and 3.4.3 demonstrate the substantial contribution of this thesis to the official test beam analysis and entered the corresponding paper published in JINST [30].

The physics performance measured here is found to be in good agreement with the nominal ones expected for the electromagnetic compartment of the HGCAL [4]. Notwithstanding the CE-E prototype tested did not entirely reflect the final structure foreseen for the HL-LHC operations, the test beam data analysis results are a cornerstone for validating the detector design and assessing its expected physics performance.

References

1. Contardo D, Klute M, Mans J, Silvestris L, Butler J (2015) Technical proposal for the phase-II upgrade of the CMS detector. Technical Report CERN-LHCC-2015-010. LHCC-P-008. CMS-TDR-15-02, Geneva
2. Butler J, Contardo D, Klute M, Mans J, Silvestris L, and on behalf of the CMS Collaboration (2015) CMS phase II upgrade scope document. Technical Report CERN-LHCC-2015-019. LHCC-G-165, CERN, Geneva
3. CMS Collaboration (2017) The phase-2 upgrade of the CMS tracker. Technical Report CERN-LHCC-2017-009. CMS-TDR-014, CERN, Geneva
4. CMS Collaboration (2017) The phase-2 upgrade of the CMS endcap calorimeter. Technical Report CERN-LHCC-2017-023. CMS-TDR-019, CERN, Geneva
5. CMS Collaboration (2017) The phase-2 upgrade of the CMS barrel calorimeters. Technical Report CERN-LHCC-2017-011. CMS-TDR-015, CERN, Geneva
6. CMS Collaboration (2017) The phase-2 upgrade of the CMS muon detectors. Technical Report CERN-LHCC-2017-012. CMS-TDR-016, CERN, Geneva
7. CMS Collaboration (2017) Technical proposal for a MIP timing detector in the CMS experiment Phase-2 upgrade. Technical Report CERN-LHCC-2017-027. LHCC-P-009, CERN, Geneva
8. CMS Collaboration (2020) The phase-2 upgrade of the CMS level-1 trigger. Technical Report CERN-LHCC-2020-004. CMS-TDR-021, CERN, Geneva
9. Lobanov A (2018) Electronics and triggering challenges for the CMS high granularity calorimeter. JINST 13(02):C02056
10. Callier S, Dulucq F, de La Taille C, Martin-Chassard G, Seguin-Moreau N (2011) SKIROC2, front end chip designed to readout the electromagnetic calorimeter at the ILC. JINST 6(12):C12040–C12040
11. CALICE Collaboration. https://twiki.cern.ch/twiki/bin/view/CALICE/CalicePapers
12. Borg J, Callier S, Coko D, Dulucq F, de La Taille C, Raux L, Sculac T, Thienpont D (2017) SKIROC2_CMS an ASIC for testing CMS HGCAL. JINST 12(02):C02019–C02019

13. Akchurin N et al (2018) First beam tests of prototype silicon modules for the CMS high granularity endcap calorimeter. JINST 13(10):P10023
14. Acar et al B (2020) Construction and commissioning of CMS CE prototype silicon modules. JINST 16(04):04002
15. H2A beamline at CERN-SPS. http://sba.web.cern.ch/sba/BeamsAndAreas/h2/H2manual.html
16. Quast T (2020) Qualification, performance validation and fast generative modelling of beam test calorimeter prototypes for the CMS calorimeter endcap upgrade, Jul 2020. Presented 28 May 2020
17. Agostinelli S et al (2003) Geant4–a simulation toolkit. Nucl Instrum Methods Phys Res Sect A: Accel Spectrom Detect Assoc Equip 506(3):250–303
18. Allison J et al (2016) Recent developments in geant4. Nucl Instrum Methods Phys Res Sect A: Accel Spectrom Detect Assoc Equip 835:186–225
19. The G4Beamline toolkit. http://www.muonsinternal.com/muons3/G4beamline
20. Spanggaard J (1998) Delay wire chambers - a users guide. Technical Report SL-Note-98-023-BI, CERN, Geneva
21. Barnyakov AYu et al (2007) Investigation and development of microchannel plate phototubes. Nuclear Instrum Methods Phys Res Sect A 572(1):404–407
22. Acar B et al (2020) The DAQ system of the 12,000 channel CMS high granularity calorimeter prototype. JINST 16(04):04001
23. Helmuth S (2005) Semiconductor detector systems. Oxford University Press, Oxford
24. Fabjan Christian W, Gianotti Fabiola (2003) Calorimetry for particle physics. Rev Mod Phys 75(4):1243–1286
25. Paganis Stathes, Psallidas Andreas, Steen Arnaud (2017) Optimizing the performance of a high-granularity silicon-pad EM calorimeter. JINST 12(06):P06013
26. Zyla PA et al (2020) Review of particle physics. PTEP 2020(8):083C01
27. Grindhammer G, Peters S (1993) The Parameterized simulation of electromagnetic showers in homogeneous and sampling calorimeters. In: International conference on Monte Carlo simulation in high-energy and nuclear physics - MC 93, 2
28. Collaboration CMS (2017) Particle-flow reconstruction and global event description with the CMS detector. JINST 12(10):P10003
29. Acosta D et al (1991) Localizing particles showering in a Spaghetti calorimeter. Nucl Instrum Method A 305:55–70
30. Acar B et al (2022) Response of a CMS HGCAL silicon-pad electromagnetic calorimeter prototype to 20–300 GeV positrons. JINST 17(05):P05022

Chapter 4
The "Golden Channel" and Its Properties

The $H \rightarrow ZZ \rightarrow 4\ell$ decay channel ($\ell = e, \mu$) is often referred to as *the golden channel* due to its many virtues. Two of its *golden* properties are the large signal-to-background ratio, with an essentially flat background under the resonant H boson peak, and the completely resolved final state, with an optimal reconstruction of the four-leptons system, made possible by the good lepton resolution of the CMS detector. Figure 4.1 shows a candidate $H \rightarrow ZZ \rightarrow 4\ell$ event, comprising of a muon-antimuon pair (red lines) and an electron-positron pair (green lines). One can see the relatively clean environment, where the $H \rightarrow ZZ \rightarrow 4\ell$ signature can be identified and reconstructed.

Initially designed to be a discovery analysis meant to identify a statistically significant fluctuation over a flat background, the $H \rightarrow ZZ \rightarrow 4\ell$ is now a precision physics analysis tuned to probe very granular phase space regions of the golden channel. The "golden journey" of this thesis starts in this chapter, where the fundamental inputs to the analysis are introduced. This chapter describes the selection of leptons and jets candidates, starting from the reconstructed PF objects introduced in Sect. 2.3. More precisely, loose selection criteria are applied on the PF electron and muon candidates, which are in turn identified and isolated by more stringent requirements to define the opposite sign lepton pairs employed in the analysis to reconstruct the H boson candidates, as schematically summarized in Fig. 4.2.

M. Bonanomi, *Response of the High Granularity Calorimeter HGCAL and Characterisation of the Higgs Boson*, Springer Theses,
https://doi.org/10.1007/978-3-031-26833-5_4

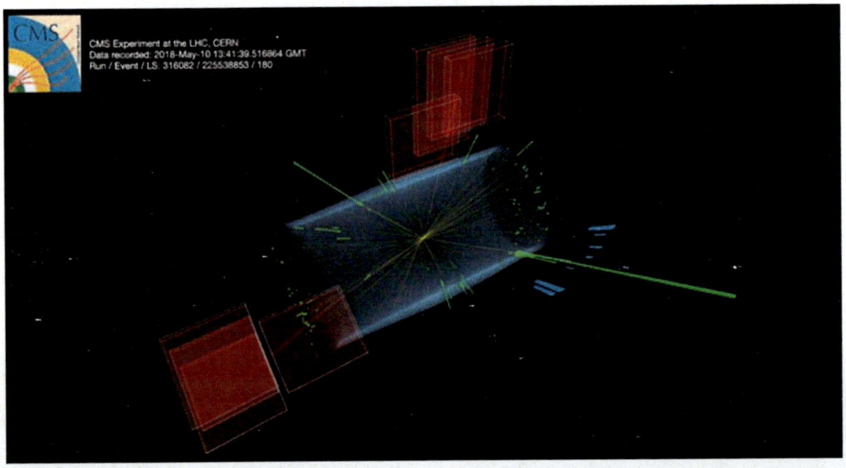

Fig. 4.1 Event recorded by the CMS detector in 2018 at a proton-proton centre of mass energy of 13 TeV. The event is a candidate for the gluon-fusion production of a Higgs boson with decays into a pair of Z bosons. One Z decays to a muon-antimuon pair (red lines) and the other decays to an electron-positron pair (green lines)

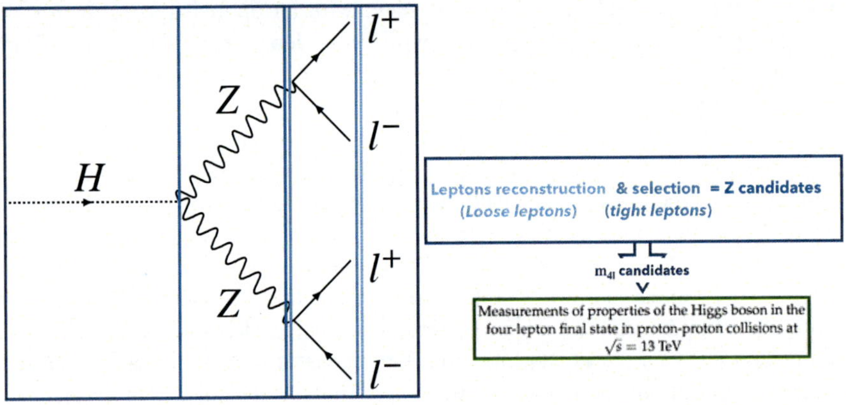

Fig. 4.2 From right to left: loose requirements are applied to select among the PF leptons candidates the electrons and muons to be considered for the analysis. More stringent criteria, further detailed in the text, are applied on these particles to define the "tight leptons" that compose the opposite sign pairs used to reconstruct the H boson candidates considered in the $H \to ZZ \to 4\ell$ analysis

4.1 Electrons

In order to select electron candidates suitable for the study of the $H \to ZZ \to 4\ell$ decay, one starts from the PF electrons reconstructed as described in Sect. 2.3. Additional isolation and identification criteria are applied to reject non-prompt electrons, i.e., not produced by a Z boson decay, and to increase the purity of the final H boson candidates selected for the analysis.

4.1.1 Electrons Selection

Loose requirements are imposed on the transverse momentum and rapidity of the PF electron candidates to preserve the reconstruction efficiency as much as possible while rejecting part of the QCD background contribution. The primary vertex of the tracks is used to introduce additional cuts:

$$p_T^e > 7 \text{ GeV}, \mid \eta^e \mid < 2.5,$$

$$d_{xy} < 0.5 \text{ cm, and } d_z < 1 \text{ cm},$$

resulting in what the analysis refers to as *loose electrons*. An additional requirement on the significance of the impact parameter (SIP) to the primary vertex is imposed to ensure that electrons are consistent with a common primary vertex, thus suppressing the presence of fake electrons originating from photon conversion:

$$\text{SIP}_{3\text{D}} = \frac{\mid \text{IP}_{3\text{D}} \mid}{\sigma_{\text{IP}_{3\text{D}}}} < 4.$$

A multivariate discriminant implementing the eXtreme Gradient Boosting (XGBoost) algorithm [1] is employed to identify and isolate the electrons. The XGBoost algorithm defines a series Boosted Decision Trees (BDT) trained to discriminate between prompt and isolated electrons arising from Z decays and fake leptons originating from jets or hadronic decays. The training of the BDT model is performed on simulated samples of Drell-Yan+jets, separately for the three data taking periods considered in the analysis (2016, 2017, and 2018). A total of 24 observables sensitive to the properties of the electron candidates are used, as summarized in Table 4.1. The BDT is trained to distinguish prompt from non-prompt electrons starting from the cluster shape, the matching between a track and the associated cluster, the track-quality information, the PF isolation sums, and the mean energy density in the event introduced to ensure PU-resilience of the algorithm.

The best set of hyperparameters of the XGBoost model is obtained by means of a grid searching algorithm based on a bayesian optimization procedure, as extensively detailed in Ref. [2].

The training is performed in six independent bins of p_T^e, η^e, with a special focus on low-energy electrons. Two p_T^e with a bin boundary at 10 GeV are defined, each comprising three η^e regions with boundaries at 0.8 and 1.479. The model's performance is assessed by looking at the corresponding Receiver Operating Characteristic (ROC) curves in each training bin and quantified by measuring the Area Under the Curve (AUC). Two exemplary ROC curves for the 2018 training in the $\mid \eta^e \mid > 1.479$ region for the $p_T^e < 10$ GeV and $p_T^e > 10$ GeV bins are shown in Fig. 4.3

Table 4.1 Observables used as inputs to the BDT model employed in the analysis for the identification and isolation of electrons

Observable type	Observable	Definition
Cluster shape	$\sigma_{i\eta i\eta}$, $\sigma_{i\varphi i\varphi}$	RMS of the energy-crystal number spectrum along η and φ
	η-ϕ width	Width of the ECAL supercluster along η and ϕ
	H/E	Ratio of the hadronic energy behind the electron supercluster to the supercluster energy
	$(E_{5\times5} - E_{5\times1})/E_{5\times5}$	Relative difference between the energy contained in the 5 × 5 crystals around the cluster seed ($E_{5\times5}$) and the energy of the crystals containing the cluster seed itself ($E_{5\times1}$)
	$R_9 = E_{3\times3}/E_{SC}$	Ratio of the sum of the seed and adjacent 3 × 3 crystals ($E_{3\times3}$) to the super cluster energy (E_{SC})
	$E_{\mathrm{PS}}/E_{\mathrm{raw}}$	Energy fraction in pre-shower. Used only for endcap training bins
Track-cluster matching	E_{SC}/p_{in}	Ratio of the supercluster energy to the inner track momentum
	E_{ele}/p_{out}	Ratio of the energy of the PF cluster to the associated outer track momentum
	$1/E_{tot} - 1/p_{in}$	Energy-momentum agreement
	$\Delta\eta_{in}$, $\Delta\varphi_{in}$	Distances in η and φ between the energy-weighted supercluster energy and inner track extrapolation
	$\Delta\eta_{seed}$	Distance between the cluster seed and the outer track extrapolation to the ECAL
Track information	$f_{brem} = 1 - p_{out}/p_{in}$	Fractional momentum loss between inner and outer tracks
	N_{KF}, N_{GSF}	Number of hits of the KF and GSF track
	χ^2_{KF}, χ^2_{GSF}	Reduced χ^2 of the KF and GSF tracks
	$N_{\text{miss. hits}}$	Number of expected but missing inner hits
	$P_{\text{conv.}}$	Conversion probability obtained from the χ^2 of the conversion vertex fit
Isolation	$\mathcal{I}_{\gamma}$	PF photon isolation p_T sum
	$\mathcal{I}_{\text{ch. had.}}$	PF charged hadrons isolation p_T sum
	$\mathcal{I}_{\text{neu. had.}}$	PF neutral hadrons isolation p_T sum
PU-resilience	ρ	Mean energy density in the event

Eventually, the electron candidates are selected for the analysis if their associated BDT score is sufficiently large. The normalized distributions for signal and background electrons derived from the training on the 2018 dataset are shown in Fig. 4.4.

The operative working points of the BDT models are chosen separately for each data taking period and bin to match, and in most of the cases outperform the isolation and identification efficiencies obtained in the Run-I analyses, where a cut-based

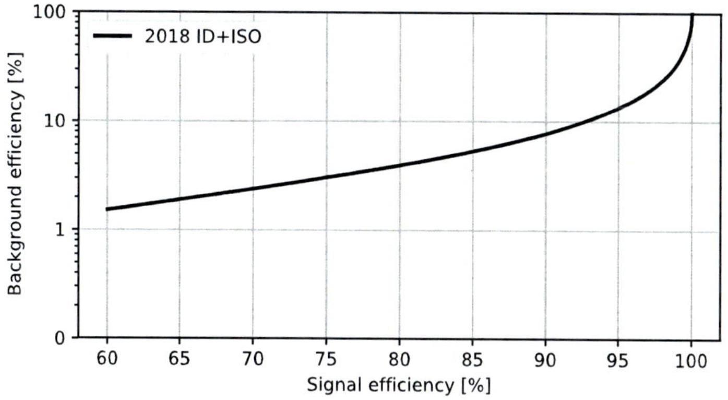

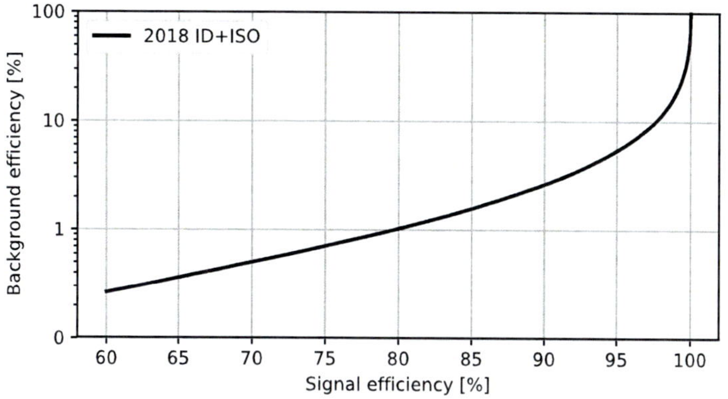

Fig. 4.3 The ROC curves for the BDT developed for electrons identification. The curves depict the performance of the algorithm on 2018 data in the two endcap regions ($|\eta^e| > 1.479$) for the low (top) and high (bottom) transverse momentum bins with $p_T^e < 10$ GeV and $p_T^e > 10$ GeV

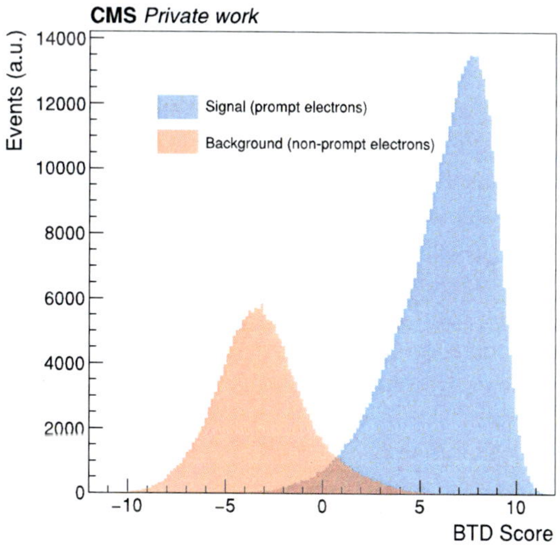

Fig. 4.4 The BDT score of signal (blue) and background (red) electrons obtained from the training on the 2018 dataset in the high p_T^e and η^e bin, corresponding to the region $5 < p_T^e < 10$ GeV, $|\eta^e| > 1.479$

Table 4.2 Minimum BDT scores required for the electron identification criteria obtained from the trainings on the 2016, 2017, and 2018 data taking periods

2016			
	$\|\eta\| < 0.8$	$0.8 < \|\eta\| < 1.479$	$\|\eta\| > 1.479$
$5 < p_T < 10$ GeV	0.9503	0.9461	0.9387
$p_T > 10$ GeV	0.3782	0.3587	−0.5745
2017			
	$\|\eta\| < 0.8$	$0.8 < \|\eta\| < 1.479$	$\|\eta\| > 1.479$
$5 < p_T < 10$ GeV	0.8521	0.8268	0.8694
$p_T > 10$ GeV	0.9825	0.9692	0.7935
2018			
	$\|\eta\| < 0.8$	$0.8 < \|\eta\| < 1.479$	$\|\eta\| > 1.479$
$5 < p_T < 10$ GeV	0.8956	0.9111	0.9401
$p_T > 10$ GeV	0.0424	0.0047	−0.6042

approach relying only on kinematic observables was employed. The WP values are presented in Table 4.2 and correspond to an overall signal efficiency of 95% for high p_T^e electrons and about 80% (70%) for low p_T^e electrons in the barrel (endcap) region.

4.1.2 *Energy Calibration*

The energy of PF electrons is extracted by combining the information from the tracker and the electromagnetic calorimeter, with a predominance of the latter, especially for electrons with large transverse momenta. Photon pairs arising from π^0 decays ensure the precise calibration of all the ECAL crystals, thus guaranteeing a good reconstruction of both electrons and photons. Dedicated corrections are computed using a simulated $Z \to e^+e^-$ sample and applied in p_T^e, $|\eta^e|$ bins of data to correct the residual misalignment between the data and the MC energy scales and to reduce the width of the dielectron mass spectrum. A matching of the $Z \to e^+e^-$ mass resolutions in data and simulation is achieved from applying pseudo-random Gaussian smearing to the MC electron energies as a function of p_T^e and $|\eta^e|$. The electron energy scale is measured in data from a fit with a Crystal Ball function to the $m_{e^+e^-}$ spectrum around the Z peak. The dielectron invariant mass and the corresponding energy scale and resolution are shown in Fig. 4.5 for the three data taking periods considered in this analysis. The agreement is overall satisfactory, even though for large $m_{e^+e^-}$ values ($m_{e^+e^-} > 95$ GeV) a shift between data and MC energy scales is observed, especially in the 2018 dataset. This effect, possible consequence of the miscalibration of some crystals or of the loss in their response due to ageing, has a marginal impact on the analysis and it is covered by the systematic uncertainties associated to the leptons identification and energy scales.

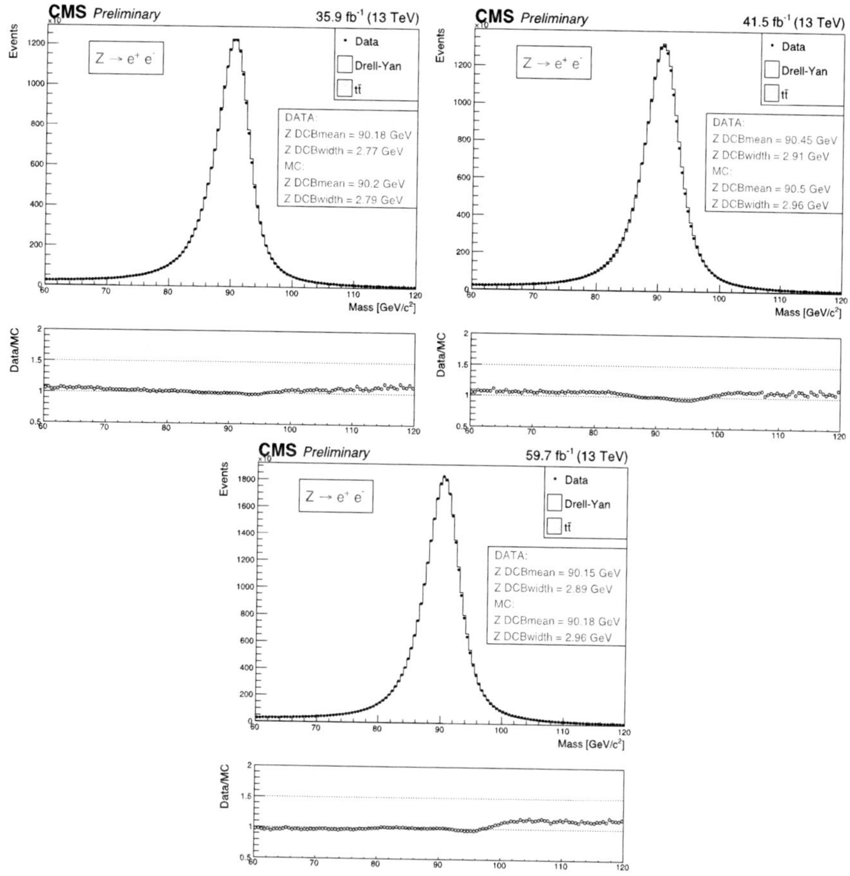

Fig. 4.5 Electron energy scale and resolution measured with $Z \rightarrow e^+e^-$ events for the 2016 (top left), 2017 (top right), 2018 (bottom) data taking periods

4.1.3 Electrons Efficiency Measurements

The selection efficiency of electrons is measured by means of a tag-and-probe (TnP) technique. For this purpose, two sets of leptons are built starting from Drell-Yan+jets samples, produced with the MadGraph aMC@NLO generator: one has to pass relatively loose criteria and it is labeled as *probes*, the other is subject to more stringent requirements, and it is referred to as *tags*. The former ensures a full coverage of the dilepton invariant mass under the resonant peak, while the latter is ideally composed of pure electrons, with a negligible background contamination. The *tag* and the *probe* electrons selected are same flavour and opposite charge leptons. Eventually, the efficiency is measured in mutually exclusive bins and is defined as the ratio of the number of passing probes (N_P) to the total number of probes:

$$\varepsilon_{sel} = \frac{N_P}{N_P + N_F},$$

being N_F the number of probes that do not pass the selection criteria.

The *tag* electrons used for the selection efficiency measurements are required to satisfy the single electron trigger path and to have transverse momentum $p_T^e >$ 30 GeV and a supercluster pseudorapidity $|\eta_{SC}^e| < 2.17$, whereas *probes* are selected among GSF candidates (cf. Sect. 2.3.2) with opposite charge to that of the tag.

A dedicated procedure is employed for the measurement of the selection efficiency in bins of low transverse momentum. More precisely, for electrons with $7 < p_T^e <$ 20 GeV, the tags are required to have a BDT score for isolation and identification larger than 0.92 and they have to satisfy the following kinematic cut:

$$\sqrt{2 \times \mathrm{PF}_{\mathrm{MET}} \times p_T^e(tag) \times (1 - \cos(\phi_{\mathrm{MET}} - \phi_{tag}))} < 45\ \mathrm{GeV}.$$

The final measurement of the electron selection efficiencies are performed in p_T^e and $|\eta_{SC}^e|$ bins using a template fit. The resonant shape of the probes is extracted from simulation and convoluted with a Gaussian. The distributions in data are then fitted with the template mentioned above and an error-function with a one-sided exponential tail. The selection efficiencies measured on the 2017 dataset are shown, as a function of the transverse momentum and pseudorapidity, in Fig. 4.6 for both Data and MC, along with the corresponding scale factors. For all three years, the efficiencies range between 95 and 99% for electrons with transverse momentum $p_T^e >$ 30 GeV. On the other hand, the efficiencies in the low p_T^e bins are much smaller and range between 75 and 85%, due to the poor statistics exploitable for a $Z \rightarrow e^+e^-$ resonance in these regions. The corresponding scale factors have uncertainties

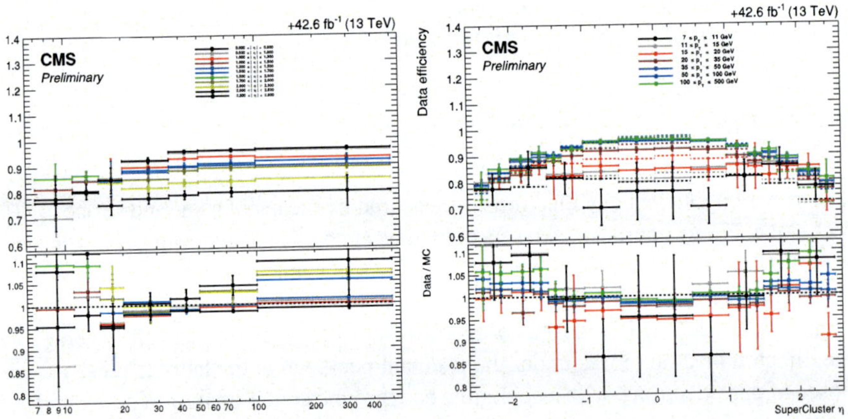

Fig. 4.6 Electron selection efficiencies measured in data as a function of the transverse momentum (left) and pseudorapidity (right) measured using the TnP technique. The Data-MC ratio is displayed in the bottom plots. Results are shown for the 2017 data-taking period [3]

that range between 4 and 8%, as shown in the bottom panels of Fig. 4.6. These uncertainties represent the most significant contribution to the absolute precision of the measurements presented in this analysis: contrary to their counterpart for the muons, where the J/Ψ resonance can be exploited for the TnP technique, the reduced statistics, and the worse resolution the $Z \to e^+e^-$ resonance ultimately affect the measurement of the selection efficiencies in the low transverse momentum region, resulting in the experimental uncertainty with the largest impact on the analysis.

4.2 Muons

Similar to electrons, muons are identified starting from the global PF candidates created as detailed in Sect. 2.3.3. The following sections present the dedicated identification and isolation criteria applied to these objects to define a set of good quality muons for the next steps of the analysis.

4.2.1 Muons Selection

The tracks of the PF muon candidates are classified according to the subdetectors used to reconstruct them. Hence, one can have *global*, *tracker*, or *standalone* muons according to whether the tracker and the muon chambers, or only one of these two submodules of the CMS experiment, are used in the reconstruction. In the context of this analysis only *global* and *tracker* muons are retained. The primary vertex of the tracks is used to introduce additional preselection cuts on the PF muon candidates:

$$p_T^\mu > 5 \text{ GeV}, \mid \eta^\mu \mid < 2.4$$

$$d_{xy} < 0.5 \text{ cm, and } d_z < 1 \text{ cm},$$

resulting in what the analysis refers to as *loose muons*. An additional requirement on the significance of the impact parameter (SIP) to the primary vertex is imposed:

$$\text{SIP}_{3\text{D}} = \frac{\mid \text{IP}_{3\text{D}} \mid}{\sigma_{\text{IP}_{3\text{D}}}} < 4\,, \tag{4.1}$$

similarly to what is done for electrons. These identification criteria aim at reducing contamination from fake muons originating from in-flight decay of hadrons and cosmic rays, therefore enhancing the selection of leptons coming from the primary vertex.

Loose muons with transverse momentum $p_T^\mu < 200$ GeV are considered for the analysis if they match the corresponding PF muon ID (cf. Sect. 2.3.3). Loose muons with larger transverse momentum are considered as identified if they pass either the

Table 4.3 The requirements for a muon to pass the Tracker High-p_{T} ID. Requirements used in the analysis to select high p_{T} muons

Description	Implementation
Muon station matching	Applied on tracker muons. Muon is matched to segments in at least two muon stations
Good p_{T} measurement	$\frac{p_{\mathrm{T}}}{\sigma_{p_{\mathrm{T}}}} < 0.3$
Vertex compatibility $(x - y)$	$d_{xy} < 2$ mm
Vertex compatibility (z)	$d_z < 5$ mm
Pixel hits	At least one pixel hit
Tracker hits	Hits in at least six tracker layers

corresponding PF muon ID or the Tracker High-p_T ID. In particular, loose muons with $p_T^{\mu} > 200$ GeV are retained for the analysis if they pass the criteria detailed in Table 4.3.

The cases where a single muon can be misidentified as two or more particles are dealt with using a procedure referred to as "ghost cleaning," where tracker muons that do not satisfy the global muon definition are removed and if two muons share 50% or more of their track segments, only the one with the best track quality is retained.

A PF-based isolation definition is employed to distinguish between prompt muons and unwanted contributions from the QCD background processes or non-prompt muons. In order to deal with the presence of an increasingly large PU rate in the three data taking periods considered in the analysis (cf. Fig. 2.4), the so-called "$\Delta\beta$ correction" is applied to take into account the energy deposit of neutral particles in the isolation cone:

$$\Delta\beta = \frac{1}{2} \sum_{\text{charged}} p_{\mathrm{T}}^{\text{PU.}} \tag{4.2}$$

Hence, the muons' relative isolation factor ($\mathcal{I}_{\mu}$) is defined as:

$$\mathcal{I}_{\mu} = \frac{1}{p_{\mathrm{T}}^{\mu}} \left(\sum_{\text{charged}} p_{\mathrm{T}} + \max\left[0, \sum_{\text{neutral}} p_{\mathrm{T}} + \sum_{\text{photons}} p_{\mathrm{T}} - \Delta\beta \right] \right), \tag{4.3}$$

where the three scalar sums run over the transverse momenta of charged hadrons, neutral hadrons, and photons originating from the primary vertex of the event and are computed within a cone of angular radius $\Delta R = 0.3$ around the muon direction. *Tight muons* are selected, starting from isolated muon candidates, only if they satisfy the isolation requirement of $\mathcal{I}_{\mu}(\Delta R = 0.3) < 0.35$, where the working point is chosen to match that of electrons and it was optimized in previous versions of the $\mathrm{H} \to \mathrm{ZZ} \to 4\ell$ analysis.

4.2.2 Energy Calibration

A precise determination of the muon energy scale and the proper correction of any discrepancy between data and simulation impacts the reconstruction efficiency and the ultimate precision of the analysis directly. Discrepancies between data and simulation can arise from imperfections in the detector modelling, from effects related to the presence of a strong magnetic field, or from energy losses along the muon track. Dedicated corrections are applied to the muon candidates, exploiting the so-called *Rochester method* [4], to cope with these situations. These corrections are derived using the Z and J/Ψ resonances in the di-muon channel as a reference and they are applied iteratively to obtain an optimal matching between data and simulation.

The measurements of the energy scale and resolution are performed similarly to what is done for electrons: a Crystal Ball fit on the $Z \rightarrow \mu\mu$ dimuon resonance is employed to extract the energy scale and resolution, as shown in Fig. 4.7 for the three data taking periods considered in the analysis.

4.2.3 Muons Efficiency Measurements

Even though the reconstruction of muon PF objects is simpler if compared to the electrons, residual discrepancies between data and simulation may be found after the application of the selection criteria detailed in Sect. 4.2.1. In order to correct for such possible discrepancies, one measures the reconstruction and identification efficiencies for both Data and MC and derives suitable scale factors (SF) to account and correct for these effects, following the same strategy detailed for electrons in Sect. 4.1.3.

The TnP technique detailed in Sect. 4.1.3 for electrons is applied also to muons, where Z and J/Ψ resonances in the di-muon channel are exploited to measure selection efficiencies in p_T^{μ} and η^{μ} bins. In particular, the presence of the J/Ψ resonance allows enhancing the quality of the measurement in bins of low transverse momentum.

Figure 4.8 shows the selection efficiencies for muons in different regions of the detector. The efficiencies are found to be larger than 98% in all the regions of the detector, independently on the transverse momentum of the muon candidates. The corresponding SFs are found to be in agreement with 1.0 in most of the bins, with sub-percent uncertainties, and maximal deviations of the order of just a few percents in bins corresponding to muons with low transverse momentum in the barrel or in the very forward regions at $|\eta^{\mu}| > 2$.

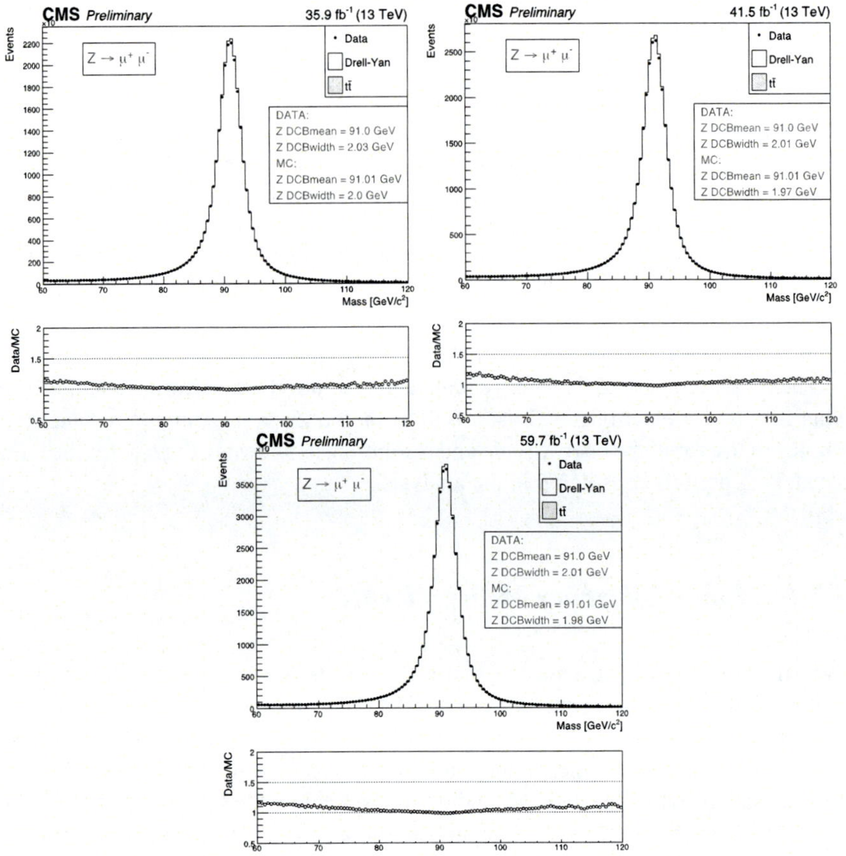

Fig. 4.7 Muon energy scale and resolution measured with $Z \rightarrow \mu^+\mu^-$ events for the 2016 (top left), 2017 (top right), 2018 (bottom) data taking periods

4.3 Final State Radiation Recovery

The $Z \rightarrow \ell^+\ell^-$ decay can radiate high energy photon with an 8% (15%) probability for dimuon (dielectron) final states. This phenomenon takes the name of final-state radiation (FSR) and it needs to be accounted for and corrected by including the photon's four momenta in the computation of the properties of the H boson candidate.

The FSR radiation recovery starts with PF photons candidates, defined as of Sect. 2.3.1, to which p_T^γ and η^γ cuts are applied:

$$p_T^\gamma > 2\ \text{GeV}, \ \mid \eta^\gamma \mid < 2.5$$

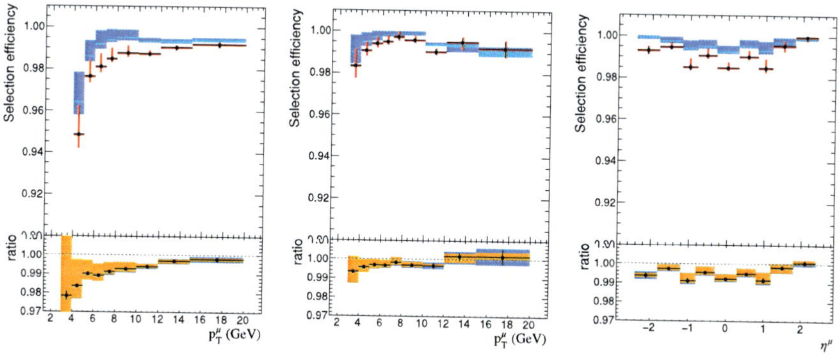

Fig. 4.8 Muons selection and identification efficiencies measured using the TnP technique on J/Ψ events with the 2018 dataset. The efficiencies are shown as a function of the transverse momentum (p_T^{μ}) for muons with $p_T^{\mu} < 20$ GeV in the barrel (left) and in the endcap (middle) regions, as well as for muons with $p_T^{\mu} > 5$ GeV as a function of the pseudorapidity η^{μ} (right). The black error bars represent the statistical uncertainty on data, while the red ones show the contribution of the systematic uncertainties. For the simulation the statistical and systematical contribution of the uncertainties are shown in the azure and orange rectangles, respectively

Photons are further selected by means of an isolation requirement, defined as:

$$\mathcal{I}_{\gamma} = \frac{1}{p_{\mathrm{T}}^{\gamma}} \left(\sum_{\text{photons}} p_{\mathrm{T}} + \sum_{\text{neutral}} p_{\mathrm{T}} + \sum_{\text{charged}} p_{\mathrm{T}} \right) \tag{4.4}$$

computed using a 0.2 GeV threshold on charged hadrons, and a 0.5 GeV threshold on neutral ones.[1] The $\mathcal{I}_{\gamma}$ isolation parameter is defined within a cone of radius $\Delta R = 0.3$.

Photons candidates matching any *loose* electron are removed from this computation and are associated directly the two electromagnetic PF candidates in a procedure referred to as *supercluster veto*.

In a next step photons are associated to the closest *loose* lepton and are discarded when:

$$\Delta R(\gamma, l)/E_{T,\gamma}^{2} < 0.012, \text{ and } \Delta R(\gamma, l) < 0.5$$

In situations where more than one PF photon candidate survives the selection criteria defined above, the one with the lowest $\Delta R(\gamma, l)/E_{T,\gamma}^{2}$ is retained.

The FSR recovery algorithm was developed in the context of the Run-I $\mathrm{H} \rightarrow \mathrm{ZZ} \rightarrow 4\ell$ analysis, where it was fundamental to ensure a good resolution in the reconstruction of the invariant mass of the four-lepton system without losing events under the resonant peak. FSR recovery plays a marginal role in the context of the full Run-II analysis presented in this thesis, as it affects only about 5.4% of the

[1] A jet veto cone of R = 0.0001 and R = 0.01 is applied to charged and neutral hadrons, respectively.

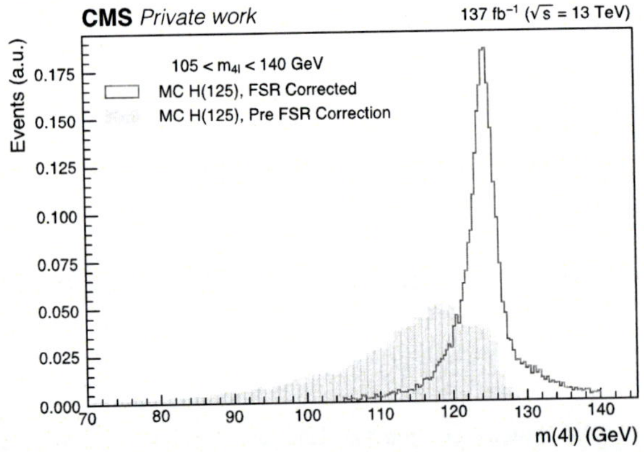

Fig. 4.9 Invariant mass of the H boson candidates for events with final state radiation. The solid histogram represents the $m_{4\ell}$ reconstructed mass is shown in solid and empty histograms before and after the FSR recovery, respectively

events considered for the analysis.[2] Nevertheless, the effect of the FSR recovery is substantial for these events and therefore it is worth considering, as shown in Fig. 4.9.

The fraction of events affected by FSR is expected to be of 6.4% , 3.6%, and 5.2% in the 4e, 4m, and 2e2m final states, respectively.

The marginal role played by the FSR recovery in the analysis must not be underestimated. Even though the impact on the final results may be negligible, proper treatment of the FSR photons is fundamental for the correct reconstruction of the leptons resolution and energy scales, which directly impact on the ultimate resolution of the reconstructed invariant mass of the $H \rightarrow 4\ell$ decay. Any residual imperfect knowledge of these two quantities must be considered as a systematic uncertainty in the analysis.

4.4 Jets

A precise reconstruction of the jets present in the event is a key handle for the discrimination of the different production mechanisms of the H boson. As briefly mentioned in Sect. 2.3 (cf. Fig. 2.25), the reconstructed hadronic jets can be used to discriminate between the VBF and the other production modes. The number of jets present in the event and their origin is used in the first step of the analysis for the categorization of the events, as described in Chap. 5. Jet-related observables are also used in the context of the measurement of $H \rightarrow ZZ \rightarrow 4\ell$ differential cross sections to probe specific phase space regions sensitive to possible BSM effects and to provide a complete characterization of the production side of the $H \rightarrow ZZ \rightarrow 4\ell$ process, as presented in Chap. 7.

The analysis presented in this thesis exploits PF jets reconstructed by means of an anti-κ_T clustering algorithm with a radius of $\Delta R = 0.4$ [5]. The jet energy and

[2] I.e. events with a four-leptons invariant mass falling within the range $105 < m_{4\ell} < 140$ GeV.

scale are corrected for following the standard CMS jet energy corrections (JECs) detailed in Sect. 2.3.4. The transverse momentum of these hadronic jets is obtained from the vectorial sum of the momenta of all the particles contained within the jet. Only jets with $p_T^j > 30$ GeV, and $| \eta^j | < 4.7$ are considered for the analysis. These are cleaned from the presence of any tight lepton and FSR photon requiring a radial separation of $\Delta R(j, \ell/\gamma) > 0.4$.

Following the central CMS recommendations, tight identification criteria are employed to define the jets considered for the analysis. More precisely, dedicated cuts are applied to control the fractions of neutral hadron and electromagnetic constituents, the number of particles in the jet, and the fractions of charged hadron and electromagnetic constituents in the central region of the detector. Additional selection criteria are adopted to mitigate spurious jets arising from nearby bunch crossings, as described in the next section.

4.4.1 Pileup Jets Identification

To mitigate the presence of additional tracks and energy deposits in the calorimeters arising from PU events, a dedicated PU jet identification algorithm is applied on jets with reconstructed transverse momentum lower than 50 GeV, following the recommendation of the CMS Jet Physics Objects Group (POG). The main reason behind this choice is that the algorithm for the PU jet ID is trained with jets with $p_T < 50$ GeV, where the highest composition of PU jets is expected. The PU jet ID algorithm is implemented via a boosted decision tree (BDT) model trained against the following features:

- the tracks' trajectories of the particles within a jet, to determine whether the corresponding jet is compatible with the primary interaction vertex;
- the jet shape topology, which is used to discriminate hard jets against PU ones;
- the objects multiplicity in the event.

Three operative working points[3] (WP) are assigned from the BDT score and can be used for the mitigation of PU jets. As an example, the performance of the BDT for the loose WP is shown in Fig. 4.10, where one can observe a global efficiency larger than 90% throughout all the p_T and η regions. The BDT is trained on 2016 Z($\rightarrow \mu\mu$)+jets events simulated with MadGraph aMC@NLO and the resulting model is applied, with a tight WP, separately to the three data taking periods considered. Signal jets, i.e. originating from the hard scattering, are defined as jets with a geometrical match to a hard-scattering generator level jet with $p_T > 8$ GeV, whereas all the other jets are classified as pileup jets.

[3] The WPs are referred to as loose, medium, and tight.

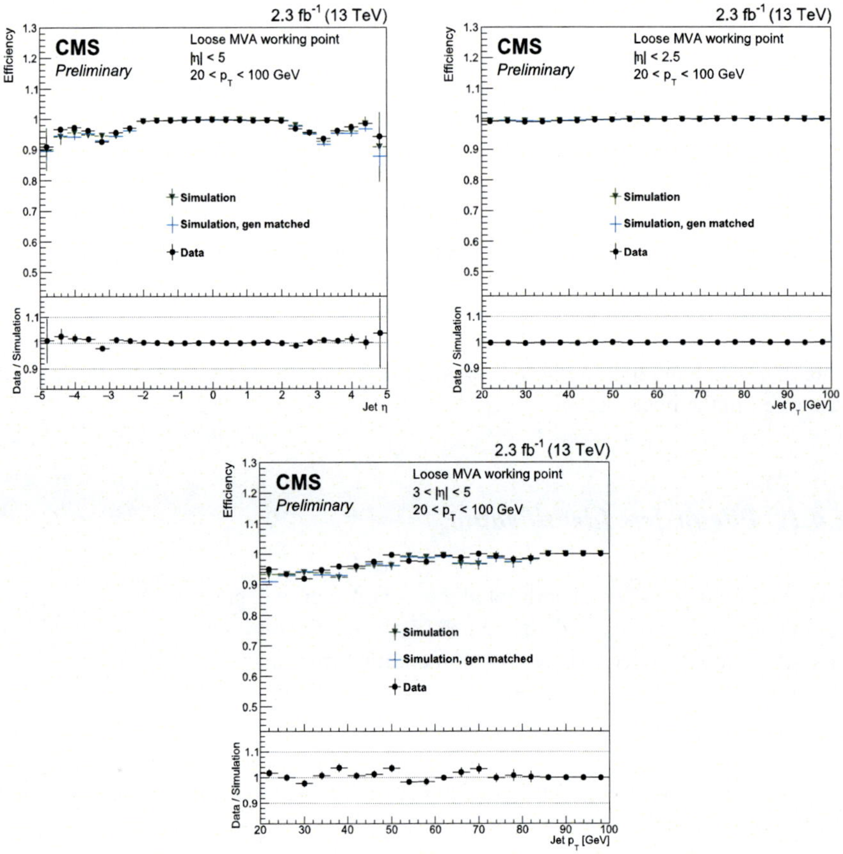

Fig. 4.10 The PU jet ID efficiency on Z($\rightarrow$ $\mu\mu$)+jets is shown for the loose working point of the BDT model described in the text. The efficiencies for Data and MC are shown as a function of the jet pseudorapidity (top left) and transverse momentum of central (top right) and forward jets (bottom). Figure taken from Ref. [6]

4.4.2 *Enhancing the Purity of Jets*

This analysis targets a comprehensive characterization of the H $\rightarrow$ ZZ $\rightarrow$ 4ℓ decay channel via the measurement of production and differential cross sections, as well as the measurement of cross sections in the Simplified Template Cross Sections (STXS) framework. In order to probe with sufficient precision all the different phase space regions and to enhance the sensitivity to the different production mechanisms of the H boson, the analysis relies on a twofold event categorization based on matrix element kinematic discriminants (cf. Sect. 5.3). The number of jets in the events, along with their properties, carries great discrimination power to distinguish among different production mechanisms and phase space regions, therefore largely affects the purity of the categories defined for the analysis.

It is fundamental to have a precise identification and isolation of the jets in the events, as well as a substantial reduction of all the spurious effects arising from problems in the data taking or flaws in the reconstruction procedure. The following presents all the *ad hoc* corrections applied in the analysis to cope with known issues that may affect the quality of the reconstructed jets.

L1 Prefiring

A significantly large fraction of high η L1 Trigger Primitives (TP) was associated with the wrong bunch crossing in 2016 and 2017 data taking periods due to the gradual timing shift of the ECAL not being properly propagated to the L1 TPs. The L1 triggering rules forbid two consecutive bunch crossings to fire. Hence the events affected by this issue could self veto when the ECAL recorded a significant energy deposit in the $2 <| \eta |< 3$ regions. This effect is observed in data but not modeled in the simulations. Hence, dedicated weights, obtained as the product of the non prefiring probability of all objects in (p_T, η) bins, are applied to 2016 and 2017 simulated events to reproduce the behaviour observed in data. The effect of the L1 prefiring weights is shown in Fig. 4.11 for ggH and VBF simulated signal samples. While the effect is negligible for the former, a 2–3% discrepancy is observed in the endcap regions for the latter, supporting the need for this weight in the analysis.

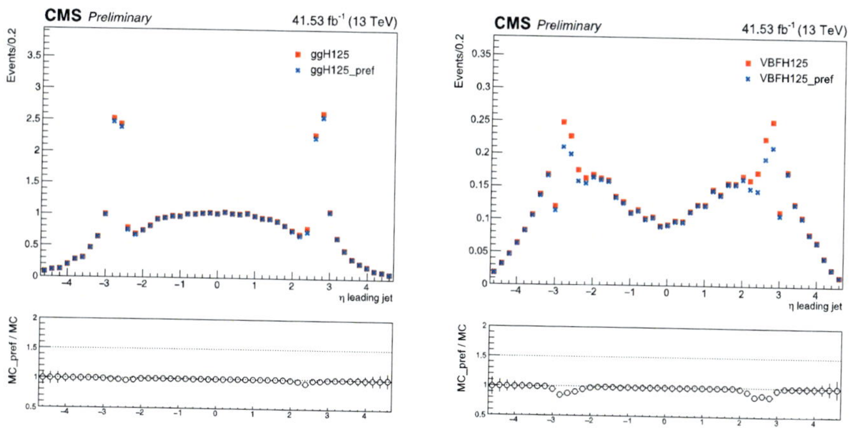

Fig. 4.11 The effect of the L1 prefiring weights on 2017 MC samples for the ggH (left) and VBF (right) production mechanisms. The samples without the L1 prefiring weight applied are depicted in the red histograms, while the blue histograms show the effect of the application of these weights. The ratio between the two histograms is shown in the bottom panels of each plots, showing a 2–3% discrepancy in the endcap regions for the VBF production mode

HEM 15/16 Failures

Two modules of the HCAL front-end electronics (HEM15 and HEM16) suffered irreparable damage in mid 2018 due to a CMS-wide power interlock. This issue was expected to cause larger multiplicities in the $-2.4 <\mid \eta \mid< -1.4$ and $-1.6 < \phi < -0.6$ regions. Dedicated studies were carried out to compare the events reconstruction before and after the HEM problem occurrence. No significant difference was found for the $\mathrm{H} \rightarrow \mathrm{ZZ} \rightarrow 4\ell$ analysis objects and therefore no particular procedure was adopted to cure the HEM15 and HEM16 failures.

4.4.3 b-Jets Tagging

The analysis relies on identifying b-quark induced jets, hereafter referred to as *b-jets*, as an additional tool for the the categorisation of the events detailed in Chap. 5. The process of discriminating b-jets against light flavour jets is often referred to as *b-tagging*. Given the wide applicability of the *b-tagging* procedure in many analyses, recommendations are provided centrally in CMS.

The $\mathrm{H} \rightarrow \mathrm{ZZ} \rightarrow 4\ell$ analysis presented in this work relies on the use of the DeepCSV algorithm [7, 8], where a Deep Neural Network (DNN) combines the information on the heavy jet's impact parameter, track quality, and secondary vertex position to identify jets originating from b-quarks. More precisely, the DeepCSV architecture consists of four fully-connected hidden layers, each with 100 nodes. The activation of each node is regulated by a rectified linear unit (ReLU) function that fires according to the input values of the given node. The last layer consists of five nodes, of which the activation is regulated by a normalized exponential function,[4] corresponding to the five jet flavour categories employed for the training of the DNN. The training of the architecture is performed simultaneously in all the vertex categories and jet flavours using jets with transverse momentum between 20 GeV and 1 TeV contained in the tracker acceptance. A mixture of tt and multijet events, with a relative ratio of 2:1:4 for b:c:udsg jets, ensures the reduction of possible dependency of the training on the heavy-flavour quark production process. The efficiency of the algorithm to correctly identify a jet of a given flavour f is defined as the ratio of the number of f-tagged jets to the total number of jets of flavour f and it is found to be consistently above 50% on simulation. The difference between the tagging efficiency in data and simulation is corrected for by applying dedicated per jet data-to-simulation scale factors ($SF_f = \varepsilon_f^{\mathrm{data}}(p_{\mathrm{T}}, \eta)/\varepsilon_f^{\mathrm{MC}}(p_{\mathrm{T}}, \eta)$) obtained from the ratio of the tagging efficiency for a f-flavour jet in data to the one in simulation. Three operative working points, referred to as "loose", "medium", and "tight", are defined

[4] The usage of a normalized exponential function instead of the ReLU function used in the other layers allows to interpret the output as a probability for a specific jet flavour category.

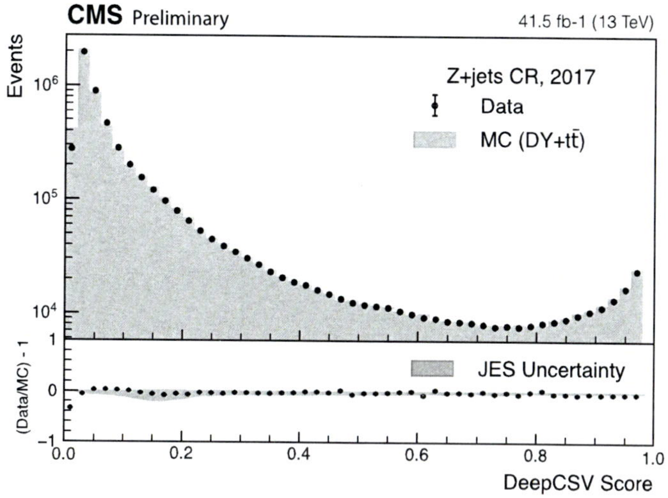

Fig. 4.12 The b-jet tagging score for the medium WP of the DeepCSV algorithm for the leading jet in the Z+jets CR. The black points represent data and the orange histogram the MC simulation. The bottom panel shows the ratio of data to simulation distributions. The effect of the uncertainty on the jet energy scale is shown by the gray band

and correspond to thresholds after which the misidentification probability is 10%, 1%, and 0.1%, respectively. The analysis exploits the DeepCSV algorithm with a medium working point, of which the corresponding score is shown in Fig. 4.12.

For this choice of WP, data to MC scale factors are applied to simulated jets, as a function of the jet transverse momentum (p_T^j), pseudorapidity (η^j) and flavour, by downgrading (upgrading) the b-tagging status of b-tagged (untagged) jets fraction.

4.5 Summary

Table 4.4 summarizes the properties of the main objects used in the $H \rightarrow ZZ \rightarrow 4\ell$ analysis. Electrons and muons, also referred to as "tight leptons", are combined together to build Z boson candidates, which are in turn used to build the H boson events that enter the statistical analysis, as extensively discussed in the next chapter.

Table 4.4 Summary of the selection requirements applied to the physics object that enter the $H \to ZZ \to 4\ell$ analysis

Electrons
$p_T^e > 7\text{GeV} \mid \eta^e \mid < 2.5$
$d_{xy} < 0.5\,\text{cm}\ d_z < 1\,\text{cm}$
$\lvert SIP_{3D}\rvert < 4$
ID and isolation with cuts on BDT score as of Table 4.2
Muons
Global or Tracker Muon
Discard Standalone Muon tracks if reconstructed in muon system only
$p_T^\mu > 5\text{GeV} \mid \eta^\mu \mid < 2.4$
$d_{xy} < 0.5\,\text{cm}\ d_z < 1\,\text{cm}$
$\lvert SIP_{3D} < 4$
PF muon ID if $p_T < 200$ GeV, PF muon ID or High-p_T muon ID (Table 4.3) if $p_T > 200$ GeV
$\mathcal{I}_{PF}^\mu < 0.35$
FSR photons
$p_T^\gamma > 2\text{GeV}\ \text{abs}\eta^\gamma < 2.4$
$\mathcal{I}_{PF}^\gamma < 1.8$
$\Delta R(\ell, \gamma) < 0.5\ \Delta R(\ell, \gamma)/(p_T^\gamma)^2 < 0.012\text{GeV}^{-2}$
Jets
$p_T^{jet} > 30\text{GeV}\ \text{abs}\eta^{jet} < 4.7$
$\Delta R(\ell/\gamma, \text{jet}) > 0.4$
Cut-based jet ID (tight WP)
Jet pileup ID (tight WP)
Deep CSV b-tagging (medium WP)

References

1. Chen T, Guestrin C (2006) XGBoost. ACM
2. Rembser J (2020) Measurement of triboson production in themultilepton final state and electron studieswith the CMS experiment at the LHC. PhD thesis, IPP Paris
3. CMS Collaboration (2021) Electron and photon reconstruction and identification with the CMS experiment at the CERN LHC. JINST 16(05):P05014

4. Bodek A, van Dyne A, Han JY, Sakumoto W, Strelnikov A (2012) Extracting muon momentum scale corrections for hadron collider experiments. Eur. Phys J C 72:2194
5. Cacciari Matteo, Salam Gavin P, Soyez Gregory (2008) The anti-k_t jet clustering algorithm. JHEP 04:063
6. Jet algorithms performance in 13 TeV data (2017) Technical Report CMS-PAS-JME-16-003, CERN, Geneva
7. CMS Collaboration (2016) Identification of b-Quark Jets with the CMS experiment. Technical report CMS-PAS-BTV-15-001, CERN, Geneva
8. Collaboration CMS (2018) Identification of heavy-flavour jets with the CMS detector in pp collisions at 13 TeV. JINST 13(05):P05011

Chapter 5
Events Selection and Categorization

The "golden journey" towards the characterization of the H boson properties in the 4ℓ decay channel continues in this chapter, where the events selection and their subsequent categorization are presented.

While in Run-I and previous versions of the $H \to ZZ \to 4\ell$ analysis the properties of the H boson were measured inclusively or, at most, per production mechanism, this analysis aims at probing fine granular phase space regions of each production mode. Often the H boson properties are assessed by measuring signal strength modifiers or fiducial cross sections. The former is defined as the ratio of the observed cross section times branching fraction "$(\sigma \cdot \mathrm{BR})^{\mathrm{obs.}}$" to the corresponding SM expectation "$(\sigma \cdot \mathrm{BR})^{\mathrm{exp.}}$", while the latter consists of measuring cross sections in a phase space region defined to minimize the model dependence of the results. However, neither of these approaches allows a complete characterization of specific phase space regions sensitive to a given production mechanism. For this reason, the measurements are also performed with a third, complementary approach: the Simplified Template Cross Section (STXS) framework. The STXS framework aims at reducing the theory dependence of the measurements whilst, at the same time, maximising the experimental sensitivity. Signal cross sections are measured for each production mechanism in dedicated phase space regions, also referred to as STXS bins, with enhanced sensitivity to either SM predictions or BSM scenarios. A more detailed description of the STXS framework, is given in Sect. 5.2.

A dedicated events categorization is employed to characterize the H boson production cross sections both in an inclusive way and in the more fine granular description of the STXS. The selected events are divided, exploiting the topology of the event and a set of matrix element discriminants, into seven mutually exclusive categories with enhanced sensitivity to the five main production mechanisms of the H boson. In the second step of the categorization of the events, these seven categories are further split into a total of twenty-two sub-categories targeting the STXS bins, as detailed in Sect. 5.3.2.

M. Bonanomi, *Response of the High Granularity Calorimeter HGCAL and Characterisation of the Higgs Boson*, Springer Theses,
https://doi.org/10.1007/978-3-031-26833-5_5

Once the events are selected and categorized, they are used as inputs for the signal extraction by means of the statistical analysis procedure described in Chap. 7.

5.1 Events Selection

The PF objects selected with the identification and isolation requirements outlined in the previous chapters are considered for the next step of the analysis, which targets the selection of events compatible with a Higgs boson candidate.

The events are required to have fired the HLT paths described in Sect. 2.2.4. Conversely to the Run-I analysis, an event is selected if any of the HLT paths is fired, independently on the reconstructed final state. This OR logic ensures the proper identification of production mechanisms with additional leptons, thus improving even further the trigger efficiency.

The events selection is triggered by a primary vertex (PV) requirement, i.e., only events with a well-reconstructed PV are retained. More in detail, each event is requested to have a small PV radius, $R_{\text{PV}} < 2$ cm, a large number of degrees-of-freedom, $N_{\text{PV}} > 4$, and collisions limited to the z-axis, $z_{\text{PV}} < 24$ cm.

The Z boson candidates are formed, in the events fulfilling the PV criteria, from same flavor and opposite sign (e^+e^-, $\mu^+\mu^-$) *tight leptons*, as defined in Chap. 4, with invariant mass in the range $12 < m_{l^+l^-} < 120$ GeV. The ZZ candidates considered for the analysis consist of pairs of non-overlapping Z, where the candidate with invariant mass closer to the nominal Z boson mass is referred to as Z_1, while the second one is referred to as Z_2. The ZZ candidates from this selection chain define three mutually exclusive channels corresponding to the three 4ℓ final states considered in the measurements: 4μ, $4e$, and $2e2\mu$.

In addition, the following criteria are imposed to enhance the sensitivity to the H boson decays and improve the signal-to-background ratio further.

- **Minimum Z_1 mass:** to be considered for the analysis, the di-lepton pair forming the Z_1 candidate must have an invariant mass larger than 40 GeV.
- **Ghost removal:** all the lepton pairs selected must have a radial distance of $\Delta R(\ell_i, \ell_j) > 0.02$.
- **Lepton p_T:** two of the four selected leptons must have $p_T > 10$ GeV and at least one $p_T > 20$ GeV, thus ensuring an optimal trigger efficiency.
- **QCD suppression:** the invariant mass of all the opposite-sign di-lepton pairs that can be built with the four leptons selected must be larger than 4 GeV ($m_{l^+l'^-} >$ 4 GeV). FSR photons are not included in the computation of the invariant mass to reproduce the scenario of low mass QCD di-lepton resonances.
- ***Smart* cut:** the alternative ZZ candidate that can be built in the 4μ and $4e$ channels is referred to as Z_aZ_b and is used to identify and to reject events with an on-shell Z accompanied by a low-mass $\ell^+\ell^-$ resonance. More precisely, candidates with $m_{Z_b} < 12$ GeV are not retained in the analysis if the invariant mass of Z_a is closer to the nominal Z boson mass than the one of Z_1 in the main ZZ candidate.

- **Invariant mass of the 4ℓ system:** the events satisfying the above criteria and with a four-lepton system invariant mass larger than 70 GeV ($m_{4\ell} > 70$ GeV) are retained in the analysis.

While in most of the cases a single ZZ candidate is selected from the requirements aforementioned, in events originating from the production of a Higgs boson in association with a pair of top quarks ($t\bar{t}$H) or an additional vector boson (VH) additional leptons may be found in the final state, thus resulting in more than one ZZ candidate identified. In such cases, the one with the largest value of the kinematic discriminant $\mathcal{D}_{bkg}^{kin}$ is eventually retained, unless the two candidates are composed of the same four leptons, and the one with m_{Z_1} closest to the nominal Z boson mass is considered for the analysis. A detailed description of the matrix element approach and of the kinematic discriminants used in the analysis is given later in Sect. 5.1.1.

5.1.1 Matrix Element Kinematic Discriminants

The invariant mass of the 4ℓ system is probably the most straightforward observable to consider when it comes to the discrimination between signal and background events: the sharp resonance peak typical of the $H \to ZZ \to 4\ell$ decay is substantially different from the flat shapes of the main background processes. Hence, it could be used for the statistical analysis. Nevertheless, with the increasing statistics collected from the experiments and with the ultimate goal of precision measurements of the H boson properties, the analyses introduce advanced methods to enhance the sensitivity to different production mechanisms and the purity of the event categories. Two complementary approaches are usually considered: machine learning algorithms and matrix elements methods, as they permit both a full disclosure of the hidden relations among the physical observables that characterize a particular decay channel and an increased discrimination power between signal and background processes.

The $H \to ZZ \to 4\ell$ analysis presented in this thesis exploits the so-called Matrix Element Likelihood Approach (MELA) both in the categorization of the events and in the signal extraction. The computations are performed via the MELA package [1, 2], which relies on the JHUGen [3] and MCFM [4] generators to compute matrix elements for signal and background events, respectively. The underlying assumption supporting the use of the MELA method is that the $H \to 4\ell$ decay can be completely characterized by means of eight independent degrees of freedom, namely:

- $m_{4\ell}$, the invariant mass of the 4ℓ system;
- m_{Z_1} and m_{Z_2}, the invariant masses of the two Z boson candidates;
- $\theta^{\star}$, identified as the angle between the z-axis and the Z_1 direction in the 4ℓ rest frame;
- Φ and Φ_1, the azimuthal angles between the planes containing the H boson and the decay products of the two Z bosons;
- θ_1 and θ_2, the angles between the Z_1 and Z_2 flight directions and the planes containing the di-lepton systems originating from the decay of these vector bosons.

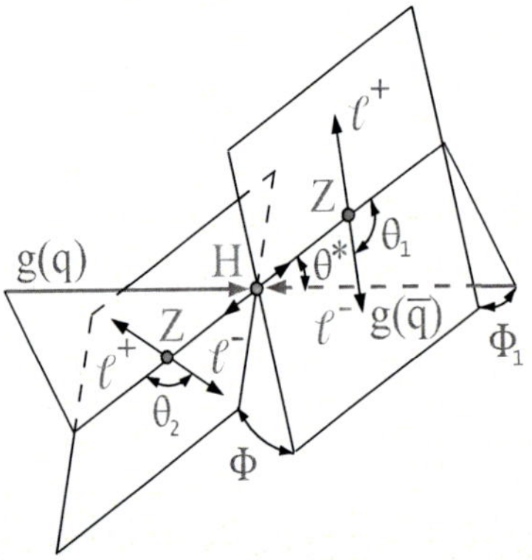

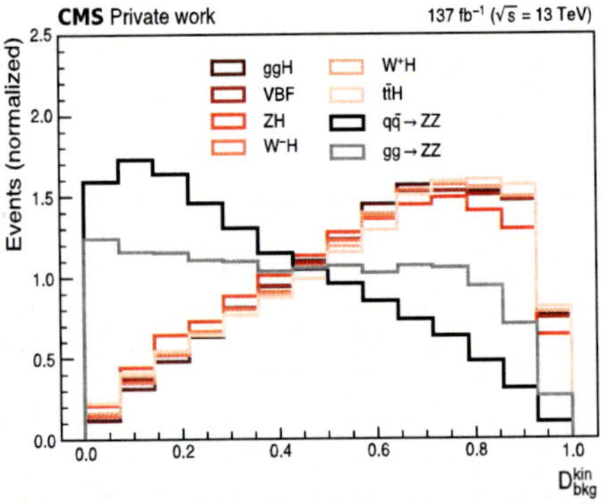

Fig. 5.1 Schematic representation of the $H \rightarrow ZZ \rightarrow 4\ell$ decay (left) with the illustration of the five angles that comprise the $\vec{\Omega}^{H\rightarrow 4\ell}$ set of kinematic observables used to build the kinematic discriminant ($\mathcal{D}^{kin}_{bkg}$) shown in the right plot for the H boson production mechanisms and the two irreducible background processes that enter the analysis

While the $\theta^{\star}$, Φ, and Φ_1 angles are defined in the H boson rest frame, θ_1 and θ_2 are defined with respect to the two Z bosons rest frames. The angular variables, of which the set[1] is hereafter referred to as $\vec{\Omega}^{H\rightarrow 4\ell}$, are depicted in the left side of Fig. 5.1, where a schematic representation of the $H \rightarrow ZZ \rightarrow 4\ell$ decay channel is given for the ggH production mode.

Hence, one can extract from the JHUGen and MCFM generators the probability[2] $\mathcal{P}_a$, as a function of the masses m_i and angular observables $\vec{\Omega}$ for a given value of $m_{4\ell}$, for an event to have been produced in a physical process a. The MELA method takes these probabilities as input to compute the *likelihood ratio* test statistics:

$$q_{a,b} = \frac{\mathcal{P}_a}{\mathcal{P}_b}, \tag{5.1}$$

which is the best discriminating variable between the two hypotheses, read *physical processes a* and b, as a consequence of the Neyman-Pearson lemma[3] [5]. The applications of Eq. (5.1) cover a large variety of scenarios: from the discrimination between signal and background processes, to the comparison between two different signal processes (e.g. production mechanisms of the H boson), up to the test of the

[1] Along with the inclusion of the m_{Z_1} and m_{Z_2} observables.

[2] There is a proportionality relation between the number of events N generated in a particular phase space under a given hypothesis, read physical process, a and the probability to obtain an event in the same phase space using a probability distribution $\mathcal{P}_a$.

[3] Which states that the likelihood ratio (q_{H_0,H_1}) is the most powerful test statistics to discriminate a null hypothesis H_0 against an alternative hypothesis H_1.

SM against BSM hypotheses. In the analysis presented in this manuscript the structure of Eq. (5.1) is slightly modified to build matrix element kinematic discriminants as:

$$\mathcal{D} = \frac{\mathcal{P}_a}{\mathcal{P}_a + \mathcal{P}_b}, \tag{5.2}$$

which still contains the full information of Eq. (5.1) but have the additional advantage of being bounded between 0 and 1, therefore facilitating any binning required for the analysis and the definition of suitable working points for the application of the discriminant.

The main kinematic discriminant of which the analysis makes extensive use targets the $gg/q\bar{q} \to 4\ell$ processes and is defined as:

$$\mathcal{D}^{\text{kin}}_{\text{bkg}} = \left[1 + \frac{\mathcal{P}^{q\bar{q}}_{\text{bkg}}(\vec{\Omega}^{\text{H}\to 4\ell}|m_{4\ell})}{\mathcal{P}^{gg}_{\text{sig}}(\vec{\Omega}^{\text{H}\to 4\ell}|m_{4\ell})} \right]^{-1}, \tag{5.3}$$

where $\mathcal{P}^{q\bar{q}}_{\text{bkg}}$ defines the probability for background events, mostly driven by the dominant $q\bar{q} \to ZZ$ process, and $\mathcal{P}^{q\bar{q}}_{\text{sig}}$ is the probability for signal events, of which the ggH production mechanisms is the predominant. The right plot of Fig. 5.1 represents the distribution of such a discriminant for all the production modes considered in the analysis, as well as for the two irreducible background processes $q\bar{q} \to ZZ$ and $gg \to ZZ$. As previously discussed, $\mathcal{D}^{\text{kin}}_{\text{bkg}}$ is used as an additional selection criteria to identify the analysis ZZ candidate whenever more than one four-lepton system is identified. $\mathcal{D}^{\text{kin}}_{\text{bkg}}$ is also embedded in the definition of the likelihood function used in the statistical analysis, as further detailed below.

The analysis results heavily rely on the use of this kind of matrix element discriminants, as the production cross sections of the SM H boson are measured by means of a two dimensional maximum likelihood fit. More precisely, the likelihood function maximized to extract the analysis results consists of the product of two one dimensional likelihoods: the first one is proportional to the unbinned distribution of the invariant mass of the four-lepton system ($m_{4\ell}$). In contrast, the second one embeds a conditional probability density function (pdf) built from a two-dimensional template of the kinematic discriminant as a function of the $m_{4\ell}$ distribution. The conditional distribution of $\mathcal{D}^{\text{kin}}_{\text{bkg}}$ is shown in Fig. 5.2 for signal and background processes in the reconstructed category of the analysis sensitive to the ggH production mechanism.

The two dimensions of this conditional template are finely binned to appreciate the strong correlation between the kinematic discriminant and $m_{4\ell}$, which brings additional power for the discrimination between the signal and background processes, as shown in the figure. More quantitatively, the signal-to-background separation improves by 20% when the full kinematic information of the $\text{H} \to \text{ZZ} \to 4\ell$ decay is included in the fits under the form of the conditional pdf just described. Even though the $\mathcal{D}^{\text{kin}}_{\text{bkg}}$ discriminant is really effective in selecting signal events against background ones, its performance can be enhanced for the VBF and VH processes, where the presence of jets in the event provides an additional handle for the discrimination

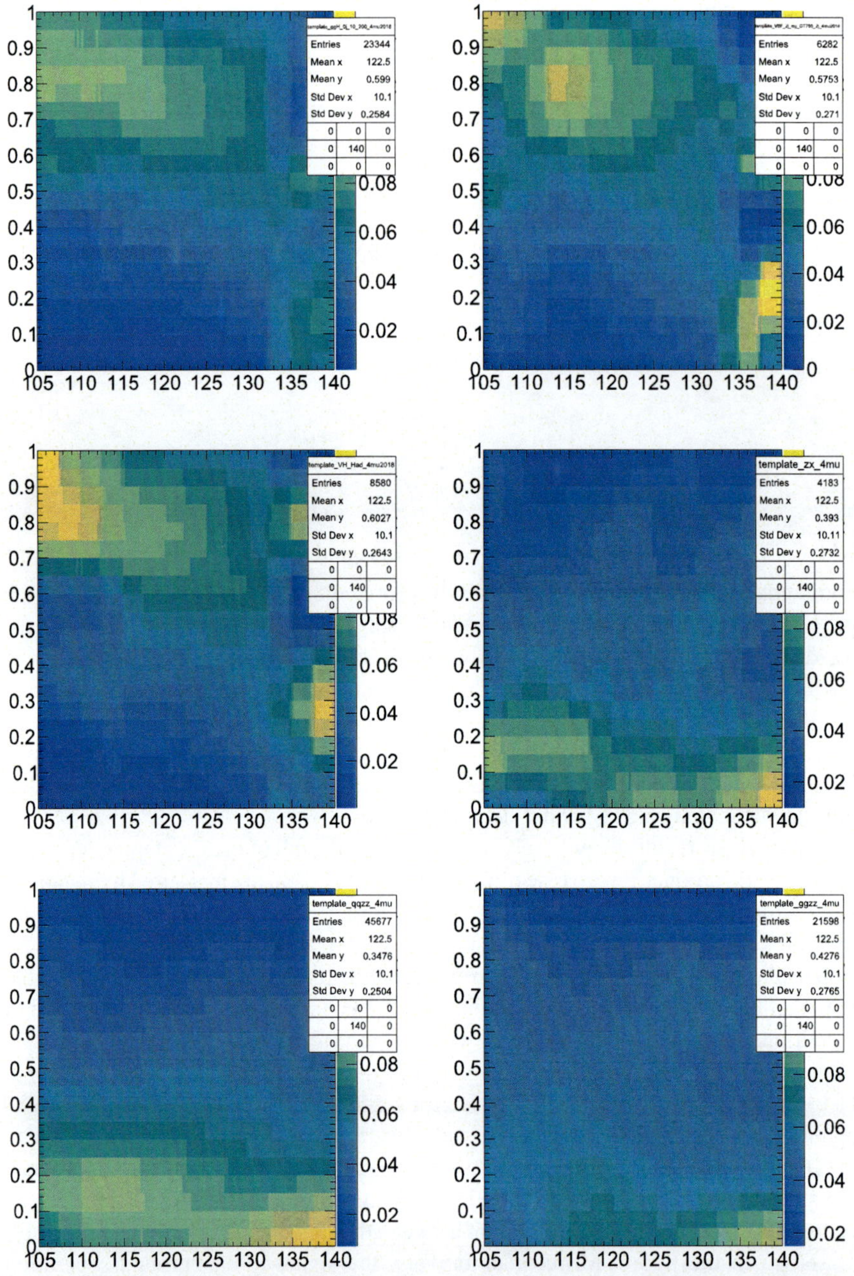

Fig. 5.2 Conditional distribution of $\mathcal{D}_{\text{bkg}}^{\text{kin}}$ in the reconstructed event category sensitive to the ggH production mode for: ggH signal (top left), VBF signal (top right), ZH signal (middle left), Z+X background (middle right), $q\bar{q} \rightarrow 4\ell$ (bottom left) and $gg \rightarrow$ZZ (bottom right) as a function of $m_{4\ell}$

of the signal against the QCD background processes $ZZ/Z\gamma^\star/\gamma^\star\gamma^\star \to 4\ell$ and the electroweak (EW) processes of vector boson scattering (VBS) or triboson (VVV) production. For this reason two additional discriminants are defined:

$$\mathcal{D}_{\text{bkg}}^{\text{VBF+dec}} = \left[1 + \frac{c^{\text{VBF}}(m_{4\ell})[\mathcal{P}_{\text{bkg}}^{\text{EW}}(\vec{\Omega}^{\text{H+jj}}|m_{4\ell}) + \mathcal{P}_{\text{bkg}}^{\text{QCD}}(\vec{\Omega}^{\text{H+jj}}|m_{4\ell})]}{\mathcal{P}_{\text{sig}}^{\text{EW}}(\vec{\Omega}^{\text{H+jj}}|m_{4\ell})}\right]^{-1} \quad (5.4)$$

$$\mathcal{D}_{\text{bkg}}^{\text{VH+dec}} = \left[1 + \frac{c^{\text{VH}}(m_{4\ell})[\mathcal{P}_{\text{bkg}}^{\text{EW}}(\vec{\Omega}^{\text{H+jj}}|m_{4\ell}) + \mathcal{P}_{\text{bkg}}^{\text{QCD}}(\vec{\Omega}^{\text{H+jj}}|m_{4\ell})]}{\mathcal{P}_{\text{sig}}^{\text{EW}}(\vec{\Omega}^{\text{H+jj}}|m_{4\ell})}\right]^{-1}, \quad (5.5)$$

where the set of observables $\vec{\Omega}^{\text{H}\to 4\ell}$ provided as input for the MELA computations is complemented here by the information of the two additional jets in the event. $\mathcal{P}_{\text{sig}}^{\text{EW}}$ is the probability for the VBF and VH signal, $\mathcal{P}_{\text{bkg}}^{\text{EW}}$ represents the probability of the EW VBS and VVV background processes, while $\mathcal{P}^{\text{QCD}}$ the one of the $ZZ/Z\gamma^\star/\gamma^\star\gamma^\star \to 4\ell$ QCD production in association with two jets. The factors $c(m_{4\ell})$ are introduced to reshape the distributions in such a way that their shapes are evenly spread in the [0, 1] interval, crossing at 0.5 for signal and background processes, thus facilitating further rebinning and implementation of selection cuts in the analysis.

The performance of these two kinematic discriminants sensitive to the VBF and VH topologies are compared to the $\mathcal{D}_{\text{bkg}}^{\text{kin}}$ discriminant in Fig. 5.3, where one can appreciate the substantial improvement in the ROC curves quantifying the separation between the signal processes and the $q\bar{q} \to 4\ell$ background. More precisely, the inclusion of these two discriminants in the analysis was to improve the measurements of the VBF and VH signal strength modifiers by about 10 to 15%.

5.2 The Simplified Template Cross Section Framework

Before moving on to the description of the event categories considered in the analysis, it is worth to spend some time introducing the theoretical pillar on which these are built: the Simplified Template Cross Section (STXS) framework.

After the discovery of the H boson, one of the main goals of the LHC Run-II physics programme has been the precision measurement of the H boson properties and couplings to probe in detail the SM predictions. In Run-I these results were often presented in the form of signal strength modifiers, an approach with two substantial limiting factors: a strong model dependence of the results and an "inclusive"[4] approach to the measurements. An alternative way of assessing the H boson properties was to perform fiducial cross section measurements, i.e., measured in a phase

[4] The term "inclusive" here refers to the fact that signal strengths are often measured either per-decay (e.g. the "inclusive" $H \to ZZ \to 4\ell$ signal strength) or inclusively per-production mode (e.g. ggH production mode).

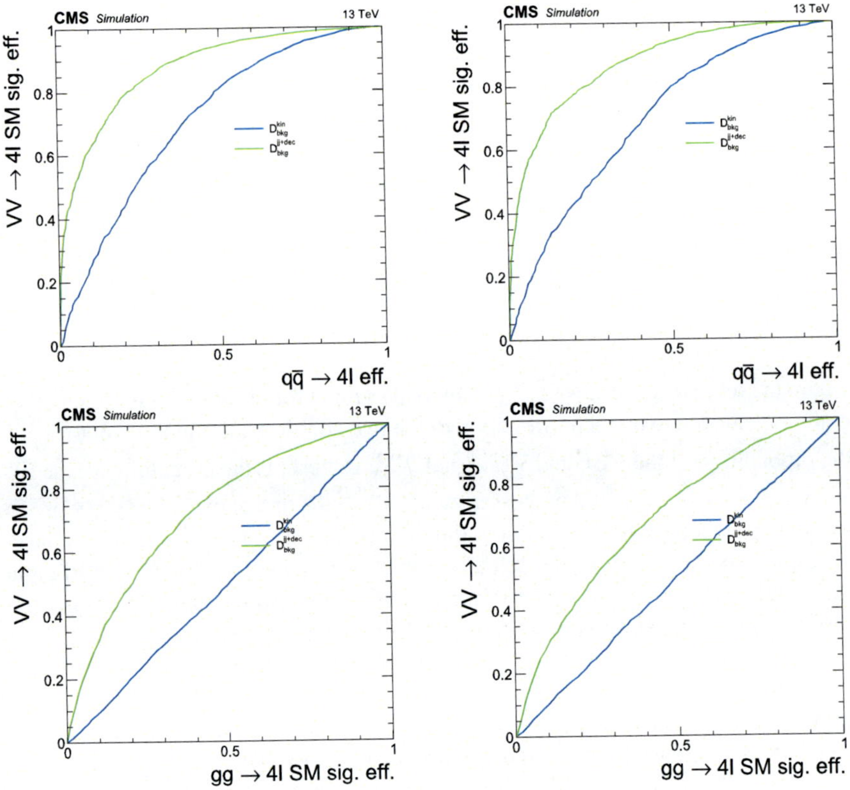

Fig. 5.3 *Top:* Comparison of $\mathcal{D}_{\text{bkg}}^{\text{VBF+dec}}$ and $\mathcal{D}_{\text{bkg}}^{\text{kin}}$in the VBF-2jet tagged category (left) and $\mathcal{D}_{\text{bkg}}^{\text{VH+dec}}$ and $\mathcal{D}_{\text{bkg}}^{\text{kin}}$in the hadronic VH-tagged category (right) between the VBF (left) or VH (right) processes and the $q\bar{q} \to 4\ell$ background. *Bottom*: Comparison of $\mathcal{D}_{\text{bkg}}^{\text{VBF+dec}}$ and $\mathcal{D}_{\text{bkg}}^{\text{kin}}$in the VBF-2jet tagged category (left) and $\mathcal{D}_{\text{bkg}}^{\text{VH+dec}}$ and $\mathcal{D}_{\text{bkg}}^{\text{kin}}$in the hadronic VH-tagged category (right) between the VBF (left) or VH (right) processes and the gluon fusion signal process

space region defined to reduce the model acceptance effects. This approach, specifically designed to maximize the theory independence of the measurements, presents two main drawbacks: it is still inclusive in the production mechanisms of the H boson and it is substantially limited by the statistics available, thus being possible only on a subset of all the decay channels. The STXS framework is designed to provide the analyses with a common ground for measurements that combine the advantages of both the signal strength modifiers and the fiducial cross sections approaches.

More precisely, the STXS framework defines mutually exclusive phase space regions, hereafter referred to as *bins*, to maximize the experimental sensitivity while simultaneously reducing the theoretical uncertainty, directly folded into the measurement. In addition, the STXS bins are defined for each production mode of the H boson separately, thus reducing the model dependence of the measurements and

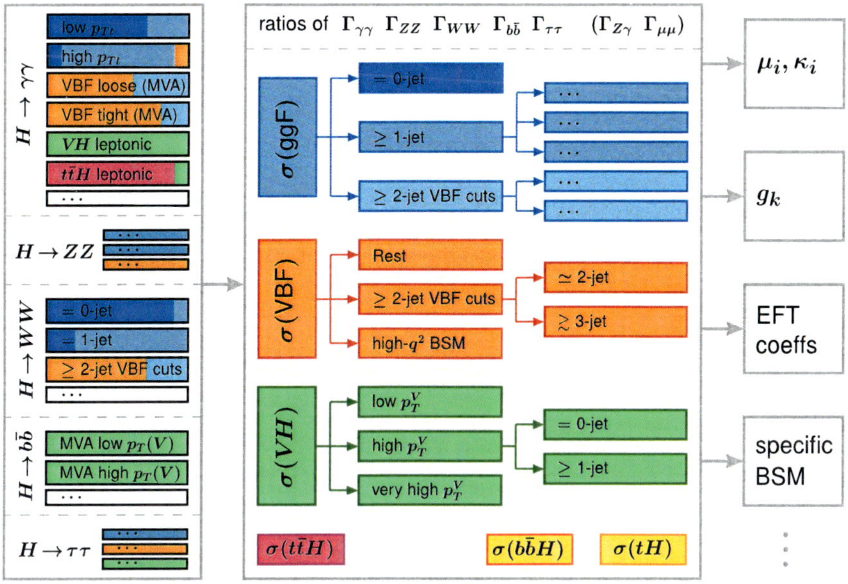

Fig. 5.4 Schematic overview of the STXS approach

facilitating their re-interpretation. A schematic view of the STXS framework is given in Fig. 5.4. As shown in the left box, the analyses design a particular events categorization to target different phase space regions of a particular production mechanism of the Higgs boson. The analysis categories are chosen to match the true STXS bins used as a reference. Cross sections in the STXS framework are then measured by means of a fit over all the analysis categories, as illustrated in the central box. Eventually, as depicted in the right box of Fig. 5.4, these results will serve as an input for interpretations that are not limited to a single theoretical model, but can range from the measurements of the SM H boson couplings, to the test of specific BSM scenarios, up to the constrain of EFT Wilson's coefficients. As mentioned above, such versatility is at the basis of the STXS approach where, conversely to the usual signal strength modifiers, the theory uncertainties on the SM predictions are not embedded in the cross section measurements, thus easing further re-interpretations and ensuring extended longevity of the results.

The original binning scheme proposed in the STXS framework is shown in Fig. 5.5 and it is referred to as Stage 0. It was designed to target the main production mechanisms of the H boson in an exclusive way and its bins are `ggH`, `qqH`, `VH-lep`, and `ttH`. All the bins are defined with a global cut on the Higgs boson rapidity at $|y_{\rm H}| < 2.5$ to reflect the lack of sensitivity of the experimental measurements beyond this range. The `qqH` bin is defined to be sensitive to the VBF production mode and to the associated productions (VH) with a hadronically decaying vector boson. The `VH-lep` production bin includes VH events with leptonic decays of the vector boson V. The $b\bar{b}$H and tH production processes are characterized by a low

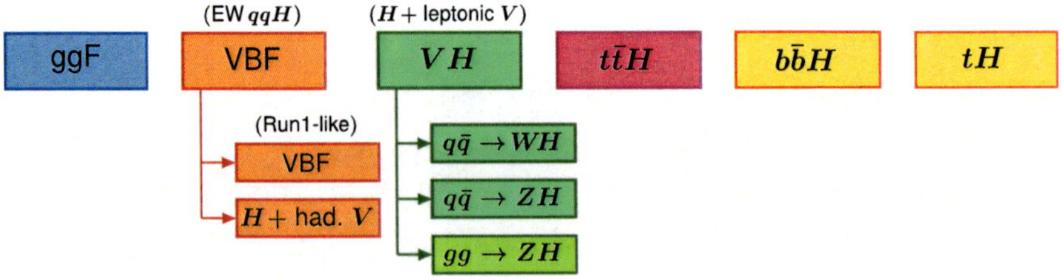

Fig. 5.5 The Stage 0 of the STXS framework

statistics and therefore considered together with the ggH and ttH production bins, respectively. Similarly, the $gg \to ZH$ process with hadronic decays of the Z boson is included in the ggH bin.

With the increasing statistics collected by the LHC experiments and with the feedbacks coming from the different analyses, the STXS framework evolved towards a more fine-grained binning, with dedicated phase space regions defined specifically to probe the presence of any BSM physics. All the bins are defined on generator-level objects in a phase space volume with a $| y_{\rm H} | < 2.5$ acceptance to reflect the Stage 0 bins. For what concerns the generator-level objects, the jets considered are clustered with the anti-$k_{\rm T}$ algorithm with a radius parameter of 0.4 and a $p_{\rm T}$ threshold of 30 GeV are selected, while no particular requirement on the transverse momentum or rapidity is imposed to leptons.

The current proposed binning is referred to as STXS Stage 1.2 and is depicted in the central panels of the four graphs shown in Fig. 5.6. Due to the low sensitivity expected in some of the bins, a specific merging scheme is employed to maximize the experimental sensitivity to each cross section measured. The merging scheme adopted in the analysis results in a total of 19 production bins, depicted in the lower panels of the graphs in Fig. 5.6 and detailed in the following.

The Stage 0 ggH production bin is further split in two bins with boundary at $p_{\rm T}^{\rm H} = 200$ GeV. The events with $p_{\rm T}^{\rm H} > 200$ GeV identify a phase space region primarily sensitive to possible BSM effects. This region is further divided into four sub-bins in the original Stage 1.2 scheme, all of which are grouped into the ggH/$p_{\rm T} > 200$ bin for the analysis in order to maximize the sensitivity. The SM-like ggH phase space region contains events with $p_{\rm T}^{\rm H} < 200$ GeV and is further split into three categories according to the jet multiplicity in the event. Events with zero or one jets are split into sub-bins as a function of $p_{\rm T}^{\rm H}$. Given the high sensitivity of the analysis to these phase space regions, the same production bins of the STXS Stage 1.2 are considered: ggH-0j/$p_{\rm T}$[0, 10], ggH-0j/$p_{\rm T}$[10, 200], ggH-1j/$p_{\rm T}$[0, 60], ggH-1j/$p_{\rm T}$[60, 120], and ggH-1j/$p_{\rm T}$[120, 200], where the number in brackets refer to the $p_{\rm T}^{\rm H}$ bin boundaries in GeV. The events with two or more jets are further classified according to the di-jet system invariant mass with a boundary at $m_{\rm jj} = 350$ GeV. Similarly to the zero and one jet bins, for events with $m_{\rm jj} < 350$ GeV the production sub-bins are defined as a function of $p_{\rm T}^{\rm H}$. Also in these cases the analysis adopts the same binning as of the original Stage

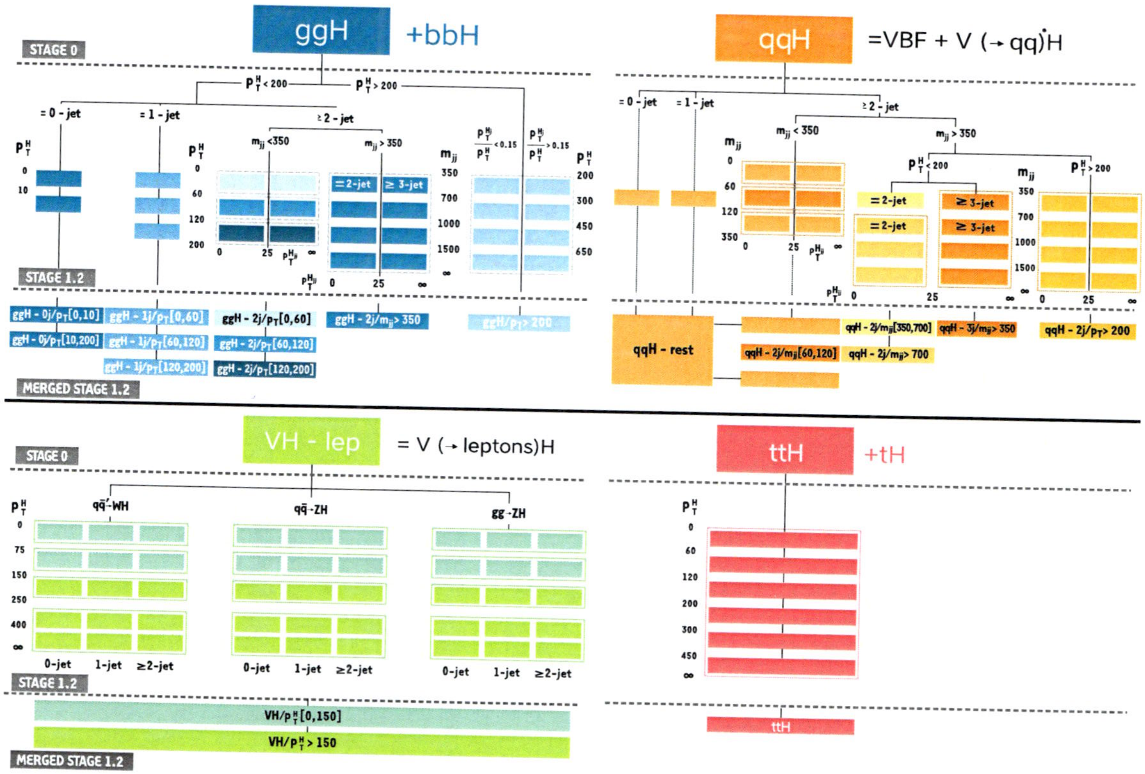

Fig. 5.6 Binning of the gluon fusion production process, the electroweak production process (combines VBF and VH with hadronic V decay), the VH production process with leptonic V decay (combining WH, ZH, and gluon fusion ZH production), and the $t\bar{t}H$ production process in the merged stage 1.2 of the STXS framework used in the $H \rightarrow ZZ \rightarrow 4\ell$ analysis

1.2: ggH-2j/p_T[0, 60], ggH-2j/p_T[60, 120], and ggH-2j/p_T[120, 200]. The events with $m_{jj} > 350$ GeV identify the ggH-like phase space region with VBF-topology. Similarly to what is done for the ggH/p_T > 200 bin, also in this case the four sub-bins defined in the Stage 1.2 scheme are merged together into a single bin called ggH-2j/m_{jj} > 350.

Starting from the inclusive Stage 0 qqH bin, further sub-bins are defined as a function of the number of jets in the event. The majority of the events produced by VBF are characterized by the presence of two or more jets, as shown in Fig. 2.25. Given the low sensitivity of the analysis to qqH-like events with zero or one jets, the corresponding Stage 1.2 bins are merged together into a single bin called qqH-rest. The qqH-like phase space region with two or more jets in the event is further split according to the di-jet invariant mass (m_{jj}). The events with $m_{jj} < 350$ GeV are divided into three bins: the first, with $m_{jj} < 60$ GeV, and the third, with $120 < m_{jj} < 350$ GeV, are merged together with the events with less than one jet in the qqH-rest bin; the second one, with $60 < m_{jj} < 120$ GeV, identifies the phase space region sensitive to VH production with a hadronic decay of the V boson, therefore is retained for the analysis and referred to as qqH-2j/m_{jj}[60, 120]. The qqH-like events with two or more jets and di-jet invariant mass larger than 350 GeV are split into two regions around $p_T^H = 200$ GeV. The region where $p_T^H > 200$ GeV defines an exclusive bin called qqH-2j/p_T > 200. For what concerns events with $p_T^H < 200$ GeV: the Stage 1.2 scheme proposes a further classification based on the transverse momentum of the H+di-jet system (p_T^{H+jj}), using 25 GeV as a bin boundary. The analysis defines two bins in the $p_T^{H+jj} < 25$ GeV region: one with $350 < m_{jj} < 700$ and the other with $m_{jj} > 700$ referred to as qqH-2j/m_{jj}[350, 700] and qqH-2j/m_{jj} > 700, respectively. The remaining events with $p_T^{H+jj} > 25$ GeV are grouped into a single bin called qqH-3j/m_{jj} > 350.

In the Stage 1.2 classification, a bin per each of the $q\bar{q} \rightarrow$ WH, gg $\rightarrow$ ZH, and $q\bar{q} \rightarrow$ ZH processes is defined. A further splitting as a function of the transverse momentum of the vector boson (V) is applied with boundaries at p_T^V =75, 150, 250, and 400 GeV, as shown in Fig. 5.6. Given the small number of expected events originating from the VH-lep production mechanisms, only two bins are considered in the reduced Stage 1.2 scheme adopted for the analysis: VH-lep/p_T^H[0, 150] and VH-lep/p_T^H > 150.

The Stage 0 ttH bin, comprising the $t\bar{t}$H, tHW, and tHq production mechanisms, is split into five bins in the Stage 1.2 with boundaries defined by the values of p_T^{H+jj}: 60, 120, 200, and 300 GeV. These production modes account for an expected yield of less than two signal events in the H $\rightarrow 4\ell$ decay channel: for this reason, the Stage 1.2 bins are merged together into a single bin, referred to as ttH.

The STXS measurements reported in this thesis comprise the first Stage 1.2 cross sections results presented in a CMS analysis using the full Run-II dataset. For completeness, the inclusive Stage 0 cross sections are also measured in the analysis. The events categorization developed to measure the cross sections in the STXS bins is presented in the next section.

5.3 Events Categorization

A comprehensive characterization of the H boson production mechanisms is possible only if the signal events are classified into mutually exclusive categories sensitive to each of them. The H $\rightarrow$ ZZ $\rightarrow$ 4ℓ analysis presented here relies on a categorization performed into two subsequent steps.

In the first step, the candidates are classified into seven mutually exclusive categories defined on the basis of the jet multiplicity in the event, the number of b-tagged jets, and the presence of additional leptons. In order to increase the purity of each category and to enhance the sensitivity to the main production modes, additional matrix element discriminants sensitive to the event topology are introduced, as described in Sect. 5.3.1. This categorization step inherits from previous versions of the H $\rightarrow$ ZZ $\rightarrow$ 4ℓ analysis that used only partial LHC Run-II datasets, where the H boson production processes could be probed only in an inclusive way. The main innovation brought to the analysis consists in the introduction of a second categorization step, where the seven categories aforementioned are further split into a total number of twenty-two reconstructed categories designed to match the merged STXS Stage 1.2 production bins closely, as detailed in Sect. 5.3.2.

5.3.1 Production Discriminants

The analysis exploits a set of observables sensitive to the different production mechanisms to probe in detail the Higgs boson production cross sections in the four-leptons decay channel and test the SM predictions in the fine granular phase space regions of the STXS framework. One such observable must be the number of jets in the event, which has high discriminating power, especially in separating the ggH process from the VBF and VH ones. To improve the discrimination of events originating from productions with associated top quarks, the number of selected b-tagged jets is considered. In events originating from the associated production with a vector boson it is highly probable to find extra leptons coming from its decay rather than from the actual H boson in the event. Therefore, also the number of such leptons is considered in the event categorization. The actual discrimination power of these three observables is depicted in Fig. 5.7, where their normalized distributions are shown for all the production mechanisms separately.

To enhance the separation of the different signal processes and therefore the ultimate sensitivity of the analysis, a set of matrix element discriminants sensitive to the production of the H boson is employed. These discriminants are the best test statistic to distinguish a null-hypothesis from an alternative one, as they encapsulate the maximal information present in an event, similarly to the $\mathcal{D}^{\text{kin}}_{\text{bkg}}$ discriminant presented in Sect. 5.1.1. Four of such matrix element discriminants are introduced

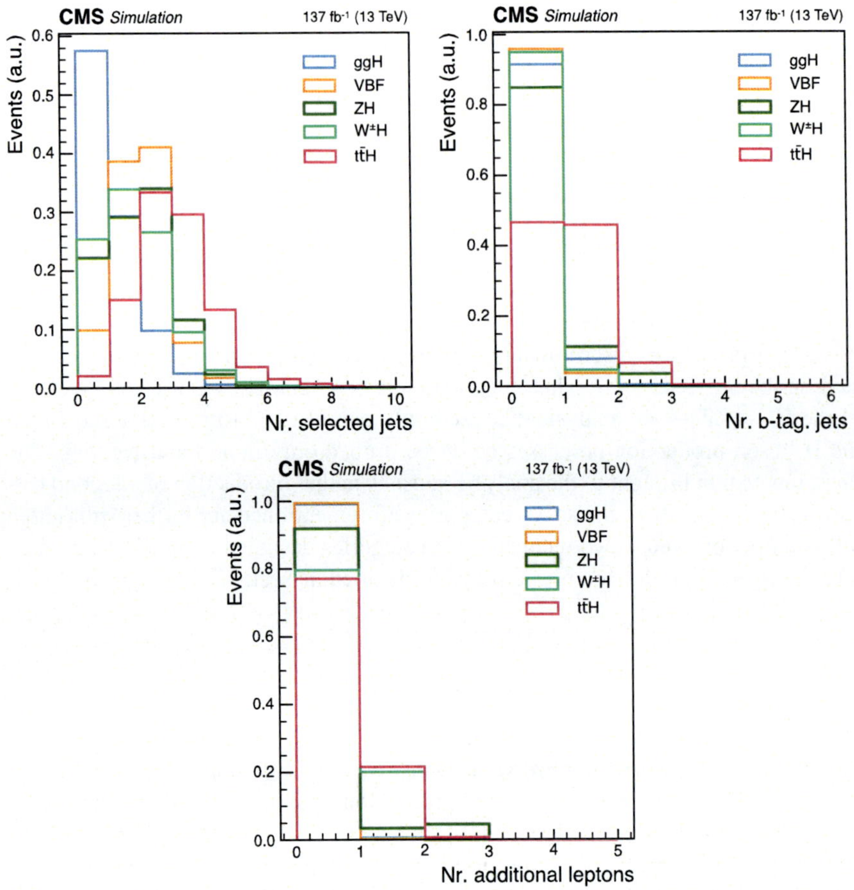

Fig. 5.7 Normalized distributions of the number of selected jets (top right), of b-tagged jets (top left), and of additional leptons (bottom) for the main production mechanisms of the H boson considered in the analysis. The distributions comprise events passing the entire set of the analysis selection criteria and are required to be in the $105 < m_{4\ell} < 140$ GeV mass range

in the analysis to distinguish the VBF and VH production mechanisms from the dominant ggH one. These are defined, in a similar fashion to the $\mathcal{D}^{\text{kin}}_{\text{bkg}}$, as follows:

$$\mathcal{D}^{\text{VBF}}_{\text{2jet}} = \left[1 + \frac{\mathcal{P}_{\text{Hjj}}(\vec{\Omega}^{\text{H+jj}}|m_{4\ell})}{\mathcal{P}_{\text{VBF}}(\vec{\Omega}^{\text{H+jj}}|m_{4\ell})}\right]^{-1} \quad \mathcal{D}^{\text{VBF}}_{\text{1jet}} = \left[1 + \frac{\mathcal{P}_{\text{Hj}}(\vec{\Omega}^{\text{H+j}}|m_{4\ell})}{\int d\eta^{\text{j}}\mathcal{P}_{\text{VBF}}(\vec{\Omega}^{\text{H+jj}}|m_{4\ell})}\right]^{-1}$$

$$\mathcal{D}^{\text{WH}}_{\text{2jet}} = \left[1 + \frac{\mathcal{P}_{\text{Hjj}}(\vec{\Omega}^{\text{H+jj}}|m_{4\ell})}{\mathcal{P}_{\text{WH}}(\vec{\Omega}^{\text{H+jj}}|m_{4\ell})}\right]^{-1} \quad \mathcal{D}^{\text{ZH}}_{\text{2jet}} = \left[1 + \frac{\mathcal{P}_{\text{Hjj}}(\vec{\Omega}^{\text{H+jj}}|m_{4\ell})}{\mathcal{P}_{\text{ZH}}(\vec{\Omega}^{\text{H+jj}}|m_{4\ell})}\right]^{-1}, \tag{5.6}$$

where the $\vec{\Omega}^{\mathrm{H+j}}$ and $\vec{\Omega}^{\mathrm{H+jj}}$ define the set of observables describing the H + j and H + jj systems, respectively. The $\mathcal{P}_{\mathrm{VBF}}$, $\mathcal{P}_{\mathrm{Hjj}}$, $\mathcal{P}_{\mathrm{Hj}}$, and $\mathcal{P}_{\mathrm{VH}}$ are the probabilities of a given event coming from the VBF process, the ggH production in association with two or one jets, and the VH ($V = W, Z$) processes, respectively. The $\int d\eta^{j}\mathcal{P}_{\mathrm{VBF}}$ term that enters the definition of $\mathcal{D}^{\mathrm{VBF}}_{\mathrm{1jet}}$ is the integral of the two-jet VBF matrix element probability over the η^{j} values of the unobserved jet, with the additional requirement of the transverse momentum of the H + jj system being zero. The $\mathcal{D}^{\mathrm{WH}}_{\mathrm{2jet}}$ and $\mathcal{D}^{\mathrm{ZH}}_{\mathrm{2jet}}$ discriminants are combined in the context of the analysis into a single discriminant, sensitive to the inclusive VH production mode, defined as the maximum between these two values $\mathcal{D}^{\mathrm{VH}}_{\mathrm{2jet}} = \max(\mathcal{D}^{\mathrm{ZH}}_{\mathrm{2jet}}, \mathcal{D}^{\mathrm{WH}}_{\mathrm{2jet}})$. The distributions of these four discriminants are shown in Fig. 5.8. The $\mathcal{D}^{\mathrm{VBF}}_{\mathrm{2jet}}$ discriminant is found to have good discrimination power between VBF events accompanied by two jets and both ggH events with two associated jets and background events. Similarly, the $\mathcal{D}^{\mathrm{VBF}}_{\mathrm{1jet}}$ discriminants allows the identification of VBF events with only one jet, even though with a more consistent contamination from ggH events. The mild signature of VH events in the detector makes them more challenging to be separated from the other processes, as shown by the distributions of $\mathcal{D}^{\mathrm{WH}}_{\mathrm{2jet}}$ and $\mathcal{D}^{\mathrm{ZH}}_{\mathrm{2jet}}$. However, part of the selection efficiency is recovered with the use of the inclusive discriminant $\mathcal{D}^{\mathrm{VH}}_{\mathrm{2jet}} = \max(\mathcal{D}^{\mathrm{ZH}}_{\mathrm{2jet}}, \mathcal{D}^{\mathrm{WH}}_{\mathrm{2jet}})$.

5.3.2 Events Categorization in the 4ℓ Analysis

The events categorization begins with the ZZ candidates, selected as described in Sect. 5.1, and divides them into seven mutually exclusive categories exploiting the kinematic observables and the matrix element discriminants defined in the previous section. The reconstructed categories are defined to maximize the sensitivity of the analysis to each production mechanism of the H boson and are described in the following.

- Events with exactly four leptons and $\mathcal{D}^{\mathrm{VBF}}_{\mathrm{2jet}} > 0.5$ enter the **VBF-2jet tagged** category if accompanied by 2 or 3 jets, of which at most 1 b-tagged one, or by 4 jets, none of which b-tagged.
- If the events meet all the above criteria but the cut on the discriminant, the event is classified as **VH-hadronic tagged** if $\mathcal{D}^{\mathrm{VH}}_{\mathrm{2jet}} > 0.5$.
- The next category targeted is the **VH-leptonic tagged**, where no more than 3 jets, none of which b-tagged, and exactly 1 additional lepton or exactly 1 pair of opposite sign and same flavor leptons are required. Events with no jets and at least one additional lepton are also considered in this category.
- Events with at least four jets, of which at least one b-tagged, and no additional leptons populate the **t$\bar{t}$H-hadronic tagged** category.
- The **t$\bar{t}$H-leptonic tagged** category groups all the events with at least one additional lepton.

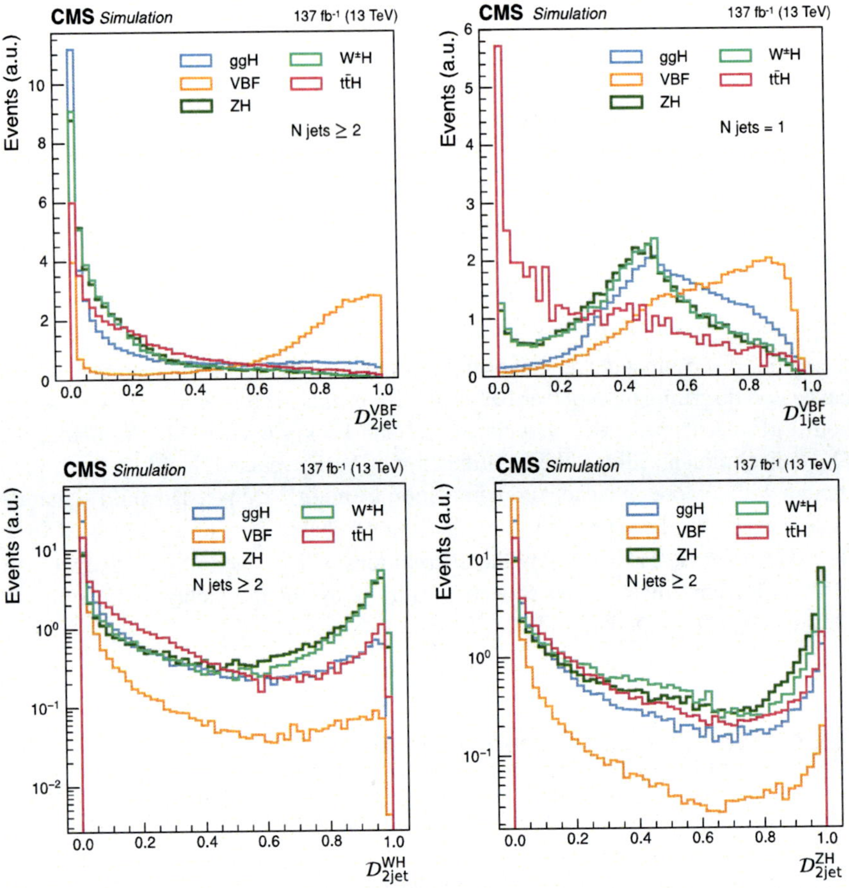

Fig. 5.8 Normalized distributions of the $\mathcal{D}_{2jet}^{VBF}$ (top left), $\mathcal{D}_{1jet}^{VBF}$ (top right), $\mathcal{D}_{2jet}^{WH}$ (bottom left), $\mathcal{D}_{2jet}^{ZH}$ (bottom right) discriminants for the main production mechanisms of the H boson considered in the analysis. The distributions comprise events passing the entire set of the analysis selection criteria and are required to be in the $105 < m_{4\ell} < 140$ GeV mass range

- The events where exactly four leptons and one jet are identified are classified as **VBF-1jet tagged** if $\mathcal{D}_{1jet}^{VBF} > 0.7$.
- The remaining events are classified as **Untagged**: a category comprising mostly events originating from gluon-fusion.

The result of this first classification step is shown in Fig. 5.9, which illustrates the fraction of signal events that compose each reconstructed category. A signal purity of about 93% is observed for the "Untagged" category, mainly sensitive to the ggH production mechanism. Similarly, the relative purity of the two reconstructed categories targeting the tt̄H process is found to be between 75% and 90%, respectively. It is also interesting to note that the "VBF-1jet tagged" category is composed of 78%

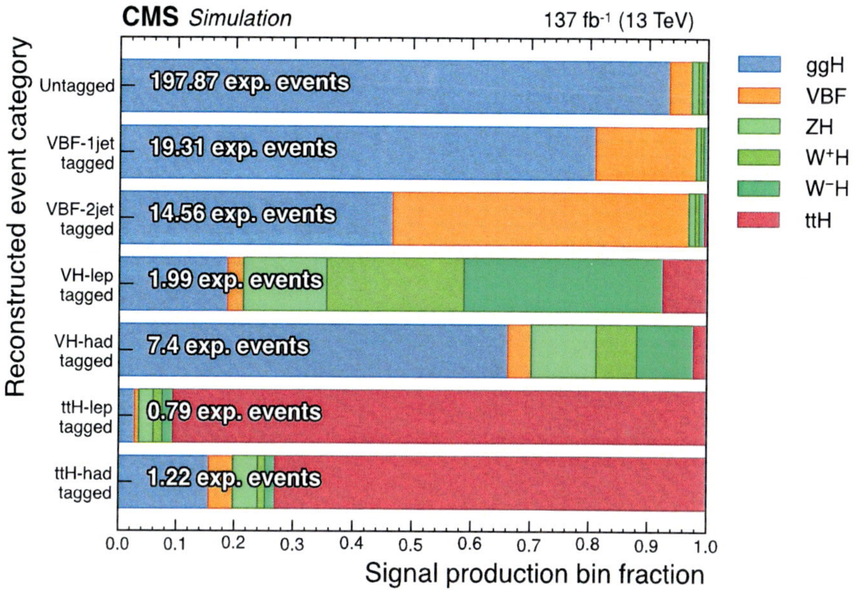

Fig. 5.9 Fraction of signal events in each of the seven event categories considered in the first step of the analysis

of events originating from gluon-fusion. This observation comes with no surprise as a consequence of the slightly less discriminating power of the $\mathcal{D}_{\text{1jet}}^{\text{VBF}}$ discriminant. Similarly, the "VH-hadronic tagged" category presents a relatively large contamination from ggH events, as expected from the distributions of $\mathcal{D}_{\text{2jet}}^{\text{WH}}$ and $\mathcal{D}_{\text{2jet}}^{\text{ZH}}$ shown in Fig. 5.8. Notwithstanding these contaminations, it is worth to mention that the results of the analysis are extracted from a simultaneous fit across all the reconstructed categories: as discussed in Chap. 7, the measurements for the VH and VBF modes are performed inclusively and therefore the higher purity of the other categories associated to these two production mechanisms allows a fairly precise determination of their corresponding cross sections.

A revised events categorization based on machine learning techniques could be envisaged for future versions of this analysis. However, the largest improvement may be expected to come from additional statistics rather than from a revised analysis strategy, as the Matrix Element Likelihood Approach employed here is already a powerful technique that ensures to capture the maximal information contained in the event.

The use of the full Run-II statistics for the analysis permits scrutinizing fine granular phase space regions of the H boson production mechanisms even in the four-leptons final state. Hence, the second step of the events categorization splits the seven mutually exclusive categories aforementioned into a total of 22 reconstructed categories, defined to closely match the merged STXS Stage 1.2 production bins (cf. Sect. 5.2). The definition of the 22 categories is presented in Table 5.1, along with

Table 5.1 Definition of the 22 reconstructed categories designed to target stage the STXS Stage 1.2 production bins. Additional kinematic requirements are imposed on reconstructed observables as detailed in the fourth column, starting from events in the seven categories resulting from the first step of the categorization. A dash indicates no requirement

Reconstructed event category	1st categorization step	Number of jets	Kinematical requirements (GeV)	Targeted production bin
Untagged-0j-$p_T^{4\ell}$[0, 10]	Untagged	0	$0 < p_T^{4\ell} < 10$	ggH-0j/p_T[0,10]
Untagged-0j-$p_T^{4\ell}$[10, 200]	Untagged	0	$10 < p_T^{4\ell} < 200$	ggH-0j/p_T[10,200]
Untagged-1j-$p_T^{4\ell}$[0, 60]	Untagged	1	$0 < p_T^{4\ell} < 60$	ggH-1j/p_T[0,60]
Untagged-1j-$p_T^{4\ell}$[60, 120]	Untagged	1	$60 < p_T^{4\ell} < 120$	ggH-1j/p_T[60,120]
Untagged-1j-$p_T^{4\ell}$[120, 200]	Untagged	1	$120 < p_T^{4\ell} < 200$	ggH-1j/p_T[120,200]
Untagged-2j-$p_T^{4\ell}$[0, 60]	Untagged	2	$0 < p_T^{4\ell} < 60$, $m_{jj} < 350$	ggH-2j/p_T[0,60]
Untagged-2j-$p_T^{4\ell}$[60, 120]	Untagged	2	$60 < p_T^{4\ell} < 120$, $m_{jj} < 350$	ggH-2j/p_T[60,120]
Untagged-2j-$p_T^{4\ell}$[120, 200]	Untagged	2	$120 < p_T^{4\ell} < 200$, $m_{jj} < 350$	ggH-2j/p_T[120,200]
Untagged-$p_T^{4\ell} > 200$	Untagged	–	$p_T^{4\ell} > 200$	ggH/$p_T > 200$
Untagged-2j-$m_{jj} > 350$	Untagged	2	$m_{jj} > 350$	ggH-2j/$m_{jj} > 350$
VBF-1jet-tagged	VBF-1jet-tagged	–	–	qqH-rest
VBF-2jet-tagged-m_{jj}[350, 700]	VBF-2jet-tagged	–	$p_T^{4\ell} < 200$, $p_T^{4\ell jj} < 25$, $350 < m_{jj} < 700$	qqH-2j/m_{jj}[350,700]
VBF-2jet-tagged-$m_{jj} > 700$	VBF-2jet-tagged	–	$p_T^{4\ell} < 200$, $p_T^{4\ell jj} < 25$, $m_{jj} > 700$	qqH-2j/$m_{jj} > 700$
VBF-3jet-tagged-$m_{jj} > 350$	VBF-2jet-tagged	–	$p_T^{4\ell} < 200$, $p_T^{4\ell jj} > 25$, $m_{jj} > 350$	qqH-3j/$m_{jj} > 350$
VBF-2jet-tagged-$p_T^{4\ell} > 200$	VBF-2jet-tagged	–	$p_T^{4\ell} > 200$, $m_{jj} > 350$	qqH-2j/$p_T > 200$
VBF-rest	VBF-2jet-tagged	–	$m_{jj} < 350$	qqH-rest
VH-hadronic-tagged-m_{jj}[60, 120]	VH-hadronic-tagged	–	$60 < m_{jj} < 120$	qqH-2j/m_{jj}[60,120]
VH-rest	VH-hadronic-tagged	–	$m_{jj} < 60$ or $m_{jj} > 120$	qqH-rest
VH-leptonic-tagged-$p_T^{4\ell}$[0, 150]	VH-leptonic-tagged	–	$p_T^{4\ell} < 150$	VH-lep/p_T^H[0,150]
VH-leptonic-tagged-$p_T^{4\ell} > 150$	VH-leptonic-tagged	–	$p_T^{4\ell} > 150$	VH-lep/$p_T^H > 150$
$t\bar{t}$H-leptonic-tagged	$t\bar{t}$H-leptonic-tagged	–	–	ttH
$t\bar{t}$H-hadronic-tagged	$t\bar{t}$H-hadronic-tagged	–	–	ttH

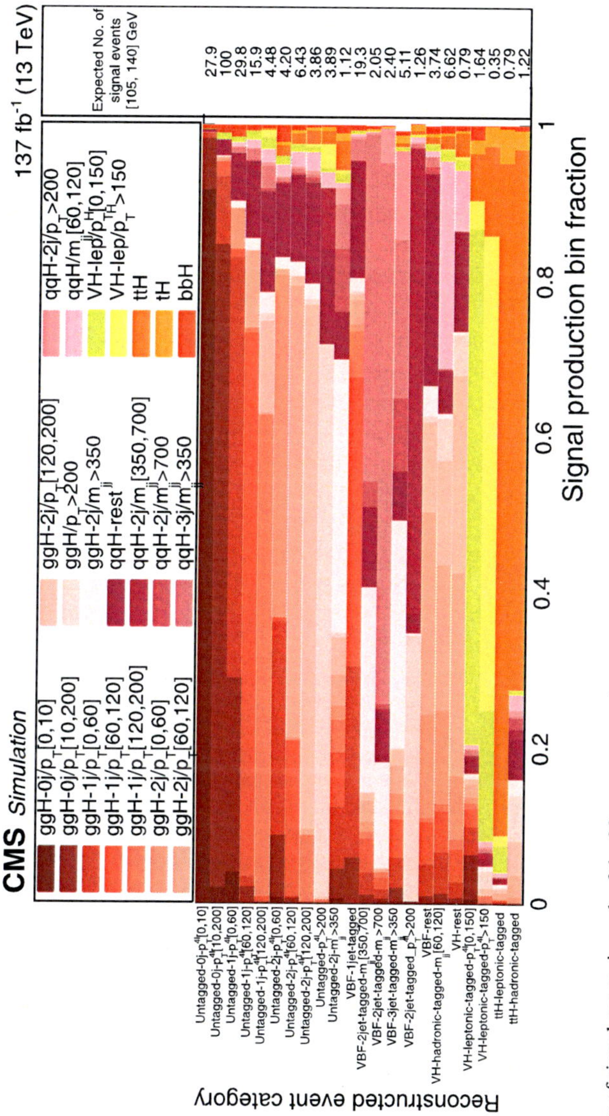

Fig. 5.10 Fraction of signal events in each of the 22 reconstructed event categories as a function of the merged STXS Stage 1.2 bins in the $105 < m_{4\ell} < 140$ GeV mass range used for the analysis

the corresponding selection cuts on the kinematic observables chosen to reproduce the generator-level ones employed in the definition of the STXS Stage 1.2 bins.

The relative signal purity obtained in the 22 reconstructed categories of the analysis is shown in Fig. 5.10, where the signal bin fractions comprising each category are illustrated. The purity of the **untagged** sub-categories targeting the ggH production mechanism is larger than 80% in the phase space regions with one or less jet, and it gets as large as 90% for the Untagged-0j-$p_{\mathrm{T}}^{4\ell}$ [10, 200] category, which drives the sensitivity of the entire analysis. Reflecting what observed in Fig. 5.9 for the first step of the categorization, the VBF-1jet-tagged and VBF-rest categories are largely populated by ggH-like events, while the reconstructed categories targeting phase space regions with two or more jets have larger contributions from VBF-like events, in accordance to what expected from the different topology of these two production mechanisms (cf. Fig. 2.25).

Even though the use of matrix element kinematic discriminants sensitive to the production processes enhances the relative purity of each category, partial correlations and cross-contamination are expected. The imperfect matching between the reconstructed categories and the true STXS bins is taken into account with the introduction of an appropriate set of nuisance parameters in the likelihood function to cope with categories migrations. In addition, it is worth to remark that a maximum likelihood fit is performed simultaneously on all the reconstructed categories to extract the STXS Stage 1.2 cross section measurements and that a conditional term proportional to the matrix element discriminants is embedded in the likelihood function to enhance the sensitivity to the topology of each production mechanism, as mentioned in Sect. 5.1.1.

References

1. Collaboration CMS (2013) Study of the mass and spin-parity of the Higgs boson candidate via its decays to Z boson pairs. Phys Rev Lett 110(8):081803
2. Collaboration CMS (2012) Observation of a new boson at a mass of 125 GeV with the CMS experiment at the LHC. Phys Lett B 716:30
3. Gritsan AV, Roskes J, Sarica U, Schulze M, Xiao M, Zhou Y (2020) New features in the JHU generator framework: constraining Higgs boson properties from on-shell and off-shell production. Phys Rev D 102(5):056022
4. Campbell JM (2020) MCFM-9.1 manual a Monte Carlo for femtobarn processes at hadron colliders
5. Neyman J, Pearson ES (1993) On the problem of the most efficient tests of statistical hypotheses. Philos Trans R Soc London Ser A, Contain Pap Math Phys Character 231(694–706):289–337

Chapter 6
Signal and Background Modelling

The "golden roadmap" that guides the analysis workflow and strategy is the Monte Carlo simulation of the phenomenology of the physical processes studied and their interaction with the CMS detector. It is fundamental to exploit the most precise generators and use the state-of-art calculations to accurately model the physics considered in the analysis and represent the experimental data properly.

There are cases in which the simulations do not reflect either the perturbative order at which theoretical calculations were performed or the experimental configuration at the time of the data taking, resulting in unwanted discrepancies that are dealt with by the application of dedicated weights to the MC events, as discussed in Sect. 6.1.

Even though in most of the cases these reweighed MC events provide an accurate description of a given physical process, as for the irreducible background and the signal processes described in Sects. 6.2.1 and 6.3, there are situations in which one has to rely on data-driven methods to estimate the contribution of a particular process. This is the case of the reducible backgrounds to the $H \rightarrow ZZ \rightarrow 4\ell$ signal, consisting of processes where in-flight decays of light mesons within jets, heavy-flavor hadrons decays, and charged hadrons overlapping with neutral-pions decays are misidentified as leptons. These backgrounds are dominated by processes in which a Z boson is accompanied by jets and therefore are referred to as "Z+X." As further detailed in Sect. 6.2.2, the MC samples available for these processes are limited in statistics, thus resulting in an unreliable estimate of the corresponding yields and distributions. For this reason, two dedicated data-driven techniques are employed in the analysis to get a more realistic assessment of this background contribution.

The $H \rightarrow ZZ \rightarrow 4\ell$ analysis presented in this thesis falls in the category of the "unbinned shape analyses", where MC simulations are used to estimate an analytical shape that parametrizes the expected distributions. The description of the signal and background distributions by means of analytical shapes is detailed in Sects. 6.2 and 6.3, respectively.

M. Bonanomi, *Response of the High Granularity Calorimeter HGCAL and Characterisation of the Higgs Boson*, Springer Theses,
https://doi.org/10.1007/978-3-031-26833-5_6

To measure the properties of the H boson in the fine granular phase space regions of the Stage 1.2 of the STXS framework, it is fundamental to achieve a reliable estimate of the signal and background shapes in each of these bins, so as to enhance the sensitivity of the analysis and to properly account for their different phenomenology. This approach results into a total of 19 analytical parametrizations of the reconstructed invariant mass $m_{4\ell}$, extracted separately per each data taking period and final state (4e, 4m, and 2e2m), which are included as probability density functions (pdfs) in the likelihood function maximized to extract the results of the analysis, as detailed in Sect. 6.4.

With such a complex scenario involved in the measurement, one can not neglect the presence of "unknowns" in the analysis. These are related to the residual mismodelling of the physical processes in the simulations, to the detector effects altering the shape of the H boson resonance, but also to the theoretical calculations of the different cross sections, as well as to the definition of the phase space regions in which the measurements are performed. All these effects enter the analysis as *systematic uncertainties* and are accounted for by introducing dedicated nuisance parameters in the statistical model, as explained in detail in Sect. 6.5.

6.1 Monte Carlo Reweighing Procedure

After the generation, showering, and hadronization of the events by means of dedicated Monte Carlo (MC) generators, the simulated particles are propagated through a detailed simulation of the detector based on the GEANT4 toolkit. However, MC generators do not always reproduce the state-of-art calculations of the theoretical cross sections, and the simulation of the detector can contain effects not observed in data and vice-versa. For these reasons, a dedicated reweighing of the simulated events is performed, as detailed in the following.

One of the most critical and challenging corrections to consider concerns the pileup (PU) profile. Projections based on simulations are used to estimate the expected PU profiles during each data taking period. These are based on a smooth PU profile and do not always match the ones recorded by the experiments, which are often the result of the convolution of different PU profiles.[1] For this reason, a dedicated weight is introduced to match the pileup profile in MC to the one observed in data, as illustrated in Fig. 6.1, where the raw and reweighed MC pileup profiles are shown.

Additional weights are applied to the simulated events to take into account the actual number of events generated and the theoretical computations of the cross section and branching fraction:

$$w_{\text{evt}} = \frac{\text{L}_{\text{int}} \cdot (\sigma \;\cdot\; \text{BR})_{\text{Th., SM}} \cdot w_{\text{PU}} \cdot \text{SF}_{4\ell} \cdot w_{\text{gen.}}}{\sum_{\text{gen. evts}} w_{\text{gen.}}}, \tag{6.1}$$

[1] One per each data taking period or section.

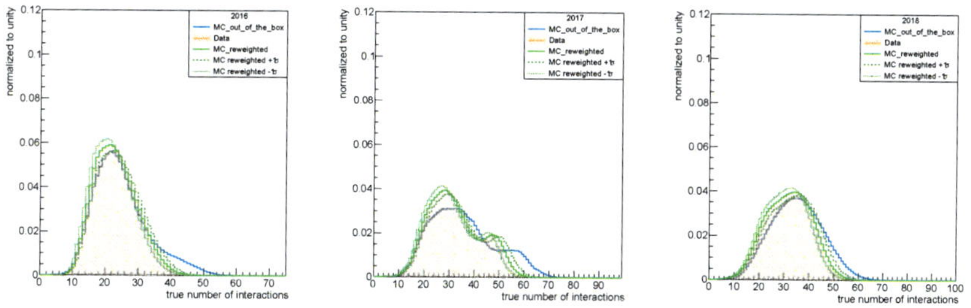

Fig. 6.1 Distribution of the number of interaction vertices reconstructed in Data (dashed orange histogram) and in MC, before (solid blue line) and after (solid green line) the application of the pileup weights described in the text. The distributions are for the three data taking periods considered in the analysis: 2016 (left), 2017 (centre), and 2018 (right)

where L_{int} refers to the integrated luminosity of the data taking period, $(\sigma \cdot BR)_{Th., SM}$ is the product between the best known theoretical prediction for the SM cross section of the generated process and its branching ratio, w_{PU} is the pileup weight introduced at the beginning of this section, $SF_{4\ell}$ is the per-event scale factor obtained from the product of the four leptons' efficiency scale factors, derived as of Sects. 4.1.3 and 4.2.3, and $w_{gen.}$ is the weight of the MC generator used. Ultimately, the product of these weights is rescaled by the sum of all the generation weights $(\sum_{gen.\ evts} w_{gen.})$ to get the expected yield for each process. The numerator of Eq. 6.1 is supplemented with an additional contribution for the 2016 and 2017 data taking periods to correct for the L1 trigger primitives prefiring issue described in Sect. 4.4.2.

The specific MC generators used to simulate the different physical processes involved in the analysis, as well as the corresponding values for $\sigma_{Th., SM} \cdot BR_{Th., SM}$ are reported in Sects. 6.2 and 6.3 for the background and signal processes, respectively.

6.2 Background Modelling

The background contributions to the $H \rightarrow ZZ \rightarrow 4\ell$ analysis are divided into irreducible and reducible ones, according to the signature of their final state. The presence of four distinct leptons, originating from the decay of two Z bosons produced either via gluon fusion ($gg \rightarrow ZZ$) or quark-antiquark annihilation ($q\bar{q} \rightarrow ZZ$), characterizes the former. The latter consists of processes where in-flight decays of light mesons within jets, heavy-flavor hadrons decays, and charged hadrons overlapping with neutral-pions decays are misidentified as leptons. These backgrounds are dominated by processes in which a Z boson is accompanied by jets and therefore are referred to as "Z+X".

While the irreducible background processes are modelled via dedicated MC simulations, the Z+X background is obtained from a dedicated data-driven estimate, as discussed in Sects. 6.2.1 and 6.2.2, respectively. The modelling of these background processes and their inclusion in the statistical analysis is described in Sect. 6.2.4.

6.2.1 Irreducible Background

The $q\bar{q} \rightarrow ZZ$ process

The complex challenge of producing a MC simulation that matches the high perturbative order of the theoretical calculations is well represented by the $q\bar{q} \rightarrow ZZ$ background. In fact, while the generation of events is performed at NLO, the cross section of this process is known with precision at NNLO [1]. The discrepancy between these two perturbative orders can be quite large, as shown in Table 6.1, where the inclusive cross sections for the $q\bar{q} \rightarrow ZZ$ process are reported at LO, NLO, and NNLO.

To account for this effect in the analysis, the process is produced with a NLO simulation using the POWHEG generator and dedicated NNLO/NLO k-factors, obtained from the ratio of the two corresponding cross sections, are used to reweigh the distributions to their state-of-art theoretical cross sections. These are applied as a function of $m(ZZ)$ and they can range between 5 and 25%, as shown in the left plot of Fig. 6.2. Additional electroweak NLO/LO k-factors are applied to take into account corrections due to the initial state quark flavor and kinematics [2]. These are computed and applied for those events with $m(ZZ) > 2m(Z)$, and their distribution is shown in the right plot of Fig. 6.2.

The $gg \rightarrow ZZ$ process

Similarly to the $q\bar{q} \rightarrow ZZ$ process, also for the $gg \rightarrow ZZ$ background a reweighing procedure is employed to cope with the difference between the state-of-art cross section computations and the perturbative order used in the generation. More precisely, the production of two vector bosons via gluon fusion ($gg \rightarrow ZZ$) is simulated at LO using the MCFM7.0 generator [3]. Since this process contributes to NNLO in pQCD, a dedicated reweighing must be applied to the simulated events.

Table 6.1 Cross sections for $q\bar{q} \rightarrow ZZ$ production at 13 TeVfor the $2\ell2\ell'$ (left column) and for the 4ℓ (right column) final states. The cross sections are computed on the inclusive phase space at LO, NLO, and NNLO

QCD order	$\sigma_{2\ell2\ell'}$ (fb)	$\sigma_{4\ell}$ (fb)
LO	$218.5^{+16\%}_{-15\%}$	$98.4^{+13\%}_{-13\%}$
NLO	$290.7^{+5\%}_{-8\%}$	$129.5^{+4\%}_{-6\%}$
NNLO	$324.0^{+2\%}_{-3\%}$	$141.2^{+2\%}_{-2\%}$

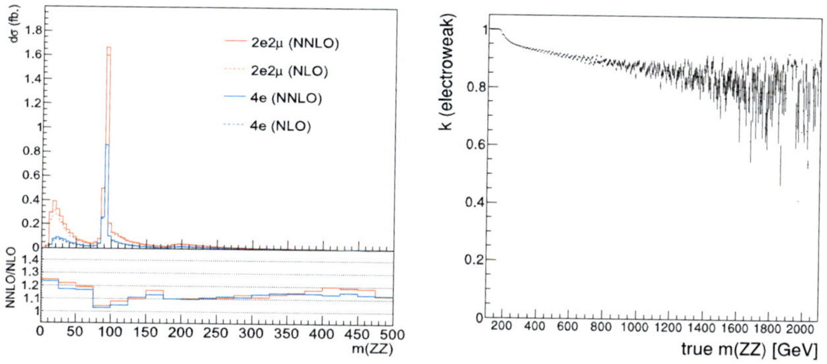

Fig. 6.2 *Left*: Distributions of m(ZZ) for the q$\bar{\text{q}}$ →ZZ background at NLO (dashed lines) and NNLO (solid lines) for the 2em (red) and 4e (blue) final states. The ratio plot shows the NNLO/NLO QCD k-factors used in the analysis. *Right*: Electroweak NLO/LO K factor for the q$\bar{\text{q}}$ →ZZ background as a function of m(ZZ) computed for events with $m(ZZ) > 2m(Z)$

Recent calculations [4] have shown that the soft collinear approximation provides a reasonable description of the background cross section and the interference term at NNLO. Moreover, one can prove that the same k-factors computed for the signal can also be used for the background [5], as their value agree at NLO [6] and at NNLO for the signal and interference terms [7].

The $gg \to H \to 2\ell 2\ell'$ cross section is computed at LO, NLO, and NNLO for a SM H boson with a width of 4.07 MeV using the HNNLO v2 MC generator [8]. Their ratio defines the NLO and NNLO k-factors for the signal and are applied to the $gg \to$ ZZ samples used for the analysis and generated at LO with the MCFM7.0 generator [3]. Figure 6.3 shows the three cross sections used for the k-factors calculation and their distributions as a function of $m_{4\ell}$.

6.2.2 *Reducible Background*

The reducible background accounts for processes such as Z+jets, $t\bar{t}$+jets, $Z\gamma$+jets, and WZ or WW + jets, which are collectively referred to as Z+X.

In contrast to what is done for the irreducible background described in Sect. 6.2.1, the usage of MC simulations to model these processes is suboptimal due to the lack of statistics in the corresponding samples mismodelling of the associated jets. Hence, two dedicated data-driven methods are adopted in the analysis to estimate the yields and shapes of the Z+X events.

In data-driven estimates one exploits events from phase space regions orthogonal to the signal region (SR) and referred to as *control regions* (CR) to extrapolate the expected background yields and shapes in the SR. The reason behind this approach is twofold: first of all, one can profit from much larger statistics, thus reducing the associated uncertainty on the final estimate of the yields; in addition, using events

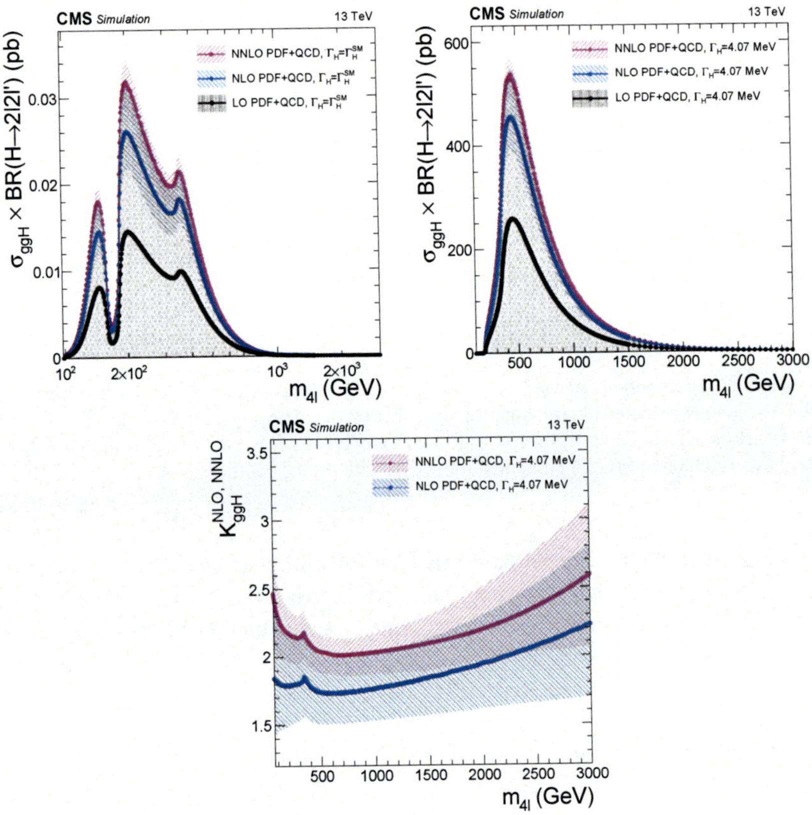

Fig. 6.3 *Top*: $gg \rightarrow H \rightarrow 2\ell 2\ell'$ cross sections at NNLO (purple), NLO (blue) and LO (black) at each H boson pole mass for the SM H boson decay width (left) and at the fixed and small decay width of 4.07 MeV (right). *Bottom*: NLO (blue) and NNLO (purple) k-factors as a function of $m_{4\ell}$ obtained from the cross sections at $\Gamma_H = 4.07$ MeV

from the CR, no bias is introduced in the statistical analysis performed on events from the SR.

The first step towards the estimate of the Z+X contribution to the SR consists of identifying of the "fake leptons," i.e., those leptons present in the final state that would mimic the topology of signal leptons. This is done in the $Z(\ell\ell) + \ell$ CR, where ℓ refers to the presence of an additional loose lepton in the event. The additional lepton ℓ is hereafter labelled according to whether it fails (F) or passes (P) the complete list of selection requirements detailed in Table 4.4. The reconstructed Z in this CR is required to have $p_T(\ell_1) > 20$ GeV and $p_T(\ell_2) > 10$ GeV. The additional lepton and the opposite sign lepton from the Z are required to have $m_{\ell^+\ell^-} > 4$ GeV. After having determined the *fake rates*, i.e., the probability of a *fake lepton* to satisfy the ZZ candidate selection criteria of Sect. 5.1, these are applied to a $Z + \ell\ell$ CR to derive the expected Z+X contribution in the SR. Two different $Z + \ell\ell$ CRs are exploited for the analysis, resulting in two complementary estimates of the Z+X contribution:

- the $2P2L_{SS}$ control region, used in the *same sign* (SS) method, is composed of a Z candidate and two same sign loose leptons;
- the 2P2F and 3P1F control regions, used in the *opposite sign* (OS) method, are characterized by the presence of two opposite sign leptons, either both failing (2P2F) or one lepton failing and the other passing (3P1F) the selection criteria.

The FSR recovery in both the $Z(\ell\ell) + \ell$ and $Z + \ell\ell$ CRs is performed with the same algorithm applied to the SR (cf. Sect. 4.3). While CRs are built ensuring a complete orthogonality to the SR, the selected events can still populate two different CRs. In these cases, it is fundamental to account and correct for these overlaps to avoid double counting and overestimates of events, as it happens for the 2P2F and 3P1F phase space regions used in the OS method.

OS estimate of fake rates

The $Z(\ell\ell) + \ell$ CR comprises all those events with a reconstructed Z boson and an additional fake lepton, hence defines the ideal phase space for the determination of the fake rates for electrons (f_e) and muons (f_μ). In order to reduce contamination from asymmetric photon conversions and from WZ and $t\bar{t}$ processes, the additional requirements on $| m_{Z_1} - m_Z |< 7$ GeV and $p_T^{\text{miss}} < 25$ GeV are applied. The fake rates are measured in the endcap and barrel regions separately, in bins of the transverse momentum of the loose lepton, after the subtraction of the WZ contribution estimated from the simulation, so as to improve the quality of the measurement.

The OS fake rates are measured, for each data taking period, separately for electrons and muons using $Z \to$ ee and $Z \to$ mm events. As an example, the fake rates measured for the 2018 data are shown in Fig. 6.4. These fake ratios are applied to the $Z + \ell\ell$ CRs to extrapolate the Z+X background contribution in the SR, as presented in the following.

Fake rates application: OS method

The 2P2F and 3P1F CRs are used to extract the final estimate of the Z+X yield in the SR. As mentioned above, these two CRs might present some overlap and therefore one should carefully avoid any double counting in the determination of the Z+X yield with the OS method.

The 2P2F CR describes signatures where only two signal leptons are found in the event, such as those originating from Z+jets and $t\bar{t}$ processes. Hence, the final contribution to the SR can be extrapolated by weighing each event by the product of the fake rates of the third (f_3) and fourth (f_4) leptons, namely:

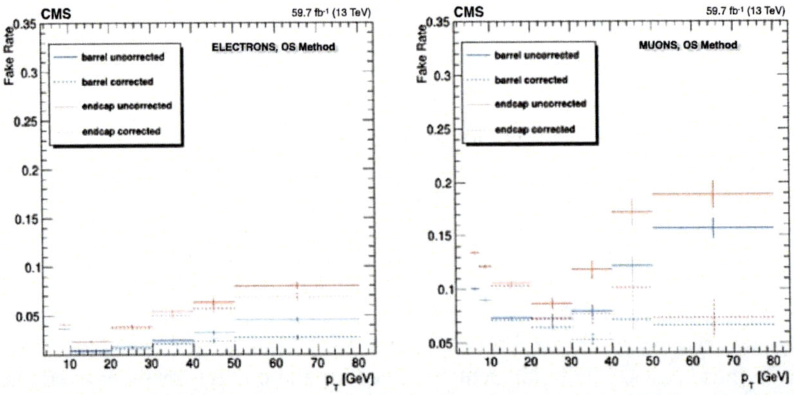

Fig. 6.4 Fake rates as a function of the probe p_T for electrons (left) and muons (right) which satisfy the loose selection criteria, measured in the $Z(\ell\ell)+\ell$ CR with 59.7 fb^{-1} data at 13 TeV. The barrel selection includes electrons (muons) up to $|\eta| = 1.479$ (1.2). The fake rates are shown before (solid lines) and after (dotted lines) removal of WZ contribution from MC

$$\mathrm{N}^{\mathrm{2P2F}}_{\mathrm{SR,\,Z+X}} = \sum_i \left(\frac{f_3^i}{1-f_3^i} \frac{f_4^i}{1-f_4^i} \right) \mathrm{N}_{\mathrm{2P2F.}}$$

Similarly, the contribution of the additional loose lepton in the 3P1F CR is taken into account by weighing each event by the corresponding fake rate:

$$\mathrm{N}^{\mathrm{3P1F}}_{\mathrm{SR,\,Z+X}} = \sum_j \left(\frac{f_4^j}{1-f_4^j} \right) \mathrm{N}_{\mathrm{3P1F}}$$

The double counting of 2P2F-like events in the 3P1F region is corrected as:

$$N^{\mathrm{bkg}}_{\mathrm{3P1F}} = \sum \left(\frac{f_i}{1-f_i} + \frac{f_j}{1-f_j} \right) N_{\mathrm{2P2F}}.$$

Combining all these contributions together, the final OS estimate of the Z+X background in the signal region can be written as:

$$N^{\mathrm{OS,\,bkg}}_{\mathrm{SR,\,Z+X}} = \left(1 - \frac{N^{ZZ}_{3P1F}}{N_{3P1F}}\right) \sum_j^{N_{3P1F}} \frac{f_a^j}{1-f_a^j} - \sum_i^{N_{2P2F}} \frac{f_3^i}{1-f_3^i} \frac{f_4^i}{1-f_4^i}$$

The number of expected Z+X events in the SR estimated with the OS method is presented in Table 6.2, for all the data taking periods and the final states considered in the analysis.

Table 6.2 Estimate of the number of ZX events in the $m_{4\ell} > 70$ GeV mass range obtained with the OS method for each final state in the three data taking periods considered in the analysis. Uncertainties include the statistical contribution, the systematics on the different processes that enter the estimate, and the systematics on the fake rates estimation

Channel	4e	4μ	$2e2\mu$	$2\mu 2e$
2016	20.1 ± 6.1	26.9 ± 8.6	25.2 ± 8.0	22.3 ± 6.8
2017	16.2 ± 5.0	32.7 ± 10.3	24.0 ± 7.7	21.3 ± 6.5
2018	25.4 ± 7.7	49.4 ± 15.4	34.2 ± 10.7	33.0 ± 10.0

SS estimate of fake rates

The measurement of fake rates with the SS method relies on the same strategy used for the OS one, with the sole difference of an extended mass range of $40 < m_{\ell^+,\ell^-} < 120\,\text{GeV}$ to properly describe the SS topology. As a result of this, more events with asymmetric photon conversion enter the $Z(\ell\ell) + \ell$ phase space, especially in the $Z + e$ case. While in the OS region the $| m_{\ell^+\ell^-} |< 7\,\text{GeV}$ cut rejects most of these events, a non-negligible fraction is left in the SS region and it has to be accounted for. Moreover, the SS region used for the fake rates measurement does not completely reflect the composition of the $2P2L_{SS}$ region. In order to correct for this effect the Z+e fake rates (f_e) are measured for different samples, corresponding to modified ranges of the $| m_{\ell^+\ell^-} - m_Z |$ and $| m_{\ell e} - m_Z |$ cuts. The fake rates and the average number of missing hits in the inner tracker ($\langle N_{\text{miss hits}}\rangle$) are measured for all these samples in several bins of transverse momentum and pseudo-rapidity of the loose electrons. The choice of the $\langle N_{\text{miss hits}}\rangle$ observable is dictated by the requirement of a variable highly sensitive to photon conversions. The expected linear relation between $\langle N_{\text{miss hits}}\rangle$ and f_e is experimentally verified, as shown in Fig. 6.5, and is used in the $2P2L_{SS}$ region to correct for the FSR effects and to measure the electron fake rates properly.

More precisely, the electron fake rates f_e in the $2P2L_{SS}$ CR are determined from the $\langle N_{\text{miss hits}}\rangle$ in each of the seven $\left(p_T^e, | \eta |^e\right)$ bins considered in the analysis, extrapolating the correct fake rate from the linear relation described above. The resulting distribution of the fake rates obtained with 2018 data for electron and muons in the p_T bins considered for the analysis is shown in Fig. 6.6.

Fake rates application: SS method

The estimate of the Z+X background from the $2P2L_{SS}$ control region is performed by weighing all the corresponding data events as follows:

$$N^{\text{OS, bkg}}_{\text{SR, Z+X}} = \left(\frac{\text{OS}}{\text{SS}}\right)^{\text{MC}} \sum_i \text{f}_3^i \times \text{f}_4^i$$

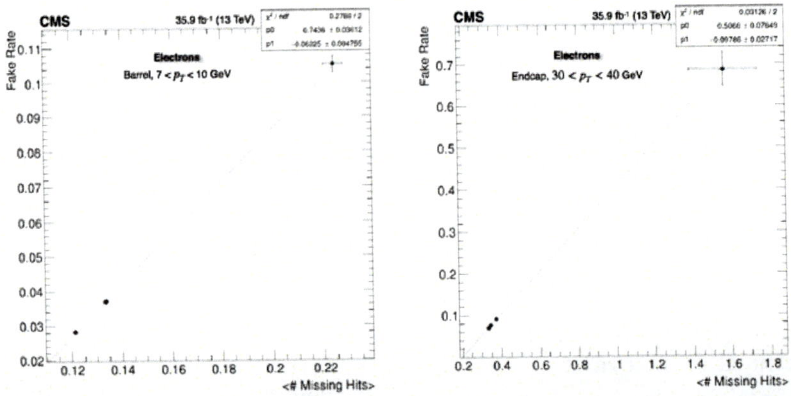

Fig. 6.5 Fake rate as a function of the fraction of loose electrons for which the track has one missing hit in the pixel detector ($\langle N_{\text{miss hits}} \rangle$) for $7 < p_T < 10$ GeV electrons in the barrel (left) and for $30 < p_T < 40$ GeV electrons in the endcaps (right). Points with error bars are extracted from the $Z(\ell\ell) + \text{e}$ CR using 35.9 fb^{-1} data at 13 TeV

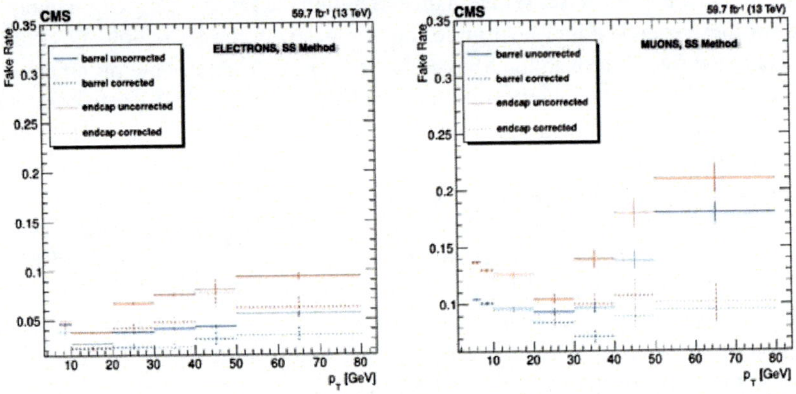

Fig. 6.6 The electron (left) and muon (right) fake rates measured in the 2P2L$_{\text{SS}}$ CR (dotted lines) correcting for $\langle N_{\text{miss hits}} \rangle$ are compared to the fake rates measured in the full SS phase space (solid lines). The fake rates are measured with 59.7 fb^{-1} data at 13 TeV

where the sum runs over all the events in the 2P2L$_{\text{SS}}$ CR, f_3 and f_4 are the fake rates for the two loose leptons, and the correction factor $\left(\frac{\text{OS}}{\text{SS}}\right)^{\text{MC}}$ accounts for the difference between the number of events in the OS and SS samples. The values of this factor, obtained from data in the three data taking periods separately, are presented in Table 6.3.

The number of expected Z+X events in the SR estimated with the SS method is presented in Table 6.4, for all the data taking periods and the final states considered in the analysis. The results agree with the prediction obtained from the OS method,

Table 6.3 The OS/SS ratios used to estimate the number of ZX events with the SS method for each final state in the three data taking periods considered in the analysis. The uncertainties on the OS/SS ratio are statistical only

Channel	4e	4μ	$2e\mu$	$2\mu 2e$
2016	1.00 ± 0.01	1.00 ± 0.03	1.00 ± 0.01	1.00 ± 0.03
2017	1.01 ± 0.01	1.04 ± 0.03	1.00 ± 0.01	1.01 ± 0.03
2018	1.01 ± 0.01	1.03 ± 0.02	1.00 ± 0.01	1.03 ± 0.02

Table 6.4 Estimate of the number of ZX events in the $m_{4\ell} > 70$ GeV mass range obtained with the SS method for each final state in the three data taking periods considered in the analysis. Uncertainties include the statistical contribution, the systematics on the different processes that enter the estimate, and the systematics on the fake rates estimation

Channel	4e	4μ	$2e2\mu$	$2\mu 2e$
2016	13.0 ± 5.4	29.7 ± 9.1	24.8 ± 7.6	16.7 ± 7.0
2017	10.9 ± 4.0	33.6 ± 10.3	26.3 ± 8.1	14.7 ± 5.5
2018	16.0 ± 5.9	52.2 ± 15.8	37.4 ± 11.4	23.3 ± 8.5

except for the 4e final state, due to the different treatment of additional electrons from photon conversions between the two methods.

Uncertainties on the Z+X estimation

The sources of uncertainty on the Z+X estimate are threefold: one is the statistical uncertainty associated with the number of events in the Z+L and Z+LL control regions and the statistical error on the $\langle N_{\text{miss hits}} \rangle$ estimate in the SS method; an additional component of systematic uncertainty is obtained from shifts by one sigma in each direction of the fake rates; the third source of uncertainty is introduced by the systematic difference among the different processes that collectively constitute the reducible background. This last contribution is estimated by measuring, in the Z+L control region, the average fake rate among the ones extracted from the individual simulated samples. The difference between this fake rate and the ones obtained reweighing the individual FRs according to the 2P2F region composition determines the last contribution to the overall Z+X uncertainty, which is found to be 38, 33, and 30% in the $4e$, $2e2\mu$, and 4μ channels, respectively.

The final systematic uncertainties associated with the Z+X prediction in each final state are reported in Table 6.5 and are included in the analysis as nuisance parameters in the likelihood function, as further discussed in Sect. 6.4.

Table 6.5 Systematic uncertainty associated to the Z+X estimate for each final state in all three data taking periods considered in the analysis

Channel	4e	4μ	$2e2\mu$
2016	41%	30%	35%
2017	38%	30%	33%
2018	37%	30%	33%

Reducible background in the STXS categories

The final Z+X estimate used for the statistical analysis consists of a weighted average of the results obtained with the SS and OS methods. The reducible background yields are extracted for all the twenty-two categories of the analysis (cf. Sect. 5.3) using the inclusive fake rates and their shapes are parametrized accordingly using Landau distributions, as described in Sect. 6.2.4.

It can be asked whether using an inclusive FR for the final Z+X estimation could have impacted on those categories targeting VBF-like phase space regions. In fact, other CMS Higgs analyses have observed that VBF events with more than 2 jets have different fake rate distributions with respect to the inclusive one. This effect is expected to be negligible in this analysis, as the Z+X estimate in each category comes with a large uncertainty of the order of 30–40%, which would ultimately cover any statistical fluctuation in the low-statistics VBF-like phase space with two or more jets. A dedicated study was carried out to demonstrate the feasibility of using an inclusive FRs for each category without deriving category-dependent fake rates. More precisely, dedicated cuts were designed to identify a CR with enhanced sensitivity to the VBF phase space requiring an angular separation of $\Delta R(l, \text{jet}) > 0.4$, the presence of two or three jets, with at most one b-tagged jet, or at least four jets, of which none b-tagged. For events with two or more jets, additional requirements on the di-jet system were introduced requiring $\Delta R_{jj} > 0.5$ and $m_{jj} > 450$ GeV. These cuts aim to reproducing the VBF-like categories in the CR, without having to rely on the computation of the dedicated production matrix element discriminant (cf. Sect. 5.1.1).

The Z+X yield in the VBF-like categories was measured after deriving dedicated FRs using the CR defined above. The expected yields are reported in Table 6.6 and compared to the prediction obtained using inclusive fake rates.

The large differences observed could sustain the initial hypothesis of the need for dedicated fake rates for these categories. For this reason, the production signal strength modifiers were measured leaving the Z+X background yields unconstrained in the VBF-like categories[2] in the maximum likelihood fits.[3] Identical results were

[2] This condition reflects a complete ignorance of the Z+X background in these categories and is implemented in the analysis assigning a 100% uncertainty to the corresponding Z+X yields.

[3] More details about the maximum likelihood fit procedure employed to extract the results of this study as well as of the rest of the analysis are given in Sect. 6.4.

Table 6.6 ZX yields in the VBF-tagged categories estimated from data using the inclusive fake rates (left) and a dedicated set fake rates (right) as described in the text

	Inclusive FR				Dedicated FR			
Category	$4l$	4μ	$4e$	$2e2\mu$	$4l$	4μ	$4e$	$2e2\mu$
VBF_1j	1.047	0.391	0.147	0.509	0.663	0.185	0.113	0.365
VBF_2j	0.882	0.295	0.097	0.490	0.054	0.011	0.016	0.028
VBF_2j_mjj_350_700_2j	0.040	0.026	0.003	0.011	0.004	0.000	0.000	0.004
VBF_2j_mjj_GT700_2j	0.016	0.013	0.002	0.001	0.091	0.018	0.039	0.034
VBF_2j_mjj_GT350_3j	0.997	0.444	0.104	0.450	0.001	0.001	0.000	0.000
VBF_GT200_2J	0.021	0.019	0.001	0.001	0.015	0.001	0.001	0.014
Inclusive VBF	3.002	1.188	0.353	1.461	0.828	0.216	0.169	0.444

obtained using the inclusive FRs, reflecting the minimal impact of the small ZX background in these categories on the final signal extraction. For this reason, the inclusive FRs were used to perform all the measurements presented in Chap. 7.

Data-driven methods for the Z+X estimate complementary to the ones presented here are envisaged for future developments of the H $\rightarrow$ ZZ $\rightarrow 4\ell$ analysis, when a larger statistics will be available and the sensitivity to the low-statistics categories will rise, thus justifying a dedicated FRs estimate.

6.2.3 *Rare Backgrounds*

The triboson production processes (ZZZ, WWZ, WZZ) and the electroweak $t\bar{t}Z$, $t\bar{t}WW$, and $t\bar{t}ZZ$ processes, collectively referred to as *rare electroweak (EW) backgrounds* contribute only to a sub-percent level to the final expected yields. Nevertheless, their inclusion in the analysis is fundamental to properly take into account the irreducible background contributions in some of the categories and achieve the most accurate description possible of the physics under study. All these processes are estimated from dedicated MC simulations. Since the yield of these processes is expected to be less than one event in all the analysis categories, the EW background events are merged to the Z+X process in the context of the background shape parametrization used for the statistical analysis. This choice is driven by the fact that any statistical fluctuation possibly introduced by merging the EWK and Z+X yields is covered by the large uncertainty[4] associated with the latter.

[4] The uncertainty on the Z+X yield range between 30 and 40% across the different categories of the analysis, as described in Sect. 6.2.2.

6.2.4 Background Line Shape

The cornerstone of any statistical analysis is the likelihood function[5] maximized to extract the parameters of interest of the measurement, may them be SM cross sections, H boson couplings, or constraints on any BSM parameter. Such a function is proportional to the product of the process yield in a given analysis category j (N^j) and its pdf ($f_N^j(x)$):

$$\mathcal{L} \propto N^j f_N^j(x).$$

In the context of an "unbinned shape analysis" such as the one presented in this thesis, the pdfs $f_N^j(x)$ enter the above expression directly as analytical parametrizations for all the reconstructed categories considered in the analysis. Needless to say, the ultimate accuracy of the analysis relies on the precise description of the $m_{4\ell}$ reconstructed distribution by means of dedicated analytical functions.

The normalization parameters N^j and the analytical expression of the pdfs $f_N^j(x)$ are derived from the dedicated MC simulations of the background processes, after the application of all the corrections and k-factors described in the previous sections. The former is taken directly from the integral of weighted $m_{4\ell}$ distributions, as described in Sect. 6.1, of each background process in the 22 analysis categories. The analytical expression of the pdfs and their parametrization is derived, separately for the three background processes, from a fit of the weighted $m_{4\ell}$ distributions using different approaches for the irreducible and reducible backgrounds.

Reducible background

The shapes of the reducible background are extracted from the distributions obtained from the SS method, as it is found to yield results in agreement with the OS one, but it comprises larger statistics than the OS CRs. The reconstructed $m_{4\ell}$ is parametrized with a Landau function in all the reconstructed categories, separately for the three data taking periods and final states considered in the analysis. The parameters of the Landau functions are extracted from a fit in the $70 < m_{4\ell} < 300$ GeV mass range and the resulting distributions are shown in Fig. 6.7 for the estimate on the 2018 dataset. The core of the Landau distribution is included within the $105 < m_{4\ell} < 140$ GeV mass range for most of the reconstructed categories, even though in some of the low populated ones it appears to be shifted towards higher masses, resulting in only the left tail of the function being included in the mass range considered for the analysis. The introduction of an additional nuisance parameter to cope with this shape uncertainty could be envisaged for future developments of the analysis, but its effect is expected to be marginal if compared the to $\mathcal{O}(30\%)$ normalization uncertainty on the small Z+X yield.

[5] A rigorous description of the likelihood function used in this analysis, along with its explicit expression is given in Sect. 6.4. The proportionality relation written here serves as overview of what is detailed in the text.

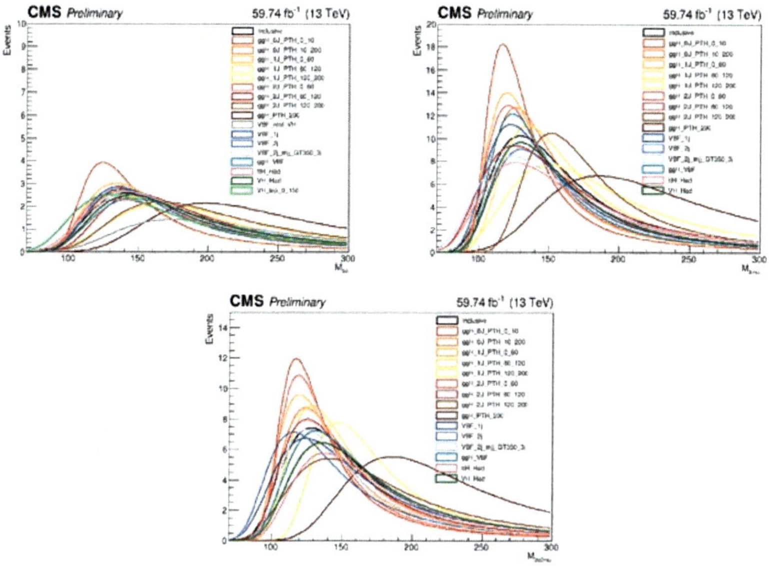

Fig. 6.7 The shape of the parametric background model for the Z+X processes in the three final states considered in the analysis: 4e (top left), 4m (top right), 2e2m (bottom). The distributions are derived for all the 22 reconstructed categories and are parametrized with a Landau function in each STXS bin

The expression of the Landau function in the $105 < m_{4\ell} < 140$ GeV mass range is included in the likelihood function as pdf for the reducible background process for each of the reconstructed categories. As an example, Fig. 6.8 shows the Z+X background shapes in the Untagged-0j-$p_{\mathrm{T}}^{4\ell}$[10, 200], VBF-3jet-tagged-$m_{\mathrm{jj}} > 350$, and VH-hadronic-tagged reconstructed categories, obtained from 2018 data in the 4e final state.

Irreducible background

Third-degree Chebyshev polynomials are used for the parametrization of the distributions for the $\mathrm{q\bar{q}} \rightarrow \mathrm{ZZ}$ and $\mathrm{gg} \rightarrow \mathrm{ZZ}$ processes in the $105 < m_{4\ell} < 140$ GeV mass range considered for the analysis.

The shapes of the $\mathrm{q\bar{q}} \rightarrow \mathrm{ZZ}$ and $\mathrm{gg} \rightarrow \mathrm{ZZ}$ processes are derived separately in all the reconstructed categories considered in the analysis. As an example, the distributions obtained from the 2018 MC samples are shown in Fig. 6.9 for the Untagged-0j-$p_{\mathrm{T}}^{4\ell}$[10, 200], VBF-1jet, and VH-leptonic-tagged-$p_{\mathrm{T}}^{4\ell}$[0, 150] categories in the 4e final state.

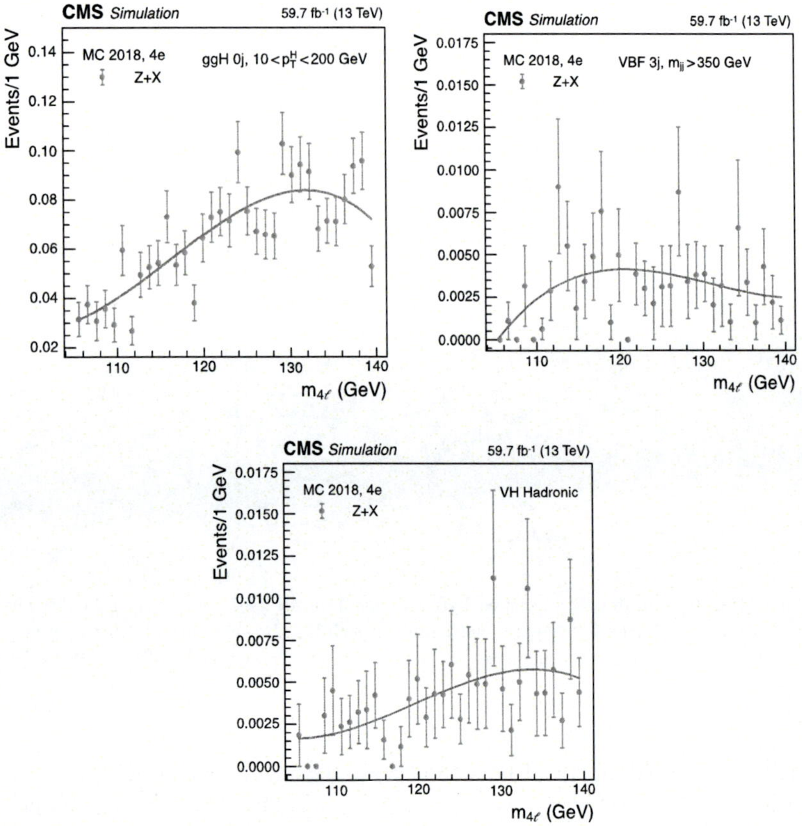

Fig. 6.8 The shape of the parametric background model for the Z+X processes in three sample categories of the analysis: Untagged-0j-$p_T^{4\ell}$[10, 200] (top left), VBF-3jet-tagged-$m_{jj} > 350$ (top right), and VH-hadronic-tagged (bottom). The points with error bars represent the Z+X events estimated from data in the 2018 data taking period and the solid lines represent the analytical shape obtained from a fit a Landau function in the $m_{4\ell} > 70$ GeV mass range, as described in the text

For all the three background processes, there are categories where the very low statistics make the fit practically unfeasible: in these cases, the inclusive shape of the specific final state, obtained from the fit of the corresponding $m_{4\ell}$ distribution, is used as pdf for the analysis. The $m_{4\ell}$ distributions for the three background processes are shown in Fig. 6.10 for the three final states separately, along with their corresponding analytical shapes obtained from the parametrizations described above.

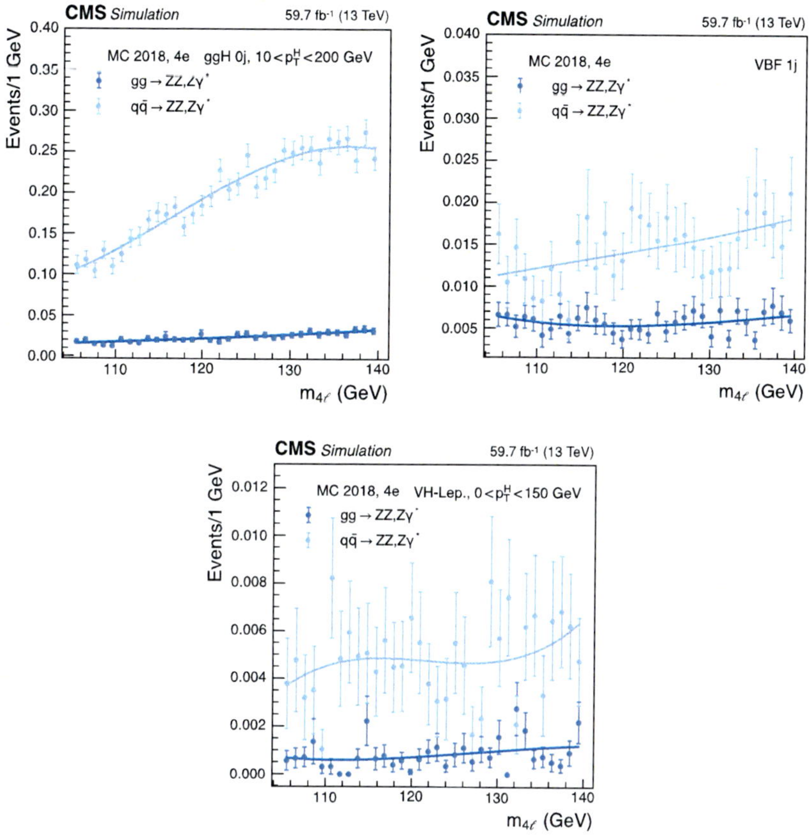

Fig. 6.9 The shape of the parametric background model for the $q\bar{q} \rightarrow ZZ$ (cyan) and $gg \rightarrow ZZ$ (blue) processes in three sample categories of the analysis: Untagged-0j-$p_T^{4\ell}$ [10, 200] (top left), VBF-1jet (top right), and VH-leptonic-tagged-$p_T^{4\ell}$ [0, 150] (bottom). The points with error bars represent weighted simulation events for the 2018 data taking period and the solid lines represent the analytical shape obtained from a fit with third order Chebyshev polynomials

6.2.5 Summary of the Background Samples Used in the Analysis

The different MC samples used for the simulation of the background processes are summarized in this section, where the details on the different corrections and k-factors are omitted, as discussed in detail in the previous sections.

The POWHEG 2.0 generator [9–11], interfaced with Pythia 8.0 [12] for parton showering, is used to produce $q\bar{q} \rightarrow ZZ$ events at NLO in pQCD. The $gg \rightarrow ZZ$ process is modelled at LO with the MCFM 7.0 [3] generator, where specific Pythia 8.0 settings are used to allow the showering only at the parton-level scale (*wimpy showers*).

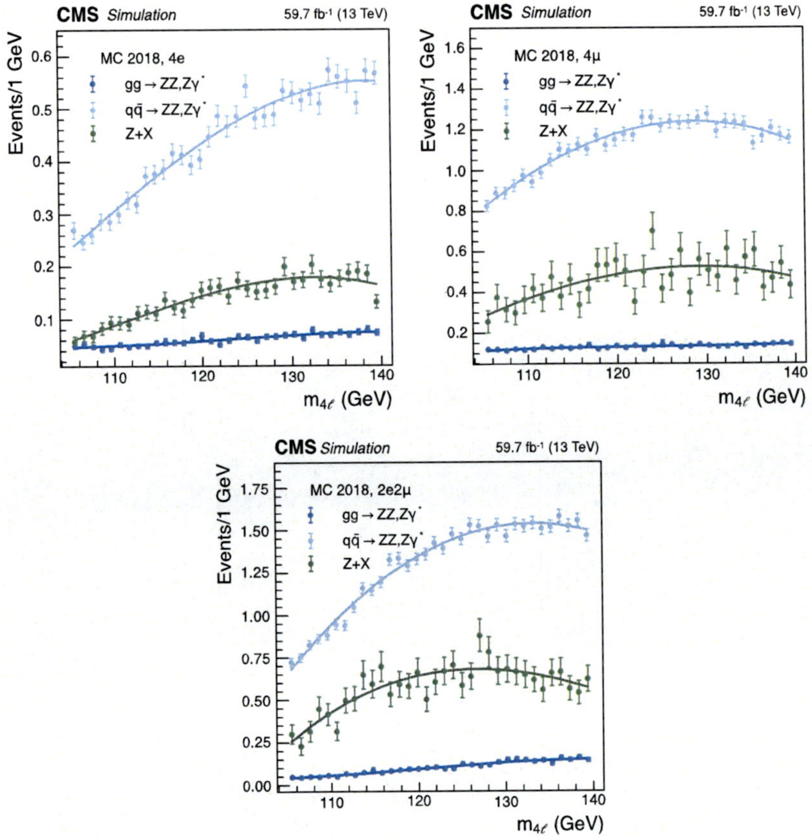

Fig. 6.10 The shape of the parametric background model for the $q\bar{q} \rightarrow ZZ$ (cyan), $gg \rightarrow ZZ$ (blue), and Z+X (green) processes in the 2018 data taking period and for the three final states considered in the analysis: 4e (top left), 4m (top right), 2e2m (bottom). The points with error bars represent the events distributions for each process and the solid lines represent the analytical shape obtained from a fit with third degree Chebyshev polynomials of the first order or a Landau function, as described in the text. The events are extracted from MC simulations and from the data-driven estimate for the $q\bar{q} \rightarrow ZZ$, $gg \rightarrow ZZ$, and Z+X, respectively

All the MC samples used to model the background processes comprising events with a Z boson accompanied by jets are produced with the MadGraph5 aMC@NLO generator [13]. As detailed in Sect. 6.2.2, the contribution of these processes, referred to as Z+X background, is estimated by means of data-driven techniques. Nevertheless, these MC simulations are still used for the optimization of these methods and for the validation of the performance of the objects reconstruction in dedicated control regions of data. The EW rare backgrounds are simulated with MadGraph5 aMC@NLO generator [13]. For all these processes, the parton showering is modelled using Pythia 8.0.

Table 6.7 Monte Carlo samples and product of theoretical cross sections and branching fractions for the background processes considered in the analysis. Parton showering is modelled with Pythia 8.0

Process	MC generator(s)	$\sigma \times BR$
$qq \rightarrow ZZ \rightarrow 4\ell$	POWHEG 2.0	1.2560 pb
$gg \rightarrow ZZ \rightarrow 4e/4\mu/4\tau$	MCFM 7.0	0.0016 pb
$gg \rightarrow ZZ \rightarrow 2e2\mu/2e2\tau/2\mu2\tau$	MCFM 7.0	0.0032 pb
$Z \rightarrow \ell\ell+$ jets	MadGraph5 aMC@NLO + FxFx	6225.4 pb
$WZ \rightarrow 3\ell\nu$	POWHEG 2.0	4.6700 pb
$t\bar{t} \rightarrow 2\ell2\nu2b$	POWHEG 2.0	87.310 pb
ZZZ	MadGraph5 aMC@NLO	0.0139 pb
WZZ	MadGraph5 aMC@NLO	0.0557 pb
WWZ	MadGraph5 aMC@NLO	0.1651 pb
$t\bar{t}+ZZ$	MadGraph5 aMC@NLO	0.0016 pb
$t\bar{t}$+WW		0.0079 pb
$t\bar{t}+Z(\rightarrow$ jets)	MadGraph5 aMC@NLO	0.2590 pb
$t\bar{t}+Z(\rightarrow \ell\ell\nu\nu)$	MadGraph5 aMC@NLO	0.2527 pb
$t\bar{t}+Z(\rightarrow \ell\ell)$	MadGraph5 aMC@NLO	0.0470 pb

Table 6.7 presents a summary of all the MC samples used in the analysis, along with the different generators used for their production and the theoretical values for their cross sections times branching fractions.

6.3 Signal Modelling

6.3.1 Overview of the Signal Samples Used in the Analysis

For the associated production of the H boson with a vector boson V ($V = W, Z$), the MINLO HVJ extension of POWHEG is used to properly take into account the presence of the additional leptons coming from the vector boson decay. All the samples are generated at NLO in pQCD using POWHEG, except for the $gg \rightarrow ZH$ production, which is produced at LO with the JHUGen generator [14]. Also the associated production with a pair of bottom quarks ($b\bar{b}H$) is simulated at LO with JHUGen. The H boson productions in association with a single top quark and an additional quark (tHq) or a W boson (tHW) are considered. These production mechanisms are modelled at LO via the MadGraph aMC@NLO generator and their inclusion in the analysis is twofold: not only it goes in the direction of having an analysis where the H boson decay into four leptons is described in the most accurate way possible, but also it allows to enhance the sensitivity to some of the STXS bins, where these modes might have a substantial contribution.

The decay of the H boson into four leptons is described using the JHUGen generator [14]. In the case of associated productions with a vector boson (VH) and with a top quark pair ($t\bar{t}$H), the H boson is allowed to decay to $H \to ZZ \to 2\ell 2X$, where X means any particle, to properly model four lepton events where two of them originate from the associated particle produced with the H boson. The normalization of these processes to the 4ℓ SM cross section is done by applying a dedicated efficiency filter (ϵ_{filter}) to the generator-level cross section.

All the samples are produced with the NNPDF 3.1 NLO set of parton distribution functions (PDFs) [15] and the full list is summarized in Table 6.8, along with the theoretical values for their cross sections times branching fractions, complemented by the product for the filter efficiency, when this is applied.

All the signal samples used for the statistical analysis are produced with a nominal value of 125 GeV for the mass of the H boson (m_H). In addition, samples at four different mass hypotheses are generated: m_H =120, 124, 126, and 130 GeV. These are used for the parametrization of the signal normalization and analytical shapes and for the evaluation of systematic uncertainties on the MC generators scales, as detailed in the next two sections.

The reconstructed invariant mass of the four-lepton system is shown in the stack histogram of Fig. 6.11 for the MC samples of all the signal and background processes for an integrated luminosity of 137 fb^{-1}, corresponding to the full Run-II statistics.

Table 6.8 Monte Carlo samples and theoretical cross section times branching fraction values for the signal processes considered in the analysis. All the samples are generated with the combination of the POWHEG 2.0, Pythia 8.0, and JHUGen 7.0 generators, indicated as "Standard" in the table. The particular settings of POWHEG 2.0 used for the productions in association with vector bosons are specified

Process	MC generator(s)	$\sigma \times BR(\times \epsilon_{\text{filter}})$
$gg \to H \to ZZ \to 4\ell$	Standard	13.34 fb
$qq \to Hqq \to ZZqq \to 4\ell qq$	Standard	1.038 fb
$q\bar{q} \to ZH \to ZZZ \to 4\ell + X$	POWHEG 2.0 (minlo HZJ)	0.618 fb
$q\bar{q} \to W^+H \to W^+ZZ \to 4\ell + X$	POWHEG 2.0 (minlo HWJ)	0.232 fb
$q\bar{q} \to W^-H \to W^-ZZ \to 4\ell + X$	POWHEG 2.0 (minlo HWJ)	0.147 fb
$gg \to ttH \to ttZZ \to 4\ell + X$	Standard	0.139 fb
$gg \to bbH \to bbZZ \to 4\ell + X$	JHUGen 7.0	0.135 fb
$gg \to ZH \to ZZZ \to 4\ell + X$	JHUGen 7.0	0.086 fb
$q\bar{q}/qg \to tHq \to tqZZ \to 4\ell + X$	MadGraph5 aMC@NLO	0.021 fb
$qg \to tHW \to tWZ \to 4\ell + X$	MadGraph5 aMC@NLO	0.015 fb

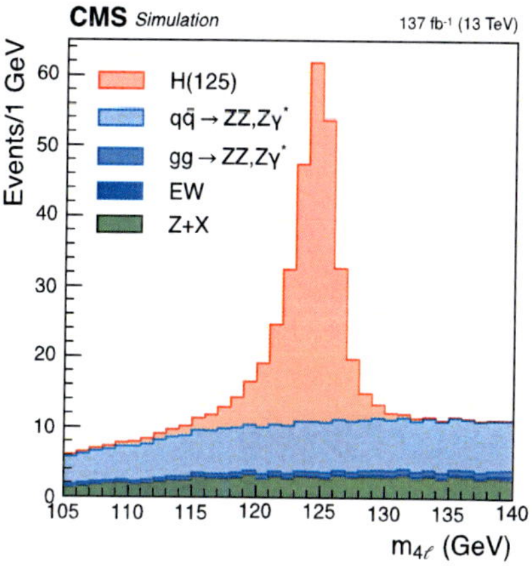

Fig. 6.11 The reconstructed distributions of the four-lepton invariant mass for signal and background processes (stacked histograms). The $m_{4\ell}$ distributions are represented in the $105 < m_{4\ell} < 140$ GeV mass range considered for the analysis with a 1 GeV bin size

The 4ℓ events are selected as of Sect. 5.1 and the statistical analysis is performed in the twenty-two categories described in Sect. 5.3, in the mass range $105 < m_{4\ell} < 140$ GeV. In this region, the background and signal shapes have substantial differences, being the former rather flat over the entire mass range, while the latter features a typical resonant shape, as illustrated in Fig. 6.11. For this reason, the parametrization of the signal line shape is somewhat more complicated than what is done for the background and described in Sect. 6.2. More precisely, the five m_H points at which the H boson samples are produced are exploited to derive a dependence of the signal pdfs' parameters and their normalization factors as a function of m_H.

6.3.2 *Normalization*

The clear resonant shape of the $m_{4\ell}$ distribution for a SM H boson shown in Fig. 6.11 is one of the main features of the $H \to ZZ \to 4\ell$ decay channel, where the final state is completely resolved and the backgrounds present a rather flat shape, resulting in the high signal-to-background ratio characteristic of the "golden" channel. In order to maximally profit from these features, the normalization and the pdf's parameters describing the analytical shape of the $m_{4\ell}$ spectrum are determined as a function of m_H.

MC simulations at five m_H points (120, 124, 125, 126 and 130 GeV) are used to derive a second-order dependence of the signal normalization as a function of m_H:

$$N_{\text{Sig}} = a_0 + a_1 \times \frac{m_H}{125.0\ \text{GeV}} + a_2^2 \times \left(\frac{m_H}{125.0\ \text{GeV}}\right)^2 \qquad (6.2)$$

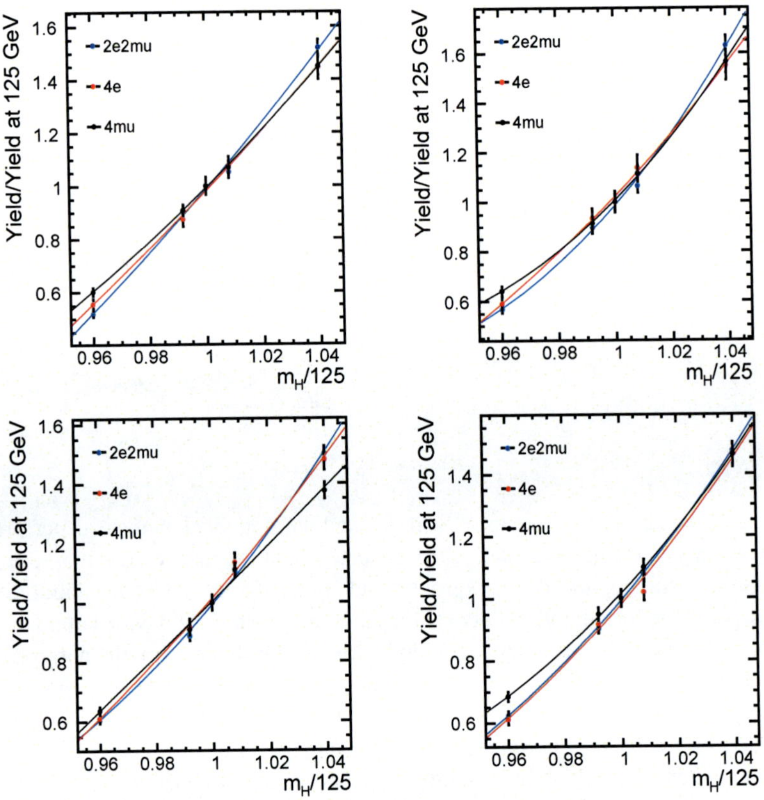

Fig. 6.12 Fits of the expected signal yields after full event selection in a $105 < m_{4\ell} < 140$ GeV mass window, for the `ggH-1j/pT[60, 120]` stage 1.2 bin in Untagged-1j-$p_T^{4\ell}$[60, 120] and VBF-1jet-tagged event categories (first row) and `ttH` bin in $t\bar{t}$H-hadronic-tagged and $t\bar{t}$H-leptonic-tagged categories (second row)

The expected signal yields are extracted from the weighted MC events (cf. Sect. 6.1), in a $105 < m_{4\ell} < 140$ GeV mass window around the Higgs boson peak for the five different m_H hypotheses. These are used to derive the a_i parameters of Eq. (6.2) by means of a polynomial fit performed independently on all the STXS Stage 1.2 bins, in the three final states, and the twenty-two categories of the analysis. The results of this parametrization are shown in Fig. 6.12 for the three final states in four different categories.

This parametrization choice allows to reduce the dependence of the statistical model on the H boson mass, which can be either fixed to its best fit value or *profiled* during the fit.

6.3.3 Line Shape

The natural width of the SM H boson is predicted to be $\Gamma_H \sim 4$ MeV, but due to detector effects and reconstruction artifacts, the reconstructed 4ℓ resonance is much larger, as observed in the red histogram of Fig. 6.11. While this effect is completely determined by the reconstruction performance, which is ultimately included in the analysis as a nuisance parameter, the actual shape of the 4ℓ resonance is affected by the leptons kinematics, resulting in a dependence on the actual H boson mass. Hence, the typical Breit–Wigner shape of the resonance is modified to take into accounts these effects and a double-sided Crystal-Ball function $f_{dCB}(m_{4\ell}\,|\,m_H)$ is used as pdf for the signal in the analysis:

$$dCB(\xi) = N_{\text{Sig}} \cdot \begin{cases} A \cdot (B + |\xi|)^{-n_L}, & \text{for } \xi < \alpha_L \\ \exp\left(-\xi^2/2\right), & \text{for } \alpha_L \le \xi \le \alpha_R \\ A \cdot (B + |\xi|)^{-n_R}, & \text{for } \xi > \alpha_R \end{cases} \tag{6.3}$$

where the independent variable $\xi = (m_{4\ell} - m_H - \Delta m_H)/\sigma_m$ is chosen to capture both the Gaussian core of the resonance, from which the dependence on the 4ℓ mass resolution σ_m, and the aforementioned systematic mass shift (Δm_H) of the peak. The two parameters α_L and α_R determine the position of the left and right boundaries outside which the Gaussian approximation of the core region is no longer valid and a power law is introduced to cope with the presence of non-Gaussian tails in the distribution. These arise from the interactions between the leptons and the detector material in the form of final state radiation emitted in the tracker and energy losses in the ECAL. The $A \cdot (B + |\xi|)^{-n_i}$ power laws used to model the tails are defined so that the n_i parameters control their prominence, while A and B ensure the continuity of the function and its first derivatives. The normalization of the pdf (N_{Sig}), corresponding to the expected number of signal events, is extracted from MC samples and it is parametrized as a function of m_H as described in the previous section.

In order to enhance the sensitivity of the analysis to the m_H dependence of the $m_{4\ell}$ shape, all the parameters of the $f_{dCB}(m_{4\ell}\,|\,m_H)$ function are expressed with a linear dependence on m_H, rather than as mere constants:

$$p_i(m_H) = a_0 + a_1 \cdot (m_H - 125 \text{ GeV})$$

Similar to what is done for the N_{Sig} normalization parameter, the H boson samples generated at five different mass points are used to obtain the expression of $p_i(m_H)$ for all the STXS Stage 1.2 bins, in the three final states, and the twenty-two categories of the analysis. An initial guess of the best fit values $p_i^0(m_H)$ is obtained from a fit of the reconstructed $m_{4\ell}$ for $f_{dCB}(m_{4\ell}\,|\,m_H = 125 \text{ GeV})$ for all the parameters, which are in turn used as input to perform a simultaneous fit over the five mass points and derive the expression of $f_{dCB}(m_{4\ell}\,|\,m_H = 125 \text{ GeV})$, $p_i(m_H)$ used for the analysis. The projection of the simultaneous fit of the $m_{4\ell}$ distributions with

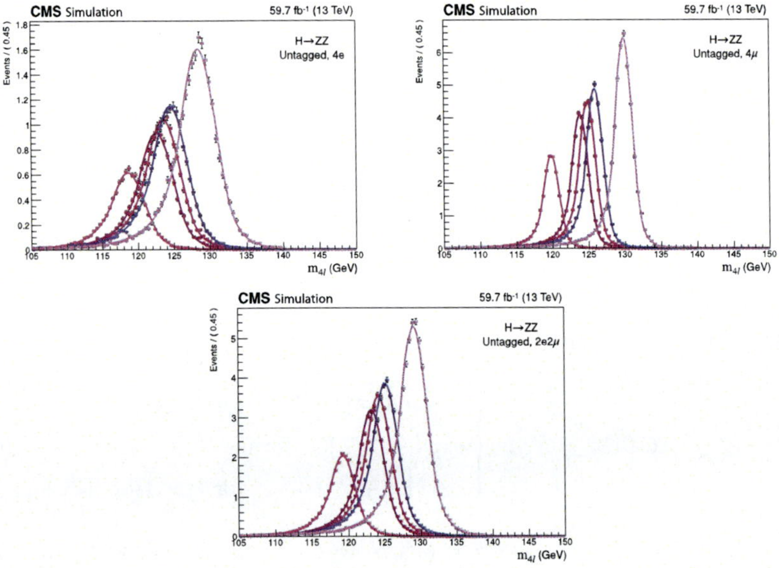

Fig. 6.13 Result of the simultaneous fit of the $m_{4\ell}$ distribution for simulated events in the Untagged reconstructed category. The projection of the simultaneous fit with the $f_{dCB}(m_{4\ell}\,|\,m_H = 125\text{ GeV})$ pdf is shown in the 4e (top left), 4m (top right), and 2em (bottom) final states, respectively

$f_{dCB}(m_{4\ell}\,|\,m_H = 125\text{ GeV})$ is shown in Fig. 6.13 for each mass hypothesis, for the three final states considered in the analysis, in the untagged category, sensitive to the ggH production mechanism.

In the categories sensitive to the associate production modes with a vector boson decaying leptonically (VH) or with a pair of top quarks ($t\bar{t}$H), a non-resonant contribution arising from one of the leptons coming from the additional vector boson being misidentified as originating from the H boson decay may distort the line shape of the $m_{4\ell}$ resonance. For this reason, an additional Landau contribution is included to Eq. (6.3), thus introducing two other parameters in the pdf definition: these are left floating in the initial fit at $m_H = 125$ GeV and they are fixed to their best fit value during the simultaneous fit. The result of the fit at $m_H = 125$ GeV is shown in Fig. 6.14 for the VH-leptonic-tagged-$p_{\mathrm{T}}^{4\ell}$ [0, 150] reconstructed category in the three final states of the analysis. While in the 4m the non-resonant contribution is softer, the presence of "wrong leptons" in the final state is much larger in the 4e and 2e2m channels, thus justifying the introduction of the additional Landau contribution in the analytical expression used to parametrize the $m_{4\ell}$ resonance.

Eventually, the resulting pdf is found to model in a satisfactory manner the expected $m_{4\ell}$ distributions of the signal processes at $m_H = 125$ GeV for all the reconstructed categories. An example is given in Fig. 6.15, where the pdf resulting from the simultaneous fit procedure described above is shown for the events in the Untagged category in the 4e and 4m final states, respectively.

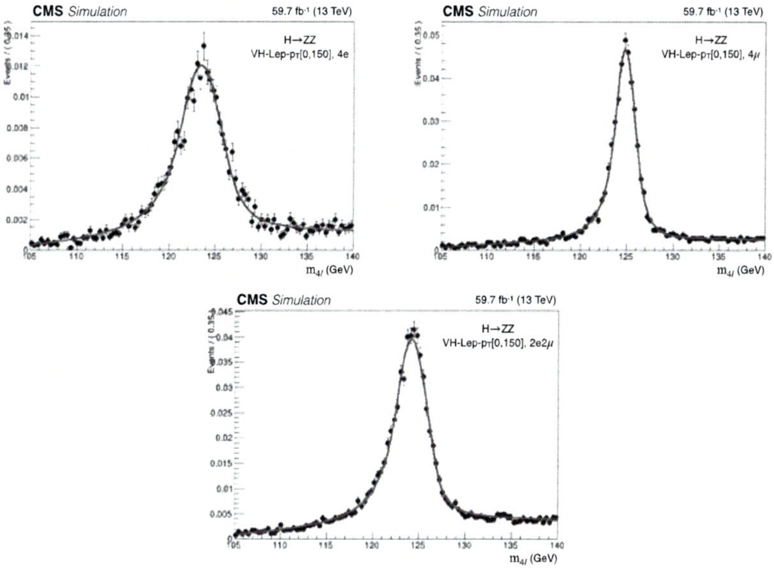

Fig. 6.14 Result of the fit of the $m_{4\ell}$ distribution for simulated events at $m_{\mathrm{H}} = 125$ GeV in the VH-leptonic-tagged-$p_{\mathrm{T}}^{4\ell}$[0, 150] reconstructed category. The pdf used for the fit comprises a Landau contribution, in addition to the standard $f_{dCB}(m_{4\ell} \mid m_H)$ double sided Crystal Ball, to cope with the presence of non-resonant events, as described in the text. The fits are shown in the 4e (top left), 4m (top right), and 2e2m (bottom) final states, respectively

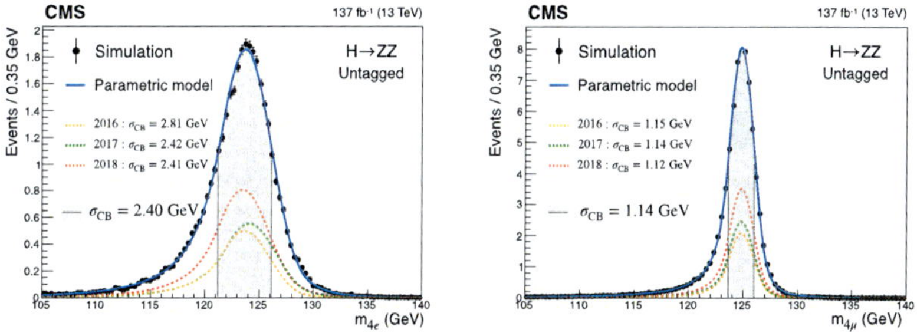

Fig. 6.15 The shape of the parametric signal model for each year of simulated data, and for the sum of all years together. The black points represent weighted simulation events of the ggH production mechanism for $m_{\mathrm{H}} = 125$ GeV and the blue line the corresponding model. Also shown is the σ_{CB} value (half the width of the narrowest interval containing 68% of the invariant mass distribution) in the gray shaded area. The contribution of the signal model from each year of data-taking is illustrated with the dotted lines. The models are shown for the 4e (left) and 4m (right) final states in the untagged event category

Having introduced the signal and background models, it is now time to dig into the details of the statistical methods used to measure the H boson properties in the $H \to 4\ell$ decay channel.

6.4 Statistical Inference in Data Analysis

After the discovery of the SM H boson by the ATLAS and CMS collaborations in 2012 with the Run-I dataset, the main focus of the Run-II has been the precision measurement of the properties of this long-sought particle. Most of the analyses developed initially to detect the H boson as a statistically significant excess over background fluctuations have become precision physics analyses that probe previously unexplored phase space regions. The $H \to ZZ \to 4\ell$ is one of such cases: the limited statistics typical of this decay channel was the major drawback in the Run-I analysis. Nonetheless, its large signal to background ratio and the completely resolved final state made it one of the cornerstones for the final claim of the discovery. Nowadays, these "golden features" are supplemented by the significant statistics available at the end of the Run-II, which opened the doors of the precision physics realm, where the different H boson production mechanisms and fine-granular phase space regions can be assessed separately. In this scenario, the analyses are moving towards an exciting era of measurements, where the statistical component of the uncertainty is becoming sub-dominant and the precision becomes ultimately limited by the systematic contribution. A detailed description of the systematic uncertainties considered in the analysis is given in the next section, while the following focuses on a general description of the statistical methods on which the measurement of the H boson properties in the "golden channel" relies.

The starting point for any analysis is constructing of a statistical model that describes the properties of the dataset $\mathcal{S}$ considered.[6] Often such a dataset comprises many events $\mathcal{S} = \{x_1, x_2, \ldots, x_n\}$, each with its own associated probability density function (pdf) $f(x_i)$. The information on the events x_i is often obtained through of auxiliary measurements (α), being them theoretical predictions or the imperfect knowledge of the detector. For this reason it is common to use *parametric models*, i.e.:

$$f(x_i) = f(x_i \mid \alpha), \; \forall x_i \in \mathcal{S}$$

Among the set of all the model parameters, one is usually interested in a few of them, which therefore take the name of *parameters of interest* (POI) and are labelled as μ. In contrast, all the additional parameters of no direct interest to the analysis are commonly referred to as *nuisance parameters*, and they are hereafter denoted as θ. The information of all the events in the dataset $\mathcal{S}$ is described by the joint pdf, called Poisson marked model in the statisticians' jargon:

[6] The notation and the structure of this section are inspired from Ref. [16].

$$\mathbf{f}(\mathcal{S} \mid \mu, \theta) = \text{Pois}(n \mid \mu) \prod_{e=1}^{n} f(x_e \mid \theta), \tag{6.4}$$

where n is the total number of events that comprise the dataset $\mathcal{S}$. In the context of this analysis $n = s + b$, where s and b represent the expected signal and background yields derived from the dedicated MC simulations. Therefore the Pois$(n \mid \mu)$ term of Eq. (6.4) becomes Pois$(n \mid \mu \cdot s(\theta) + b(\theta))$, with an explicit dependence on the nuisance parameters θ. In the high energy physics (HEP) analyses, the POI μ is often a *signal strength modifier*, defined as the ratio between the observed cross section and the corresponding SM prediction : $\mu = \sigma_{obs}/\sigma_{\text{SM}}$. This quantity is taken to scale the SM H boson expected cross section in each analysis bin by the same amount μ.

Often analyses measure the POI after subdividing the events of the dataset $\mathcal{S}$ into mutually exclusive channels c. Hence, the Poisson marked model of Eq. (6.4) can be rewritten as:

$$\mathbf{f}(\mathcal{S} \mid \mu, \theta) = \prod_{c} \left[\text{Pois}(n \mid \mu \cdot s(\theta) + b(\theta)) \prod_{e=1}^{n} f(x_e \mid \theta) \right],$$

where the Pois$(n \mid \mu \cdot s(\theta) + b(\theta))$ term refers to the product of Poisson probabilities of observing n_i events in i bins:

$$\prod_{i} \frac{(\mu s_i + b_i)^{n_i}}{n_i!} e^{-\mu s_i - b_i}.$$

For unbinned shape analyses, as the $\text{H} \to 4\ell$ presented in this thesis, one has to take into account the analytical expression of the signal and background pdfs. Therefore the analysis model becomes:

$$\prod_{i} (\mu S \cdot f_s(x_i) + B \cdot f_b(x_i)) \cdot e^{-(\mu S + B)} \tag{6.5}$$

There is a missing ingredient in Eq. (6.4), i.e., the constraint terms coming from the nuisance parameters. These are introduced in the statistical analysis as nuisance parameters (θ), with associated pdfs $\rho(\theta \mid \tilde{\theta})$, in the likelihood. As a consequence of the Bayes theorem, the systematic error pdf $\rho(\theta \mid \tilde{\theta})$ can be interpreted as the posterior obtained from an auxiliary measurement $\tilde{\theta}$ of θ:

$$\rho(\theta \mid \tilde{\theta}) \sim p(\tilde{\theta} \mid \theta) \pi_{\theta}(\theta), \tag{6.6}$$

where the $\pi_{\theta}(\theta)$ functions represent the prior of the θ measurements and it is often selected to be uniform in the context of HEP analyses. This contextualisation is the fundamental step that allows to use $p(\tilde{\theta} \mid \theta)$, which has a pure frequentist interpretation, as the pdf associated to the nuisance parameters θ. A detailed description of the systematic uncertainties that enter the analysis, along with their corresponding pdf, is given in the next section.

Taking into account all these details, the $H \to ZZ \to 4\ell$ analysis model can be expressed as:

$$\mathbf{f}(\mathcal{S} \mid \mu, \theta) = \prod_{c} \left[\text{Pois}(n \mid \mu S(\theta) \cdot f_s(x_i) + B(\theta) \cdot f_b(x_i)) \right] \prod_{i} p_i(\tilde{\theta}_i \mid \theta_i),$$

which has the advantage of being completely factorized in all its terms, a property that turns out to be extremely useful in the statistical analysis, where the results are presented in the form *maximum likelihood estimators* (MLE) of the POI(s).

Having built the marked Poisson model for the analysis, it is straightforward to obtain the likelihood function $\mathcal{L}(\mathcal{S}, \tilde{\theta} \mid \mu, \theta)$, as it has the same expression of $\mathbf{f}(\mathcal{S} \mid \mu, \theta)$, with $\mathcal{S}$ fixed:

$$\mathcal{L}(\mathcal{S}, \tilde{\theta} \mid \mu, \theta) \equiv \prod_{c} \mathcal{L}_c \left(\mathcal{S} \mid \mu S(\theta) \cdot f_s(x_i) + B(\theta) \cdot f_b(x_i) \right) \cdot \prod_{i} p_i(\tilde{\theta}_i \mid \theta_i), \tag{6.7}$$

and the MLE is the value[7] of μ that maximizes the likelihood function $\mathcal{L}$.

More precisely, the results of the analysis are extracted using a *profile likelihood ratio test statistics*, or *negative log-likelihood function*, defined as:

$$\tilde{q}_\mu = -2\Delta\text{L} = -2 \ln \frac{\mathcal{L}(\mathcal{S} \mid \mu, \hat{\theta}_\mu)}{\mathcal{L}(\mathcal{S} \mid \hat{\mu}, \hat{\theta})} \tag{6.8}$$

which is the most powerful test statistics according to the Neyman-Pearson lemma, similarly to the MELA discriminants introduced in Sect. 5.1.1. The $(\hat{\mu}, \hat{\theta})$ parameters are the MLE of the POIs and NPs, respectively. The $\hat{\theta}_\mu$ term is referred to as *conditional maximum likelihood estimate* of the nuisance parameters and it corresponds to the MLE of θ at a fixed value of μ. The same test statistics can be used both in those cases where a single POI is measured, such as the inclusive signal strength of the $H \to 4\ell$ decay, and when multiple POIs are constrained simultaneously, for example in the measurement of the signal strength modifiers of the different H boson production modes. By the Wilks theorem [17], the distribution of the negative log-likelihood $-2\Delta\text{L}$, for a model with n POIs, approaches a χ^2 distribution with n degrees of freedom, and therefore can be used to extract confidence level (CL) intervals on the measured value of μ. More precisely, the 68 and 95% CL intervals on μ are obtained for $-2\Delta\text{L} < 1$ and $-2\Delta\text{L} < 3.84$, respectively.

The dataset $\mathcal{S}$ that appears in the previous equations refers either to the actual data collected by the experiment or to some pseudo-data: the former is used to extract the *observed* results of the analysis, corresponding to the nominal value of the POIs, while the latter to measure the *expected* ones, which provide insight on

[7] More precisely, if the set of POIs and NPs is referred to as $\alpha = (\mu, \theta)$, the MLE is the value of α that maximizes $\mathcal{L}$.

the sensitivity of the analysis and a benchmark against which to compare the final results. In principle, one would have to generate a large number of pseudo-experiment (*toys*), use their outcome to build the likelihood ratio test statistics, and eventually use this to extract the expected results. However, it has been shown that in the limit of sufficient statistics a good approximation is provided by the Asimov data set [18], i.e. a representative set of data, generated under the nominal set of nuisance parameters, where statistical fluctuations are suppressed. With the increasing complexity of the analyses and the large number of POIs being measured, nowadays it is common to determine the expected results by running MLE fits on Asimov datasets and this same approach is used in the statistical inference procedure used in this thesis.

Further details on the statistical models used for the different measurements performed in the $H \to 4\ell$ analysis are detailed in the following.

6.4.1 "Golden" Statistical Inference

Signal strength and STXS cross sections

The results of the analysis, presented in Chap. 7, are obtained by means of maximum likelihood fits in the $105 < m_{4\ell} < 140$ GeV mass region. The signal strength modifiers measured are forced to be greater or equal than zero to reflect the large signal to background ratio characteristic of the $H \to 4\ell$ channel and to avoid the unphysical situations of negative pdfs in the analysis model.

The fits are performed exploiting two of the most powerful observables to discriminate between signal and background processes in the four-lepton final state: the invariant mass and the decay matrix element discriminants defined in Sect. 5.1.1. The invariant mass of the four lepton system ($m_{4\ell}$) and the discriminants (here collectively referred to as $\mathcal{D}$) are combined in the statistical model of the analysis with the product of two one-dimensional pdfs:

$$f(m_{4\ell}, \mathcal{D}) = \mathcal{P}(m_{4\ell})\mathcal{P}(\mathcal{D}|m_{4\ell}). \tag{6.9}$$

The first term ($\mathcal{P}(m_{4\ell})$) defines the probability density function of the four-lepton resonance, parametrized as described in Sects. 6.3 and 6.2.4 for signal and background processes, respectively. The second term ($\mathcal{P}(\mathcal{D}|m_{4\ell})$) refers to the conditional probability of measuring a value of the kinematic discriminant $\mathcal{D}$, given a specific value of the four-lepton invariant mass $m_{4\ell}$. This contribution is built as a two-dimensional normalized template with a 1 GeV bin width, where the granular binning is chosen to appreciate better the correlation between $m_{4\ell}$ and $\mathcal{D}$ (cf. Fig. 5.2), thus enhancing the discrimination between signal and background events during the fit. The decay-only kinematic discriminant ($\mathcal{D}^{\text{kin}}_{\text{bkg}}$) is used in most of the analysis categories, except for the VBF-2jet-tagged and VH-hadronic-tagged sub-categories. Here, the $\mathcal{D}^{\text{VBF+dec}}_{\text{bkg}}$ and $\mathcal{D}^{\text{VH+dec}}_{\text{bkg}}$ discriminants are used to exploit the increased separation power achieved by introducing in their definition the information on the number of jets in the event (cf. Sect. 5.1.1). The simulation samples of the signal processes are used to create templates for all the nineteen STXS production bins, to each of which an independent POI is assigned in the statistical analysis. Similarly, the dedicated MC samples are used to build the templates for the irreducible backgrounds, whereas the data-driven estimate is used for the Z+X contribution. Eventually, a total of $22 \times 3 = 66$ pdfs is considered in the analysis, one for every category in the three final states (4μ, 4e, $2e2\mu$), and Eq. (6.7) can be written explicitly as:

$$\mathcal{L}(\vec{\mu}) = \frac{1}{N} \prod_{\text{events}} \left[\left(\sum_{i=1}^{19} \mu_i S_i^{jk} f_S^{ijk}(m_{4\ell}, \mathcal{D}) + B^{jk} f_B^{jk}(m_{4\ell}, \mathcal{D}) \right) e^{-\sum_i \mu_i S_i^{jk} + B^{jk}} \cdot \prod_l p_l(\tilde{\theta}_l \mid \theta_l) \right], \tag{6.10}$$

where μ_i identifies the signal strength modifier associated to the ith production bin. By extension[8] from Eq. (6.7), S_i^{jk} and B^{jk} represent the expected rates for signal and background processes in the jth reconstructed category and for the final state k, and $f_S^{ijk}(m_{4\ell}, \mathcal{D})$ and $f_B^{jk}(m_{4\ell}, \mathcal{D})$ are their corresponding pdfs.

The advantage of such a complex model is that it can be easily extended to all the measurements of signal strength modifiers and STXS cross sections presented in Chap. 7. In fact, merging together all the categories that target a specific production mechanism, and associating a new POI to each of them, one can measure both the STXS Stage 0 cross sections and the signal strength modifiers of each production mode. Further details on the explicit form of Eq. (6.10) for these two measurements are given explicitly in Chap. 7, when the corresponding results are presented.

Fiducial and differential cross sections

Fiducial cross section measurements provide a complementary approach to the signal strength modifiers to study the properties of the SM H boson and to possibly detect hints of new BSM physics. In fact, while new physics can appear in the form of statistically significant discrepancies between the observed signal strength values and their corresponding SM prediction, this approach has two main drawbacks:

- Signal strength measurements are performed at the reconstruction level. Hence, detector effects, such as the limited resolution and residual imperfections in the simulation used to derive the line shape, directly impact the results of the analysis. The analysis strategy is designed so that these effects are minimized with the events' selection and accounted for by appropriate nuisance parameters, but they can not be completely removed when dealing with reconstructed objects.

[8] For simplicity the explicit dependence on the nuisance parameters θ is omitted here.

- They have a strong underlying model dependence. Signal strength modifiers are defined as constants that scale the cross sections with respect to the corresponding SM prediction. Hence, it is almost impossible to interpret these results a posteriori in any BSM scenario.

The STXS framework introduced in Sect. 5.2 is designed to partially solve the second issue, as the theoretical uncertainties on the STXS bins definition are not included in the measurement of the cross sections, but only enter the interpretation stage, as schematically illustrated in Fig. 5.4. In addition, the Stage 1.2 bins measured in this analysis define a set of fine-granular phase space regions, some of which are specifically designed to test the BSM-like phase space. Notwithstanding this, the measurements are performed starting from reconstructed objects, and even though the bins are designed to maximize the experimental sensitivity, the acceptance of the analysis and the reconstruction effects may still influence the final results, having a non-negligible effect on the kinematic distributions.

An alternative and complementary approach consists of measuring generator level cross sections in a *fiducial* phase space region defined to match as closely as possible the experimental acceptance. The ultimate goal of these measurements, which can be performed both inclusively and in differential bins of kinematic observables, is twofold: reducing the model dependence of the results and resolving the detector effects.

The set of requirements used in the definition of the fiducial phase space is presented in Table 6.9. Two substantial improvements have been introduced in this version of the analysis regarding its Run-I counterpart. First of all, the leptons considered are not taken at Born level,[9] but rather at generator-level and are corrected for the presence of any photon within a cone of $\Delta R < 0.3$ around them: for this reason they are referred to as *dressed leptons* hereafter. The second improvement consists in the inclusion of an isolation criteria in the definition of the fiducial volume. More precisely, the scalar sum of the transverse momentum of all the stable particles within a cone of $\Delta R < 0.3$ is required to be smaller than 35% of the lepton's p_T. These two requirements enhance the overlap between the fiducial volume and the experimental acceptance, reducing the results' model dependence.

The generator level ZZ candidates are obtained from the same topological selection applied on reconstruction level objects and detailed in Sect. 5.1. The sole difference between the fiducial cross section measurements and the rest of the results lies in the identification of the ZZ candidate in those cases where more than one ZZ pair is reconstructed. While this situation is dealt with by taking the candidate with the largest $\mathcal{D}^{\text{kin}}_{\text{bkg}}$ value in the standard analysis, in the fiducial measurements, a more simplistic approach is adopted and the Z_2 candidate[10] with the largest scalar sum of

[9] Meaning that the generator level leptons are not, as the name says, the LHE ones corresponding to the *MC truth*.

[10] In a ZZ candidate, Z_1 is the reconstructed vector boson with invariant mass close to the nominal Z mass.

Table 6.9 Summary of requirements used in the definition of the fiducial phase space for the $H \to 4\ell$ cross section measurements

Requirements for the $H \to 4\ell$ fiducial phase space	
Lepton kinematics and isolation	
Leading lepton p_T	$p_T > 20$ GeV
Next-to-leading lepton p_T	$p_T > 10$ GeV
Additional electrons (muons) p_T	$p_T > 7(5)$ GeV
Pseudorapidity of electrons (muons)	abs$\eta < 2.5$ (2.4)
Sum of scalar p_T of all stable particles within $\Delta R < 0.3$ from lepton	$< 0.35 p_T$
Event topology	
Existence of at least two same-flavor OS lepton pairs, where leptons satisfy criteria above	
Inv. mass of the Z_1 candidate	$40 < m_{Z_1} < 120$ GeV
Inv. mass of the Z_2 candidate	$12 < m_{Z_2} < 120$ GeV
Distance between selected four leptons	$\Delta R(\ell_i, \ell_j) > 0.02$ for any $i \neq j$
Inv. mass of any opposite sign lepton pair	$m_{\ell^+\ell'^-} > 4$ GeV
Inv. mass of the selected four leptons	$105 < m_{4\ell} < 140$ GeV

transverse momenta is retained. This choice is motivated by studies performed in the Run-I analysis which showed how it leads to increased model independence of the results.

The fiducial cross section for the $pp \to H \to 4\ell$ process (σ_{fid}), both inclusive and in differential bins of kinematic observables, is extracted by means of a maximum likelihood fit on the observed $m_{4\ell}$ distribution, $N^{f,i}_{\text{exp}}(m_{4\ell})$, defined per each final state $f = (4e, 4m, 2e2m)$ and bin i as:

$$\begin{aligned} N^{f,i}_{\text{exp}}(m_{4\ell}) &= N^{f,i}_{\text{fid}}(m_{4\ell}) + N^{f,i}_{\text{nonfid}}(m_{4\ell}) + N^{f,i}_{\text{nonres}}(m_{4\ell}) + N^{f,i}_{\text{bkg}}(m_{4\ell}) \\ &= \sum_j \epsilon^f_{i,j} \left(1 + f^{f,i}_{\text{nonfid}}\right) \sigma^{f,j}_{\text{fid}} \mathcal{L} \mathcal{P}_{\text{res}}(m_{4\ell}) \\ &+ N^{f,i}_{\text{nonres}} \mathcal{P}_{\text{nonres}}(m_{4\ell}) + N^{f,i}_{\text{bkg}} \mathcal{P}_{\text{bkg}}(m_{4\ell}). \end{aligned} \tag{6.11}$$

This quantity represents the $\text{Pois}(n \mid \mu \cdot s(\theta) + b(\theta))$ term in Eq. 6.7, and $\sigma^{f,j}_{\text{fid}}$ is the POI of the measurement.

The $\mathcal{P}(m_{4\ell})$ terms in Eq. (6.11) define the following pdfs.

- $\mathcal{P}_{\text{res}}(m_{4\ell})$ is the resonant signal pdf, defined as detailed in Sect. 6.3, with the sole difference that here the parametrization is taken inclusively on all the production mechanisms.
- $\mathcal{P}_{\text{nonres}}(m_{4\ell})$ corresponds to the pdf of the so called *non-resonant* term, associated to the VH and $t\bar{t}H$ production modes, where one of the leptons coming from the decay of the additional vector boson is misidentified as originating from the H

boson decay. This contribution is modeled by a Landau distribution and is treated as background in the measurement.

- $\mathcal{P}_{\text{bkg}}(m_{4\ell})$ defines the contribution of the background processes, modelled as described in Sect. 6.2.

The $f_{\text{nonfid}}^{f,i}$ term represents and additional resonant contribution arising from events considered in the analysis but not coming from the fiducial phase space. This fraction is estimated using simulation for each signal process considered and it originates manly from the detector effects responsible.

One of the main advantages of the fiducial cross sections approach is that the results are presented at the generator-level, where the detector effects are not included. To resolve them, one has to account for the possible migrations from the jth reconstruction-level bin to the ith generator-level one and correct for them. This procedure goes under the name of *unfolding* and relies on the use of the detector response matrix $\epsilon_{i,j}^{f}$ to solve an inversion problem of the form:

$$\vec{x}^{f}_{\text{true},i,j} = \left(\epsilon^{f}_{i,j}\right)^{-1}\left(\vec{x}^{f}_{\text{reco},i,j} - \vec{b}\right). \tag{6.12}$$

From a mathematical point of view, the unfolding procedure presents many pitfalls coming from the resolution of the least-squares problem:

$$\chi^2 = \left(\vec{x}_{\text{reco}} - \vec{b} - \mathbf{R}\vec{x}_{\text{true}}\right)^{\text{T}} \Sigma^{-1} \left(\vec{x}_{\text{reco}} - \vec{b} - \mathbf{R}\vec{x}_{\text{true}}\right) \tag{6.13}$$

One can circumvent this issue combining Eqs. (6.11) and (6.7) to perform a *likelihood-based unfolding*, where the signal extraction and the unfolding of detectors effect are performed simultaneously. In the analysis, this is possible thanks to the use of the CMS `combine` statistical package [19], which allows access to the full likelihood information and, therefore, to implement a likelihood-based unfolding. This approach has several advantages with respect to the standard one.

- Equation (6.12) requires that both the generator and reconstruction level distributions are estimated for the signal to be unfolded, after having subtracted background processes. This step is automatically taken into account in a likelihood based approach, where the generator level distribution is inferred starting from the reconstruction level pdf and the detector response matrix.
- Neither Eq. (6.12) nor Eq. (6.13) include the effect of systematic uncertainties, which can be easily included in the likelihood function.
- Systematic uncertainties can be *profiled* in the fit, thus allowing the results to be extracted using Eq. (6.8).

6.5 Systematic Uncertainties

A correct assessment of all the systematic uncertainties is fundamental for the ultimate precision achievable in the analysis. As discussed above, the systematic uncertainties are included in the likelihood function as nuisance parameters represented by the $p(\tilde{\theta} \mid \theta)$ pdfs in Eq. (6.7). One of the most common expressions for the pdf of the systematic errors is the log-normal distribution:

$$p(\tilde{\theta} \mid \theta) = \frac{1}{\sqrt{2\pi}\ln(\kappa)} \exp\left(-\frac{\left(\tilde{\theta} - \theta\right)^2 (\ln(\kappa))^2}{2}\right)\frac{1}{\theta}, \tag{6.14}$$

which solves the continuity problems typical of the truncated Gaussian pdf, while at the same time going at zero at $\theta = 0$. While log-normal uncertainties act on the normalization of a specific process considered in the analysis, some systematic uncertainties can also have an effect on the actual shape of the distributions. This is common for uncertainties associated with the pdfs' analytical expression or the energy scale and resolution corrections. In all these cases, the $p(\tilde{\theta} \mid \theta)$ pdfs are obtained from variations of ± 1 standard deviation of the corresponding nuisance in MC samples. In this analysis the systematic uncertainties are associated to log-normal pdfs, unless specified otherwise.

The systematic uncertainties are divided into experimental and theoretical contributions, as detailed below. Large part of the work of this thesis has been dedicated to thorough studies of the different systematic uncertainties considered, with particular regard to the ones associated to the STXS Stage 1.2 bins, being one of the main novelties introduced in the $H \to ZZ \to 4\ell$ analysis in the context of this thesis.

6.5.1 Experimental Uncertainties

The experimental sources of systematic uncertainties comprise all those effects originating from the detector imperfections, object reconstruction procedures, and the modelling of signal and background processes. The sources of experimental uncertainties considered in the analysis are listed below.

Luminosity measurement: the number of events recorded by the CMS detector is directly proportional to the luminosity (cf. Eq. (2.1)). This quantity is measured separately for each data taking period, with an associated uncertainty that ranges between 2.3 and 2.5% [20–22]. Ultimately, the Run-II integrated luminosity is reported with an uncertainty of 1.8%, implemented in the analysis following the partial correlation scheme recommended centrally by CMS. These nuisance parameters affect all the channels and categories of the analysis, both for signal and background processes.

Lepton reconstruction and selection efficiency: comprising of all the effects from the trigger efficiency to the identification and isolation requirements, these uncertainties range from 1 to 2.3% in the 4μ channel and from 11 to 15.5% in the $4e$ one. The large value for electrons is driven by the worse reconstruction performance at low p_T with respect to muons. In fact, while for the latter a low energy J/Ψ resonance can be exploited, for the former only the di-electron Z resonance can be exploited, resulting in the larger uncertainty on the electrons' reconstruction and selection efficiency.

Lepton momentum and scale resolution: measured from dedicated studies on a di-lepton Z resonance both in data and simulation. The scale contribution is found to have an effect of 0.04 and 0.3% in the 4μ and $4e$ channels, respectively. The resolution uncertainty is 20% in both channels. These two effects can alter the signal line shape and therefore are considered in the analysis as *shape uncertainties*. More precisely, the corresponding nuisance parameters included in the likelihood function are described by a double-sided Crystal Ball (cf. Eq. (6.3)) pdf obtained from $\pm 1\sigma$ variations of the lepton momentum and scale resolution.

Jet energy scale and resolution: these effects are estimated in a similar way to the lepton momentum and scale resolution. Their effect on the actual signal line-shape is found to be negligible, but they can cause possible migrations between event categories. For this reason, they are included in the likelihood function as log-normal uncertainties extracted from the ratio between the nominal yield in each category and the yield obtained from systematic variations of these nuisances. The scale uncertainties range between 1 to 5% at high and low jet p_T, respectively. The effect of the jet energy resolution is found to be within 1 and 2%. Since many categories of the analysis rely on the definition of the jets in the event, the effect of these nuisances is measured in all the analysis categories separately. In most of the cases, these uncertainties are found to have a negligible effect on the final results, while a more significant impact is observed in some of the VH and VBF-like phase space regions, as well as in the measurement of differential fiducial cross sections for jet-related observables.

B-tagging efficiency: studied for all the categories as it may cause migration effects, its impact is found to be negligible in all the measurements. This systematic uncertainty ranges between 1 to 3% at high and low jet p_T, respectively;

Z+X background estimation: this systematic uncertainty is described in detail in Sect. 6.2.2. It is found to be the largest uncertainty entering the analysis, with values ranging between 30 and 45% depending on the category and final state. Nonetheless, it has a marginal impact on all the measurements, as it affects only the normalization of the modest reducible background Z+X.

The determination of these uncertainties is performed independently on the three data taking periods considered. In their combination, a full correlation is considered for the nuisance parameters associated with the reconstruction and selection efficiencies, the lepton momentum and resolution, and the b-tagging efficiency. The luminosity uncertainty comes with a partial correlation scheme across the three years, which can always be written in the form of completely correlated or uncorrelated

Table 6.10 Summary of the experimental systematic uncertainties in the H → 4ℓ measurements of 2016, 2017 and 2018 data

Summary of relative systematic uncertainties			
Common experimental uncertainties			
	2016	2017	2018
Luminosity	2.6%	2.3%	2.5%
Lepton identification/reconstruction efficiencies	1.2–15.5%	1.1–12%	0.7–11%
Jet energy scale and resolution	1–20%	1–20%	1–20%
Signal related uncertainties			
Lepton energy scale	0.04–0.3%	0.04–0.3%	0.04–0.3%
Lepton energy resolution	20%	20%	20%
Background related uncertainties			
Reducible background (Z+X)	31–42%	31–38%	31–37%

contributions, thus facilitating their factorization in the likelihood function. All the other nuisance parameters are treated are uncorrelated. A summary of the systematic experimental uncertainties considered in the analysis is presented in Table 6.10.

6.5.2 *Theoretical Uncertainties*

The systematic uncertainty's theoretical component is associated with the ultimate knowledge on the cross section and phenomenology of a specific physics process. This contribution represents the main hurdle to the final performance of the analysis, due to the severe challenge of improving just by a few percent the determination of these parameters. To assess these aspects, the results of an analysis are often presented with a breakdown of the total uncertainty into its statistical and systematical components, with the latter being further divided into its experimental and theoretical contributions.

The sources of theoretical uncertainties considered in the analysis are detailed below.

Generator parameters: as discussed in detail in the previous chapters, MC simulations are extensively used at different stages of the analysis. Various parameters have to be tuned for the generation of each physical processes: among these are the renormalization (μ_R) and factorization (μ_F) scales and the choice of the PDF set used for the production. The uncertainty on the former is determined from simultaneous variations of the μ_R and μ_F parameters between 0.5 and 2. Their ratio is forced to be between 0.5 and 2 to prevent unphysical effects. The uncertainty on the PDF set is determined by comparing the results obtained from variations of the default NNPDF set [23, 24].

Irreducible background modelling: the k-factors introduced in Sect. 6.2.1 to rescale the irreducible background simulations to the most up-to-date calculations come with associated uncertainty. More specifically, a 10% uncertainty is introduced for the gg $\rightarrow$ ZZ k-factor, while a 0.1% average uncertainty affects the k-factor for the electroweak corrections of the $q\bar{q} \rightarrow ZZ$ process.

Branching fractions: a systematic uncertainty of 2% is associated to the branching ratio of the H $\rightarrow 4\ell$ decay, which is set to it SM value in all the measurements.

Pythia scales: the uncertainty on the modelling of the underlying event and of the hadronization is assessed by varying the initial and final state radiation scales in PYTHIA between 0.25 and 4 times their nominal value. These uncertainties are responsible for migration effects and range between 0.1 and 2% across the different categories.

The STXS uncertainty scheme

As outlined in Sect. 5.2, the main goal of the STXS framework is to define mutually exclusive phase space regions where the experimental sensitivity is maximized and at the same time the dependence of the measurements on the theoretical predictions is reduced. A precise estimate of the theoretical uncertainties associated with the definition of different STXS bin boundaries is needed to account for the possible event migrations around them. With the evolving of the STXS framework and the adoption of the Stage 1.2 scheme, this task was demanded directly to the analyzers. Consistent discussions also involving theoreticians are performed in the context of the LHC Higgs Working Group (LHCHWG) to prevent a complete divergence between the ATLAS and CMS approaches. At the moment of writing this thesis discussions on the topic are still ongoing and while the general principles are in place and well agreed upon, more work is needed to converge on the ultimate uncertainty scheme for the STXS Stage 1.2 measurements.

Two different sets of uncertainties are considered when dealing with the STXS framework: the acceptance uncertainties in the STXS bins and the theoretical uncertainties accounting for migrations around the bin boundaries. The former is equivalent to the QCD uncertainties of μ_R and μ_F mentioned above. In the initial stages of this analysis, inclusive QCD uncertainties were associated with the Higgs boson's different production mechanisms. However, with the analysis targeting the measurement of the STXS Stage 1.2 cross sections, these were dropped in favour of a more accurate set of systematic uncertainties reflecting the actual STXS bins. The QCD acceptance uncertainties are computed by means of scale variations, keeping the inclusive cross section fixed. More precisely, the up- and down-variations[11] are computed for

[11] If the nuisance parameter is referred to as θ, these approximately correspond to $\pm 1\sigma$ intervals around $\hat{\theta}$, the mean value of θ.

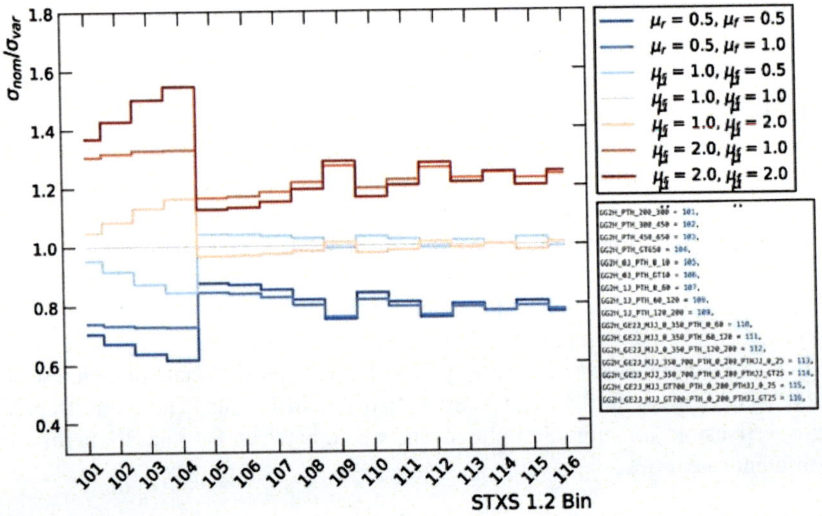

Fig. 6.16 Ratio between the nominal cross section and the one corresponding to up and down variations of the QCD scales μ_R and μ_F for the ggH-like bins of the STXS Stage 1.2 scheme. These values are used as event weights to extract the STXS acceptance uncertainties as described in the text

a choice of the QCD scales of $\mu_R = \mu_F = 2.0$ and $\mu_R = \mu_F = 0.5$, respectively. Hence, the nuisance parameters in each STXS bin are computed by multiplying these weights for the ratio of the nominal cross section in that bin to the cross section at the modified QCD scale:

$$w_i^{up/dn} = \frac{\sigma\left[\text{STXS bin } i, \mu_R = \mu_F = 1.0\right]}{\sigma\left[\text{STXS bin } i, \mu_R = \mu_F = j\right]}, \quad j = 0.5, 2.0 \text{ for } \mathit{up} \text{ and } \mathit{down} \text{ variations.}$$

Eventually, one considers as a systematic uncertainty in the fit the ratio between the number of reconstructed events for nominal and variated QCD scales, therefore taking into account all the experimental acceptance effects related to the STXS categorization. As an example, the values of w_i for the different ggH STXS Stage 1.2 bins are shown in Fig. 6.16.

Therefore the associated nuisance parameters are computed in all the analysis categories as:

$$\text{NP}^{up/dn} = \frac{\sigma_{\text{nominal}}}{\sigma(w_{\text{QCD}}^{up/dn} \times w_i^{up/dn})}, \tag{6.15}$$

where $w_{\text{QCD}}^{up/dn}$ correspond to the μ_R and μ_F weights aforementioned. These uncertainties are introduced in the analysis to cope with the presence of possible migrations among the different reconstructed categories and their effect ranges between 1 to 30%.

The second set of systematic uncertainties associated to the STXS framework is hereafter referred to as theoretical uncertainties (THU) on the migration of events between the different STXS bin boundaries. Conversely from the acceptance uncertainties described above, this set of nuisance parameters is not used in the measurements of the STXS cross sections, but enters only a posteriori interpretations of the results. In the STXS results presented in Chap. 7 these uncertainties are associated to the SM cross section prediction in each bin. The Yellow Report 4 (YR4) [25] presents the computation of these uncertainties at $\sqrt{s}$ =13 TeV for the inclusive H boson production mechanisms, corresponding to the STXS Stage 0. However, values for the more granular Stage 1.2 were not computed at the time of the YR4, and their calculation has been demanded to the analyzers. A nuisance parameter is computed to reflect the possible migration effects around the kinematic cuts that define each STXS bin boundary. Similarly to what is done for the acceptance uncertainties described above, also in this case the renormalization (μ_R) and factorization (μ_F) are varied by a factor 2 and 0.5 with respect to their nominal value, while keeping the inclusive cross section unchanged and the maximal variations are assigned to nuisance parameters. In the case of a bin divided into two sub-bins with nominal cross sections σ_A and σ_B, this approach consists in defining:

$$\begin{cases} \sigma_A^j = \sigma_A + \Delta^j \\ \sigma_B^j = \sigma_B + \Delta^j \end{cases}$$

where the parameter Δ^j defines the jth scale variation, of which the maximal value is taken as the final nuisance parameter Δ, therefore:

$$\begin{cases} \sigma_A^{up/down} = (\sigma_A \pm \Delta)\,/\sigma_A \\ \sigma_B^{up/down} = (\sigma_B \pm \Delta)\,/\sigma_B \end{cases}$$

While in most of the cases the bin boundaries are defined by means of cuts on kinematic observables and therefore standard QCD scale variations can be used, there are some bins where particular attention should be paid and simple scale variations can not be used. This is the case of the phase space regions defined by cuts on the number of jets in the event. However, since these bins remained unchanged from the early stages of the STXS framework, one can exploit the theoretical calculations performed at the time of the Stage 1.0 to define the Δ parameters for the jet-related bins. All the THU uncertainties are implemented as step functions around the corresponding STXS bin boundaries. In the case of multiple bins, the envelope of all the curves is taken as the final nuisance. Figure 6.17 shows an example for the ggH THU systematic uncertainties included in the analysis.

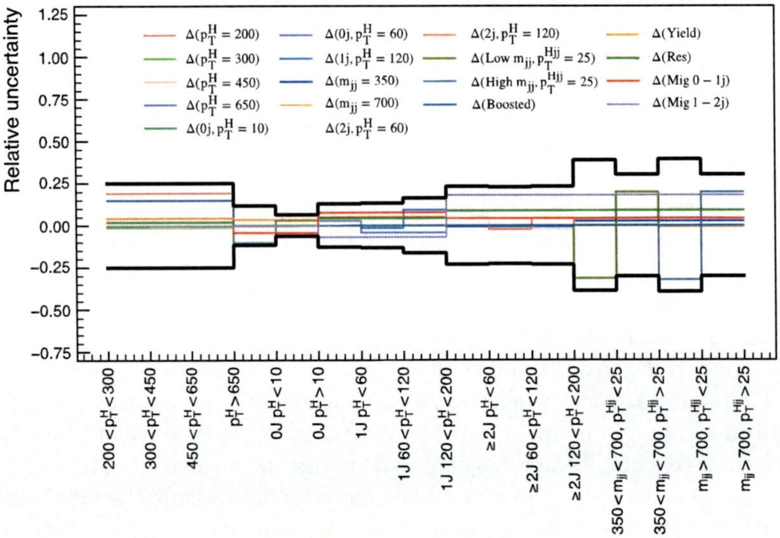

Fig. 6.17 Relative uncertainty for the different nuisance parameters in all the ggH bins of the STXS Stage 1.2. A nuisance parameter Δ is assigned to all the cut values defining a bin boundary, as shown in Fig. 5.6, following the procedure detailed in the text

More precisely, a total of ten nuisance parameters (NPs) associated with the ggH STXS bins is included in the analysis. Some of the NPs shown in Fig. 6.17 are merged to reflect the STXS Stage 1.2 merging scheme presented in Sect. 5.2. These NPs account for the uncertainties in the inclusive cross section predictions in the phase space regions defined by cuts on the number of jets in the event: these comprise both the migration between the 0 and 1-jet bin boundary and the one around the 1 and $\geq$2 jets bins. The uncertainty associated with the prediction of the ggH phase space regions with VBF topology are included in the analysis for both the 2 and $\geq$3 jets cases. Additional uncertainties are included to cope with the migration around the p_T^H bin boundaries, specifically around the 10, 60, and 120 GeV cut values. One additional NP is considered for the ggH-like bins to cope with the uncertainty on the p_T^H distribution due to missing higher-order finite top quark mass corrections. Some of the NPs shown in Fig. 6.17 are related to cuts on the m_{jj} or p_T^{Hjj} observables: these are not explicitly considered in the analysis as the corresponding true STXS bins are not defined due to the lack of statistics. Nevertheless, their effect is well covered by the VBF-topology NPs described above. All the THU uncertainties not considered in the analysis because of the merging scheme employed automatically become acceptance uncertainties, to account for possible migrations in the merged categories.

A set of THU NPs is defined also for the STXS bins associated to the VBF and VH production mechanisms to cope with the uncertainties on the p_T^H, m_{jj}, and p_T^{Hjj} distributions. Six NPs are introduced to model the uncertainties around the m_{jj} bin boundaries at 60, 120, 350, 700, 1000, and 1500 GeV, and two for the migrations around $p_T^H = 200$ GeV and $p_T^{Hjj} = 25$ GeV. Two additional nuisances are included for the treatment of migrations in the jet phase space regions, similarly to what is done for the ggH bins.

To conclude on this topic is worth stressing the profound difference between the two sets of STXS-related systematic uncertainties. While the acceptance uncertainties quantify the possible effects of migrations among the different categories of the analysis, the THU set of NPs addresses the uncertainty on the theoretical predictions associated with the definition of the STXS bins at the generator-level. This conceptual difference is reflected in the way these uncertainties enter the analysis: when signal strength modifiers are measured, both the sets of systematics are included; on the other hand, when dealing with the STXS cross sections measurements, the THU uncertainties only enter the interpretation stage and are associated to the SM prediction of the cross section in each bin during the likelihood fit.

6.5.3 Impacts of the Systematic Uncertainties on the Results

As mentioned above, one of the most compelling aspects of many LHC analyses is the ultimate precision being limited by the systematic component of the uncertainty, as for the $H \rightarrow 4\ell$ and ggH signal strength modifiers presented in the next Chapter. Therefore it is crucial to quantify the *impact* of each nuisance parameter on the measurements performed in an analysis.

The *impact* of a nuisance parameter on the POI of the analysis is assessed quantifying the shift induced on the POI by a variation of a NP by ± 1 standard deviation ($\pm 1\sigma$) with respect to its post-fit value. More precisely, the numerator of the profiled likelihood test statistics defined in Eq. (6.8) is modified by freezing a given NP to its $\pm 1\sigma$ variation, i.e. $\hat{\theta}_\mu^{\pm 1\sigma}$, while leaving all the other NPs profiled ($\hat{\theta}_\mu$) in the fit.

The relative impact of the dominant sources of systematic uncertainties on the inclusive signal strength modifier and on the STXS Stage 0 cross sections is presented in Fig. 6.18. The final precision of the results is dominated by the systematic uncertainties on the lepton reconstruction and selection efficiency, as well as on the experimental uncertainty on the measurement of the luminosity. For what concerns the systematic theoretical uncertainties, the precision on the knowledge of the SM cross sections and the modelling of the hadronization and the underlying events are found to have the largest contribution.

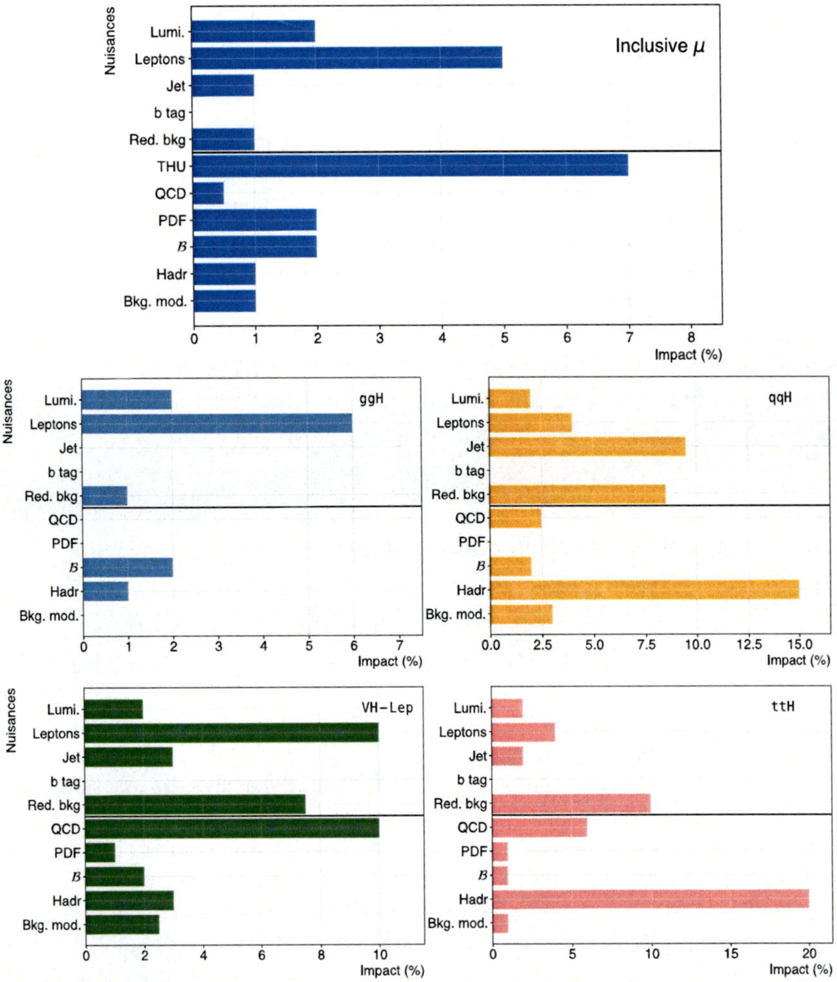

Fig. 6.18 The relative impact of the dominant sources of systematic uncertainties on the inclusive signal strength modifier μ and on the STXS Stage 0 cross sections presented in Chap. 7. All the impacts are shown in percent (rounded to the nearest 0.5%) and are combined assuming no correlation among the different NPs. The dominant experimental sources of uncertainty are found to be the integrated luminosity uncertainty (Lumi.), the lepton reconstruction and selection efficiency, leptons scale and resolution (Leptons), jet energy scale and resolution (Jet), b-tagging efficiency (B-tag), and reducible background estimation uncertainty (Red. bkg). The dominant sources of theoretical uncertainty are the ggH, VBF, and VH cross section theoretical predictions(THU), the renormalization and factorization scales (QCD), the choice of the PDF set (PDF), the branching fraction of $H \to 4\ell$ ($\mathcal{B}$), the modeling of hadronization and the underlying event (Hadr), and the background modeling (Bkg. mod.)

References

1. Grazzini M, Kallweit S, Rathlev D (2015) ZZ production at the LHC: fiducial cross sections and distributions in NNLO QCD. Phys Lett B 750:407–410
2. Bierweiler A, Kasprzik T, Kühn JH (2013) Vector-boson pair production at the LHC to $\mathcal{O}(\alpha^3)$ accuracy. JHEP 12:071
3. Campbell JM (2020) MCFM-9.1 manual a Monte Carlo for femtobarn processes at hadron colliders
4. Bonvini M, Caola F, Forte S, Melnikov K, Ridolfi G (2013) Signal-background interference effects for $gg \to H \to W^+W^-$ beyond leading order. Phys Rev D 88(3):034032
5. Passarino G (2014) Higgs CAT. Eur Phys J C 74(5)
6. Melnikov K, Dowling M (2015) Production of two z-bosons in gluon fusion in the heavy top quark approximation. Phys Lett B 744:43–47
7. Li CS, Li HT, Shao DY, Wang J (2015) Soft gluon resummation in the signal-background interference process of $gg(\to h^\star) \to ZZ$. JHEP 08:065
8. Grazzini M, Sargsyan H (2013) Heavy-quark mass effects in Higgs boson production at the LHC. J High Energy Phys 2013(9)
9. Nason P (2004) A new method for combining NLO QCD with shower Monte Carlo algorithms. JHEP 11:040
10. Frixione S, Nason P, Oleari C (2007) Matching NLO QCD computations with parton shower simulations: the POWHEG method. JHEP 11:070
11. Alioli S, Nason P, Oleari C, Re E (2010) A general framework for implementing NLO calculations in shower Monte Carlo programs: the POWHEG BOX. JHEP 06:043
12. Sjöstrand T, Ask S, Christiansen JR, Corke R, Desai N, Ilten P, Mrenna S, Prestel S, Rasmussen CO, Skands PZ (2015) An introduction to PYTHIA 8.2. Comput Phys Commun 191:159–177
13. Alwall J, Frederix R, Frixione S, Hirschi V, Maltoni F, Mattelaer O, Shao HS, Stelzer T, Torrielli P, Zaro M (2014) The automated computation of tree-level and next-to-leading order differential cross sections, and their matching to parton shower simulations. JHEP 07:079
14. Gritsan AV, Roskes J, Sarica U, Schulze M, Xiao M, Zhou Y (2020) New features in the JHU generator framework: constraining Higgs boson properties from on-shell and off-shell production. Phys Rev D 102(5):056022
15. Ball RD, Bertone V, Carrazza S, Deans CS, Del Debbio L, Forte S, Guffanti A, Hartland NP, Latorre JI, Rojo J, Ubiali M (2015) Parton distributions for the LHC run II. J High Energy Phys 2015(4)
16. Cranmer K (2014) Practical statistics for the LHC. In 2011 European school of high-energy physics
17. Wilks SS (1938) The large-sample distribution of the likelihood ratio for testing composite hypotheses. Ann Math Stat 9(1):60–62
18. Cowan G, Cranmer K, Gross E, Vitells O (2011) Asymptotic formulae for likelihood-based tests of new physics. Eur Phys J C 71:1554 [Erratum: Eur Phys J C 73:2501 (2013)]
19. CMS Collaboration. Manual of combine tool, https://cms-analysis.github.io/HiggsAnalysis-CombinedLimit
20. CMS Collaboration (2017) CMS luminosity measurements for the 2016 data taking period. Technical report CMS-PAS-LUM-17-001, CERN, Geneva
21. CMS Collaboration (2018) CMS luminosity measurement for the 2017 data-taking period at $\sqrt{s} = 13$ TeV. Technical report CMS-PAS-LUM-17-004, CERN, Geneva
22. CMS Collaboration (2019) CMS luminosity measurement for the 2018 data-taking period at $\sqrt{s} = 13$ TeV. Technical report CMS-PAS-LUM-18-002, CERN, Geneva
23. Botje et al M (2011) The PDF4LHC working group interim recommendations. CERN CDS, 1
24. Alekhin et al S (2011) The PDF4LHC working group interim report. 1
25. de Florian et al D (2016) Handbook of LHC Higgs cross sections: 4. deciphering the nature of the Higgs sector. 2/2017, 10 2016

Chapter 7
The Higgs Boson Properties in the "Golden Channel"

This "golden journey" comes to an end in this chapter, where the results of the analysis are presented. However, before unveiling the outcome of the measurements of the H boson properties in the $H \to 4\ell$ channel, it is worth taking a step back to have a global overview of the picture assembled throughout the previous chapters.

The information from the various CMS sub-detectors is combined in a Particle-Flow (PF) based reconstruction of the different particles originating from each collision. This analysis relies mainly on two kinds of *PF objects* (Chap. 4): leptons and jets, selected based on dedicated identification and isolation criteria. The latter is used to enhance the discrimination power against the different production modes of the H boson, while the former is disposed into opposite sign lepton pairs, required to satisfy the selection criteria detailed in Sect. 5.1 to be associated with the ZZ candidates eventually considered for the analysis. All the selected events are classified into seven mutually exclusive categories, designed to maximize the sensitivity of the analysis to the different production mechanisms of the H boson. For this purpose, the information on the number of jets in the event and a set of dedicated matrix element discriminants are used. To probe fine-granular phase space regions of the $H \to 4\ell$ decay, a subsequent categorization step is performed to produce analysis categories that closely match the Simplified Template Cross Sections (STXS) framework Stage 1.2 bins. Hence, a total of twenty-two categories is defined (Chap. 5) starting from the seven initial ones. A probability density function (pdf) is built for both signal and background events using the corresponding simulated samples and from the parametrization of the $m_{4\ell}$ reconstructed distribution with dedicated analytical shapes as described in Sects. 6.3 and 6.2, respectively. These pdfs are combined with matrix element discriminants sensitive to the decay information to create a 2D template for each category and final state (4μ, $4e$, and $2e2\mu$) (Chap. 6). This information is used, along with the set of nuisance parameters defined in Sect. 6.5, to build the statistical model of the analysis. The properties of the H boson in the 4ℓ decay channel are

M. Bonanomi, *Response of the High Granularity Calorimeter HGCAL and Characterisation of the Higgs Boson*, Springer Theses,
https://doi.org/10.1007/978-3-031-26833-5_7

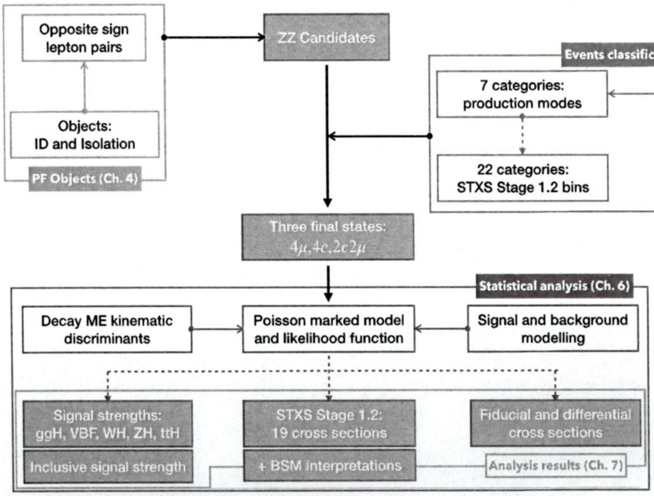

Fig. 7.1 Schematic representation of the $H \to ZZ \to 4\ell$ analysis workflow: opposite sign lepton pairs are build starting from the PF objects after the application of dedicated isolation and identification requirements (Chap. 4). The ZZ candidates are divided into seven mutually exclusive categories designed to target the main production modes. A further split of the categories is performed, for a total of 22 categories designed to closely match the true STXS Stage 1.2 bins (Chap. 5). Dedicated parametrization of signal and background processes are derived and used, along with a dedicated set of matrix element discriminants, to build the statistical model of the analysis (Chap. 6). A comprehensive characterization of the H boson properties is obtained measuring signal strength modifiers, fiducial and differential cross sections, and cross sections in the STXS framework by means of a maximum likelihood fit. Additional BSM interpretations such as constraints of the H boson anomalous couplings to vector bosons and fermions, and a measurement of the H boson self-coupling are also performed (This chapter)

measured using a maximum likelihood fit and this chapter presents the corresponding findings. The "map for the golden search" is shown in Fig. 7.1, where a schematic representation of the analysis workflow is given.

7.1 Events and Discriminants

The first step of the analysis consists of identifying all the ZZ candidates starting from the set of all the opposite sign lepton pairs selected. The distribution of the four-lepton invariant mass reconstructed in data is shown Fig. 7.2, along with the signal and background expectations extracted from simulation. A good agreement between data and simulation is observed throughout the entire $m_{4\ell}$ range.

A good statistical agreement between the events selected in the data and the simulation is also observed in the $m_{4\ell}$ distributions for the three final states considered in the analysis (4μ, 4e, and 2e2μ), as shown in Fig. 7.3.

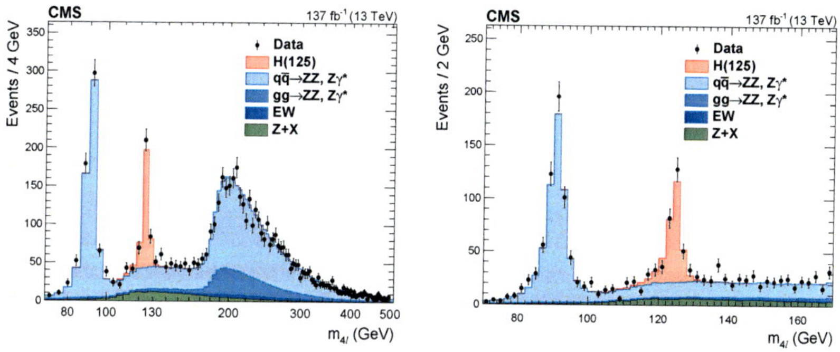

Fig. 7.2 Distribution of the reconstructed four-lepton invariant mass, $m_{4\ell}$, up to 500 GeV with 4 GeV bin size (left) and in the low-mass range, $m_{4\ell} < 170$ GeV, with 2 GeV bin size (right) for an integrated luminosity of 137 fb^{-1}. The points with error bars represent the observed data, while the stacked histograms correspond to the signal and background processes. The expected yields of the different H boson production mechanisms with $m_{\rm H} = 125$ GeV, denoted as H(125), and those of the ZZ and rare electroweak backgrounds are normalized to the SM expectations. The irreducible Z+X background yield is normalized to the estimate from the data

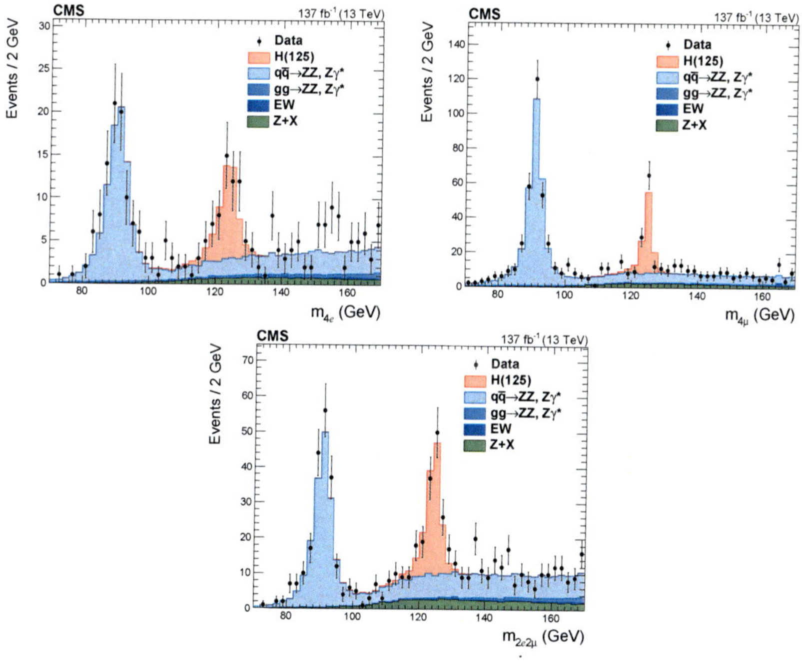

Fig. 7.3 The reconstructed four-lepton invariant mass distribution is shown separately for the three final states considered in the analysis, for an integrated luminosity of 137 fb^{-1}: 4e (upper left), 4μ (upper right), and 2e2μ (lower). The points with error bars represent the observed data, while the stacked histograms correspond to the signal and background processes. The expected yields of the different H boson production mechanisms with $m_{\rm H} = 125$ GeV, denoted as H(125), and those of the ZZ and rare electroweak backgrounds are normalized to the SM expectations. The irreducible Z+X background yield is normalized to the estimate from the data

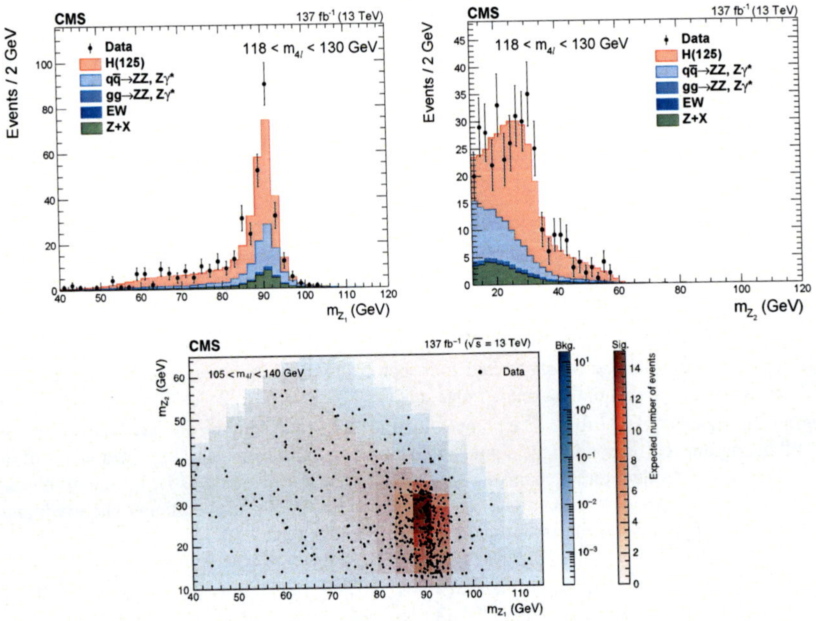

Fig. 7.4 *Top*: Distributions of the reconstructed masses of the Z_1 (left) and Z_2 (right) candidates in the $118 < m_{4\ell} < 130\,\text{GeV}$ mass region for an integrated luminosity of $137\ \text{fb}^{-1}$. The points with error bars represent the observed data, while the stacked histograms correspond to the signal and background processes. The expected yields of the different H boson production mechanisms with $m_H = 125\,\text{GeV}$, denoted as H(125), and those of the ZZ and rare electroweak backgrounds are normalized to the SM expectations. The irreducible Z+X background yield is normalized to the estimate from the data. *Bottom*: 2D distribution of m_{Z_1} and m_{Z_2} in the $105 < m_{4\ell} < 140\,\text{GeV}$ mass region for an integrated luminosity of $137\ \text{fb}^{-1}$. The black points represent the data. The blue and red color scales indicate the expected total number of background and SM H boson events for $m_H = 125\,\text{GeV}$, respectively

As mentioned in Sect. 5.1.1, the $H \to ZZ \to 4\ell$ decay is completely determined by eight degrees of freedom, two of which are the masses of the Z_1 and Z_2 candidates in the event. Hence, the soundness of the event selection can be corroborated by the good agreement between the data and MC distributions of Z_1 and Z_2 reconstructed masses presented in Fig. 7.4.

In the next step of the analysis, selected events are classified into seven mutually exclusive categories using a set of matrix element discriminants sensitive to the production of the H boson. The distributions of the $\mathcal{D}_{2\text{jet}}^{\text{VBF}}$, $\mathcal{D}_{1\text{jet}}^{\text{VBF}}$, $\mathcal{D}_{2\text{jet}}^{\text{VH}} = \max(\mathcal{D}_{2\text{jet}}^{\text{WH}}, \mathcal{D}_{2\text{jet}}^{\text{ZH}})$ discriminants (cf. Chap. 5) are shown in Fig. 7.5, for the events passing the selection criteria of the analysis, in the $118 < m_{4\ell} < 130$ GeV mass range around H boson peak.

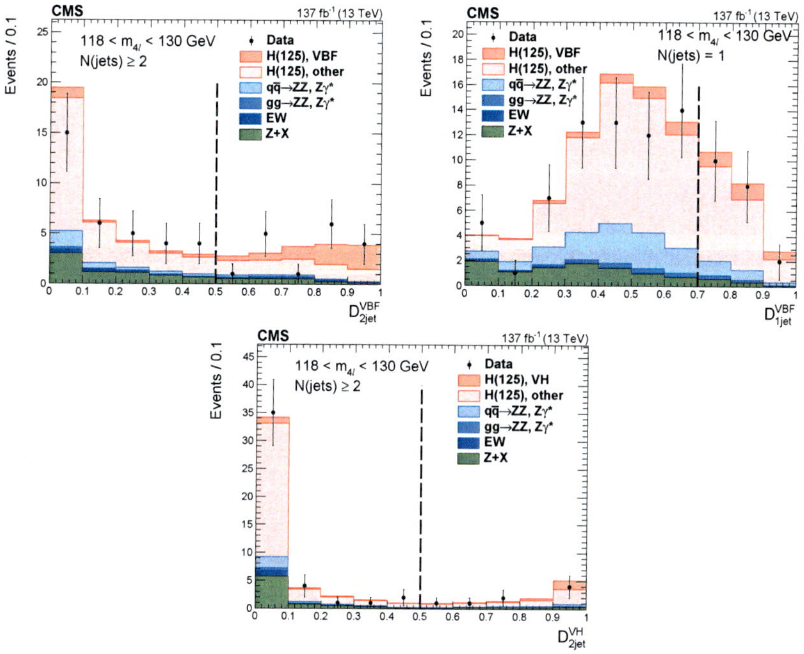

Fig. 7.5 Distributions of the categorization discriminants $\mathcal{D}_{2\text{jet}}^{\text{VBF}}$ (upper left), $\mathcal{D}_{1\text{jet}}^{\text{VBF}}$ (upper right), $\mathcal{D}_{2\text{jet}}^{\text{VH}}$ (lower) = max($\mathcal{D}_{2\text{jet}}^{\text{WH}}$,$\mathcal{D}_{2\text{jet}}^{\text{ZH}}$) in the mass region $118 < m_{4\ell} < 130\,\text{GeV}$ for an integrated luminosity of 137 fb^{-1}. The points with error bars represent the observed data, while the stacked histograms correspond to the signal and background processes. The expected yields of the different H boson production mechanisms with $m_{\text{H}} = 125\,\text{GeV}$, denoted as H(125), and those of the ZZ and rare electroweak backgrounds are normalized to the SM expectations. The irreducible Z+X background yield is normalized to the estimate from the data. The vertical dashed lines denote the operative working points used for the events categorization. For all the discriminants the SM H boson signal is shown for the production mode targeted and the other mechanisms separately

These matrix element discriminants are designed to enclose the full information of the event and to enhance the discrimination power between signal and background processes, as extensively discussed in Sect. 5.3. The seven categories sensitive to the main production mechanisms of the H boson are further split into a total of twenty-two categories, defined to closely match the merged STXS Stage 1.2 bins considered in the analysis. The expected signal and background yields at 137 fb^{-1} and the corresponding number of observed events in data, in the $105 < m_{4\ell} < 140$ GeV mass range, are reported in Table 7.1 for all the 22 categories, while Fig. 7.6 gives a visual representation of the expected composition of each category.

The results of the analysis are extracted by means of a maximum likelihood fit, performed simultaneously across all the event categories reconstructed and the three final states and data taking periods considered. As detailed in Sect. 6.4, the likelihood function employed for the statistical analysis comprises two terms: the

Table 7.1 Expected number of signal and background events and number of observed candidates after full analysis selection in the mass range $105 < m_{4\ell} < 140$ GeV and for an integrated luminosity of 137 fb^{-1}. The expected and observed yields are given for each production mechanism in the 22 categories considered for the analysis. The uncertainties listed are statistical only. The expected yields are estimated from MC simulation for the signal processes at $m_H = 125$ GeV, as well as for ZZ and rare electroweak backgrounds. The Z+X yields are estimated from data

Reconstructed event	Signal							Background				Expected		Observed
Category	ggH	VBF	WH	ZH	$t\bar{t}H$	$b\bar{b}H$	tH	$q\bar{q} \to ZZ$	$gg \to ZZ$	EW	Z+X	Signal	Total	
Untagged-0j-$p_T^{4\ell}$ [0, 10]	27.7	0.09	0.03	0.03	0.00	0.15	0.00	71.5	3.06	0.01	3.21	27.9 ± 0.1	106 ± 0	114
Untagged-0j-$p_T^{4\ell}$ [10, 200]	96.2	1.69	0.60	0.77	0.01	1.01	0.00	98.1	11.6	0.35	37.8	100 ± 0	248 ± 1	278
Untagged-1j-$p_T^{4\ell}$ [0, 60]	26.8	1.51	0.56	0.48	0.01	0.45	0.01	25.3	3.02	0.64	14.2	29.8 ± 0.1	72.9 ± 0.4	74
Untagged-1j-$p_T^{4\ell}$ [60, 120]	13.5	1.31	0.51	0.41	0.02	0.11	0.01	7.81	0.82	0.62	7.95	15.9 ± 0.1	33.1 ± 0.3	20
Untagged-1j-$p_T^{4\ell}$ [120, 200]	3.51	0.60	0.17	0.17	0.01	0.02	0.00	1.15	0.19	0.25	1.63	4.48 ± 0.05	7.69 ± 0.16	11
Untagged-2j-$p_T^{4\ell}$ [0, 60]	3.45	0.29	0.15	0.14	0.08	0.09	0.02	2.14	0.32	0.63	4.75	4.20 ± 0.06	12.1 ± 0.2	14
Untagged-2j-$p_T^{4\ell}$ [60, 120]	5.26	0.56	0.24	0.19	0.12	0.04	0.03	2.19	0.30	0.72	4.14	6.43 ± 0.06	13.8 ± 0.2	15
Untagged-2j-$p_T^{4\ell}$ [120, 200]	3.07	0.40	0.16	0.13	0.07	0.01	0.02	0.75	0.14	0.34	1.19	3.86 ± 0.05	6.28 ± 0.14	7
Untagged-$p_T^{4\ell} > 200$	2.79	0.62	0.21	0.17	0.07	0.01	0.02	0.43	0.21	0.21	0.73	3.89 ± 0.04	5.47 ± 0.11	3
Untagged-2j-$m_{jj} > 350$	0.77	0.16	0.06	0.04	0.05	0.01	0.01	0.34	0.06	0.31	1.71	1.12 ± 0.02	3.54 ± 0.14	3
VBF-1jet-tagged	15.5	3.29	0.22	0.16	0.00	0.13	0.01	6.85	1.53	0.20	2.44	19.3 ± 0.1	30.3 ± 0.2	27
VBF-2jet-tagged-m_{jj}[350, 700]	0.83	1.19	0.01	0.01	0.00	0.01	0.00	0.19	0.07	0.11	0.14	2.05 ± 0.03	2.55 ± 0.05	2
VBF-2jet-tagged-$m_{jj} > 700$	0.43	1.96	0.00	0.00	0.00	0.00	0.00	0.07	0.05	0.12	0.03	2.40 ± 0.02	2.67 ± 0.03	1
VBF-3jet-tagged-$m_{jj} > 350$	2.52	2.35	0.06	0.06	0.03	0.03	0.05	0.62	0.21	0.64	2.43	5.11 ± 0.05	9.01 ± 0.17	12
VBF-2jet-tagged-$p_T^{4\ell} > 200$	0.44	0.79	0.01	0.01	0.01	0.00	0.01	0.03	0.03	0.04	0.06	1.26 ± 0.02	1.42 ± 0.03	0
VBF-rest	2.48	0.94	0.13	0.09	0.04	0.04	0.01	0.98	0.20	0.39	2.18	3.74 ± 0.05	7.49 ± 0.17	5
VH-hadronic-tagged-m_{jj}[0, 150]	4.11	0.25	1.09	0.96	0.13	0.06	0.02	1.69	0.22	0.52	2.93	6.62 ± 0.06	12.0 ± 0.2	12
VH-rest	0.57	0.03	0.09	0.06	0.03	0.01	0.00	0.16	0.02	0.06	0.33	0.79 ± 0.02	1.36 ± 0.06	0
VH-leptonic-tagged-$p_T^{4\ell}$ [0, 150]	0.33	0.04	0.85	0.26	0.10	0.03	0.03	2.16	0.36	0.19	1.11	1.64 ± 0.02	5.47 ± 0.13	10
VH-leptonic-tagged-$p_T^{4\ell} > 150$	0.02	0.01	0.21	0.06	0.04	0.00	0.01	0.05	0.01	0.03	0.08	0.35 ± 0.01	0.52 ± 0.03	0
$t\bar{t}H$-leptonic-tagged	0.02	0.01	0.02	0.02	0.68	0.00	0.03	0.08	0.01	0.23	0.21	0.79 ± 0.01	1.32 ± 0.07	0
$t\bar{t}H$-hadronic-tagged	0.18	0.05	0.03	0.05	0.86	0.01	0.03	0.03	0.01	0.82	1.06	1.22 ± 0.01	3.15 ± 0.14	2

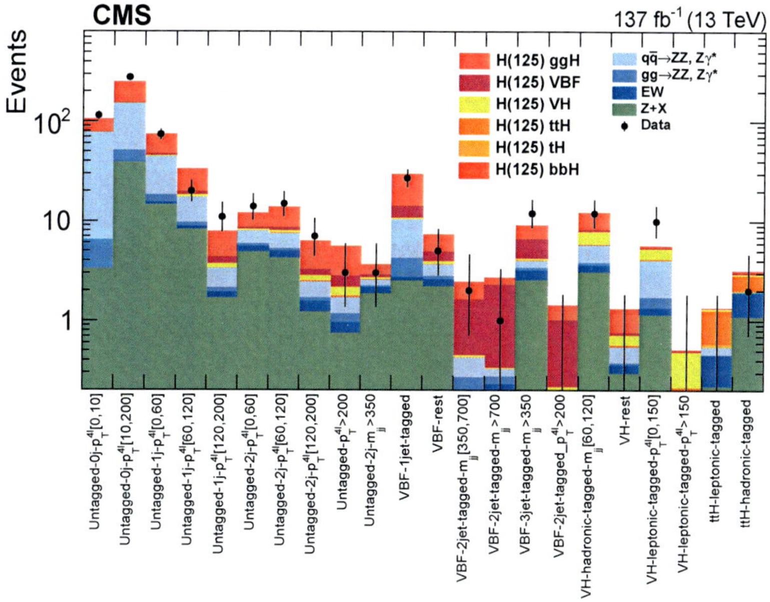

Fig. 7.6 Expected and observed number of events in the 22 reconstructed event categories considered for the analysis. The yields are extracted in the mass region $105 < m_{4\ell} < 140\,\text{GeV}$ for an integrated luminosity of 137 fb^{-1}. Points with error bars represent the data and stacked histograms represent the expected numbers of the signal and background events. The yields of the different H boson production mechanisms with $m_\text{H} = 125\,\text{GeV}$, denoted as H(125), and those of the ZZ and rare electroweak backgrounds are normalized to the SM expectations. The irreducible Z+X background yield is normalized to the estimate from the data

first one proportional to the analytical parametrization of the $m_{4\ell}$ distribution in each category, and the second one defined as a conditional template of $\mathcal{D}$ as a function of $m_{4\ell}$. This second term is introduced to enhance the sensitivity of the analysis exploiting the strong correlation between the decay matrix element discriminants and the reconstructed $m_{4\ell}$ for signal and background processes. The distribution of the $\mathcal{D}_\text{bkg}^\text{kin}$, $\mathcal{D}_\text{bkg}^\text{VBF+dec}$ and $\mathcal{D}_\text{bkg}^\text{VH+dec}$ discriminants, used to construct the $\mathcal{P}(\mathcal{D}|m_{4\ell})$ term in Eq. (6.9), are shown in Fig. 7.7, while Fig. 7.8 illustrates the correlation between the kinematic discriminants and the reconstructed invariant mass of the four-lepton system.

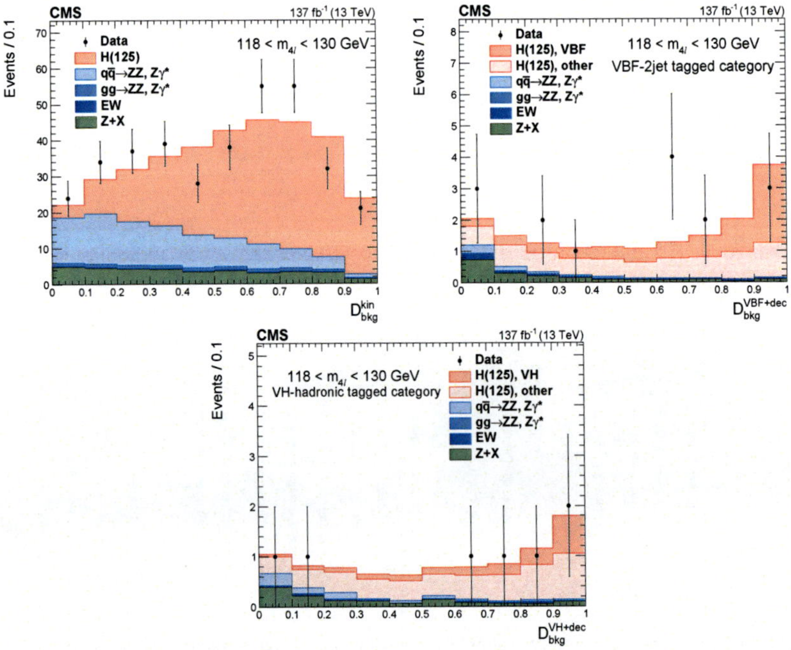

Fig. 7.7 Distributions of the $\mathcal{D}_{\text{bkg}}^{\text{kin}}$ (upper left), $\mathcal{D}_{\text{bkg}}^{\text{VBF+dec}}$ (upper right), $\mathcal{D}_{\text{bkg}}^{\text{VH+dec}}$ (lower) kinematic discriminants in the mass region $118 < m_{4\ell} < 130\,\text{GeV}$ for an integrated luminosity of 137 fb^{-1}. The points with error bars represent the observed data, while the stacked histograms correspond to the signal and background processes. The expected yields of the different H boson production mechanisms with $m_{\text{H}} = 125\,\text{GeV}$, denoted as H(125), and those of the ZZ and rare electroweak backgrounds are normalized to the SM expectations. The irreducible Z+X background yield is normalized to the estimate from the data. In the $\mathcal{D}_{\text{bkg}}^{\text{VBF+dec}}$ and $\mathcal{D}_{\text{bkg}}^{\text{VH+dec}}$ distributions the SM H boson signal is separated into two components to highlight the production mode which is targeted by the specific discriminant

7.2 Measurement of the H Boson Properties in the Four-Lepton Final State

7.2.1 Signal Strength Modifier

The flexibility of the statistical model defined for the analysis (cf. Eq. (6.10)) gives access to a large plethora of results, which ensure a complete characterization of the H boson properties in the 4ℓ final state.

All the measurements presented in the following are performed by means of a simultaneous fit across all the analysis categories, following the strategy described in Sect. 6.4. With slight modifications of Eq. (6.10) one can measure:

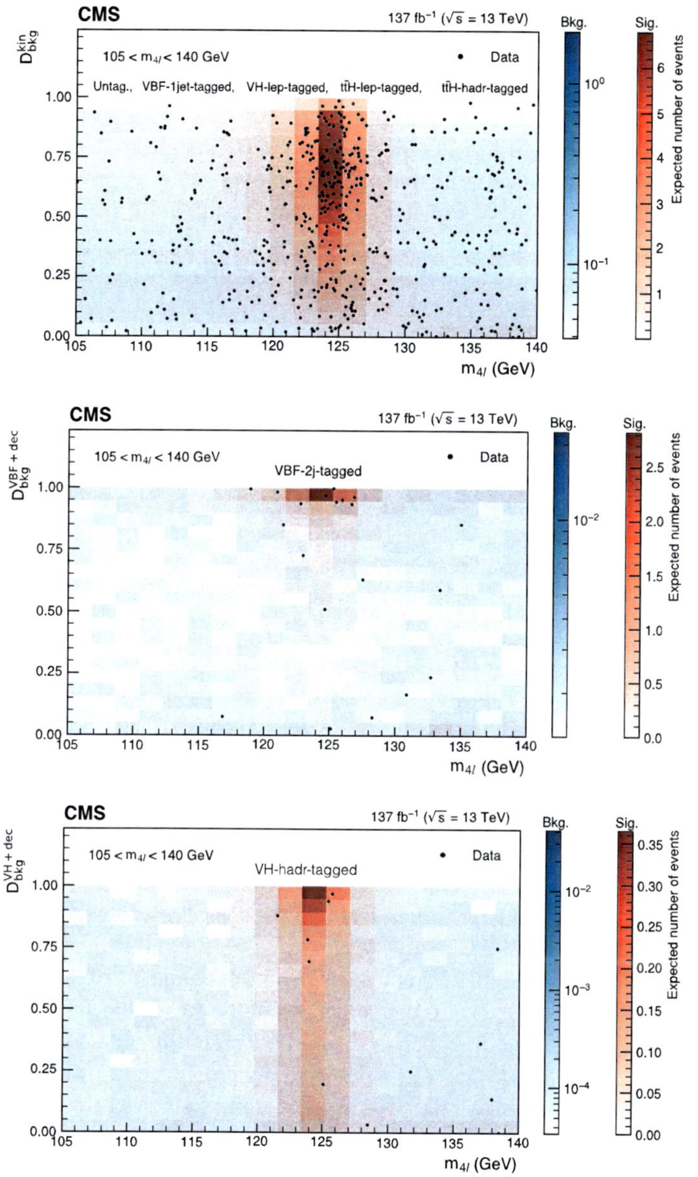

Fig. 7.8 Distributions of the $\mathcal{D}_{\text{bkg}}^{\text{kin}}$ (upper), $\mathcal{D}_{\text{bkg}}^{\text{VBF+dec}}$ (middle) and $\mathcal{D}_{\text{bkg}}^{\text{VH+dec}}$ (lower) kinematic discriminants as a function of $m_{4\ell}$ shown in the mass region $105 < m_{4\ell} < 140$ GeV for an integrated luminosity of 137 fb^{-1}. The black points represent the data from the categories listed in the legend. The blue and red color scales indicate the expected total number of background and SM H boson events for $m_{\text{H}} = 125$ GeV, respectively

- Inclusive signal strength modifier, defined as the ratio between the observed cross section in the $H \rightarrow 4\ell$ decay channel and the corresponding SM prediction (cf. Table 6.8). This is achieved by merging the different μ_i of Eq. (6.10) together into a single scaling parameter, thus obtaining the same expression as of Eq. (6.5) with $\mathcal{L} \propto (\mu S \cdot f_s(x_i) + B \cdot f_b(x_i))$;
- Signal strength modifiers for the five main production mechanism of the H boson, each defined as a scaling parameter to the expected yield of the corresponding signal process. In this approach the 19 POIs of the analysis are merged together according to which production mode they are sensitive to. Therefore one gets a simplified version of Eq. (6.10) where:

$$\mathcal{L} \propto \Big(\mu_{\mathrm{ggH,b\bar{b}H}} \cdot (s_{\mathrm{ggH,b\bar{b}H}}) + \mu_{\mathrm{VBF}} \cdot (s_{\mathrm{VBF}}) + \\ + \mu_{\mathrm{WH}} \cdot (s_{\mathrm{WH}}) + \mu_{\mathrm{ZH}} \cdot (s_{\mathrm{ZH}}) + \mu_{\mathrm{t\bar{t}H,tH}} \cdot (s_{\mathrm{t\bar{t}H,tH}}) + b\Big)$$

All the fits are performed at the fixed value of the H boson mass of $m_{\mathrm{H}} = 125.38$ GeV, corresponding to the most precise measurement to date, obtained by the CMS Collaboration from the combination of the HZZ and $\mathrm{H}\gamma\gamma$ decay channels [1].

The inclusive signal strength modifier is measured to be:

$$\mu = 0.94^{+0.12}_{-0.11} = 0.94^{+0.07}_{-0.07}\,(\mathrm{stat})^{+0.09}_{-0.08}\,(\mathrm{syst}) = 0.94^{+0.07}_{-0.07}\,(\mathrm{stat})^{+0.07}_{-0.06}\,(\mathrm{theo})^{+0.06}_{-0.05}\,(\mathrm{exp}).$$

The systematic contribution of the uncertainty is obtained by subtracting in quadrature the total error and the one obtained from the fit with all the nuisance parameters frozen to their best fit value. The additional breakdown of the systematics into experimental and theoretical components is computed similarly, from the subtraction in quadrature of the systematic error and the one corresponding to a fit where all the theoretical uncertainties are frozen.[1] The profile likelihood ratio as a function of μ is shown in Fig. 7.9 for the observed data and the expected scan for a SM Higgs boson derived from an Asimov dataset with $\mu = 1.0$. It is compelling to observe the relative proportion of the statistical and systematical components of the uncertainty, which equally contribute to the ultimate precision of the measurement. However, the inclusive signal strength modifier per-se is somehow reductive, as it does not capture features of specific phase space regions but it simply provides a global insight of the agreement between the data and the SM predictions. Nevertheless, this result clearly highlights the turning point of many LHC analyses: with the large statistics available at the end of Run-II, many analysis are entering the *precision physics realm*, with the ultimate sensitivity limited by the systematic component of the uncertainty.

1

$$\sigma_{\mathrm{syst}} = \sqrt{\sigma^2_{\mathrm{tot}} - \sigma^2_{\mathrm{syst\,frozen}}}$$

$$\sigma_{\mathrm{theory}} = \sqrt{\sigma^2_{\mathrm{tot}} - \sigma^2_{\mathrm{theory\,frozen}}} \quad \sigma_{\mathrm{exp}} = \sqrt{\sigma^2_{\mathrm{theory\,frozen}} - \sigma^2_{\mathrm{stat}}}.$$

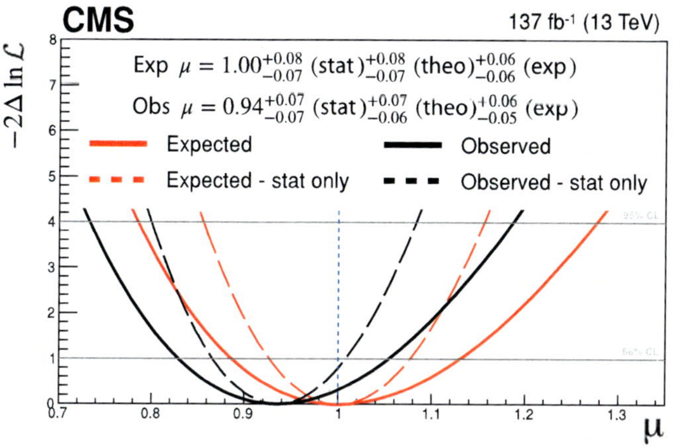

Fig. 7.9 Profile likelihood ratio scans as a function of μ for the observed data (black) and the expected scan for a SM Higgs boson derived from an Asimov dataset with $\mu = 1.0$ (red). The results are obtained for a fixed value of $m_{\mathrm{H}} = 125.38$ GeV. The scans are shown with (solid lines) and without (dashed lines) the inclusion of the systematic contribution of the uncertainties

A more detailed characterization of the H boson properties in the 4ℓ decay channel can be achieved from the measurement of the signal strength modifiers of the main production mechanisms, presented in the top plot of Fig. 7.10. It is worth mentioning that with the large statistics available for the analysis, the WH and ZH signal strength modifiers could be measured separately for the first time in the $\mathrm{H} \to 4\ell$ decay channel. The limited sensitivity of the analysis to these two production modes results in quite loose precision and, especially for the ZH case, only exclusion upper limits, rather than confidence intervals, can be set. These two contributions can be merged together to measure the inclusive VH signal strength modifier, as shown in the bottom plot of Fig. 7.10. The WH and ZH combined statistics results in a more accurate limit on μ_{VH}, where the best fit value is driven by the excess observed in the μ_{WH} measurement.

The numerical values of the different μ_i, along with the corresponding statistical and systematic uncertainties, are reported in Table 7.2. The ggH signal strength modifier $\mu_{\mathrm{ggH,b\bar{b}H}}$ is measured with competitive statistical and systematic contributions to the total uncertainty, pointing towards the latter being the limiting factor for the ultimate precision of Legacy[2] analyses in the near future. The second point worth raising comes from the observation of the systematic component of the μ_{ZH}, μ_{WH}, and $\mu_{\mathrm{t\bar{t}H,tH}}$ signal strengths measurements: its value is reported to be 0.0 as the POIs of the analysis are all lower bounded at $\mu = 0$, in order to prevent from unphysical cases where the signal to background ratio would become negative. Hence, these results are rather to be interpreted as exclusion upper limits on the signal strength modifiers of these three processes.

[2] The full Run-II analyses are often referred to as Legacy analyses.

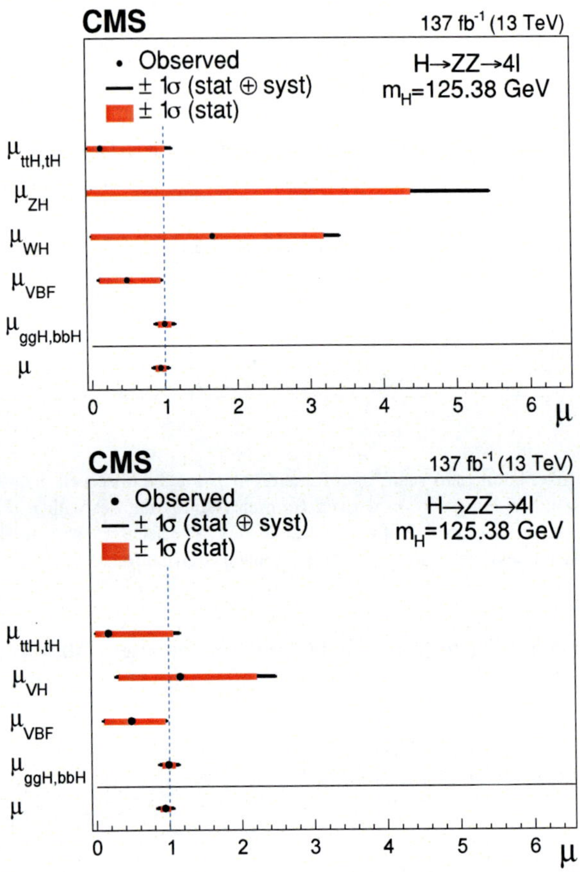

Fig. 7.10 *Top*: Results of likelihood scans for the signal strength modifiers corresponding to the five main SM H boson production mechanisms at $m_{\mathrm{H}} = 125.38\,\mathrm{GeV}$, compared to the SM prediction shown as a vertical dashed line. *Bottom*: WH and ZH production mechanisms are merged together and their inclusive contribution is measured with a single signal strength modifier μ_{VH}. The black lines indicate the one standard deviation confidence intervals including both statistical and systematic sources, while the thick red lines indicate the statistical component of the uncertainties

The presence of an excess in the WH signal strength modifier and of the ZH being fitted at 0.0 is consistent with the value observed for the inclusive VH signal strength modifier:

$$\mu_{\mathrm{VH}}^{\mathrm{obs}} = 1.15^{+1.06}_{-0.84}\,(\mathrm{stat})^{+0.79}_{-0.26}\,(\mathrm{syst})$$

which is found to be in agreement with the SM prediction of:

$$\mu_{\mathrm{VH}}^{\mathrm{exp}} = 1.00^{+1.05}_{-0.80}\,(\mathrm{stat})^{+0.88}_{-0.35}\,(\mathrm{syst}).$$

Table 7.2 Measured signal strength modifiers for the main production mechanisms of the H boson at $m_H = 125.38$ GeV. The ± 1 standard deviation uncertainties are reported with a breakdown of the statistical and systematic contributions

	Expected	Observed
$\mu_{t\bar{t}H,tH}$	$1.00^{+1.23}_{-0.77}$ (stat)$^{+0.51}_{-0.06}$ (syst)	$0.17^{+0.88}_{-0.17}$ (stat)$^{+0.42}_{-0.00}$ (syst)
μ_{WH}	$1.00^{+1.83}_{-1.00}$ (stat)$^{+0.75}_{-0.00}$ (syst)	$1.66^{+1.52}_{-1.66}$ (stat)$^{+0.85}_{-0.00}$ (syst)
μ_{ZH}	$1.00^{+4.79}_{-1.00}$ (stat)$^{+6.76}_{-0.00}$ (syst)	$0.00^{+4.38}_{-0.00}$ (stat)$^{+3.24}_{-0.00}$ (syst)
μ_{VBF}	$1.00^{+0.53}_{-0.44}$ (stat)$^{+0.18}_{-0.12}$ (syst)	$0.48^{+0.46}_{-0.37}$ (stat)$^{+0.14}_{-0.10}$ (syst)
$\mu_{ggH,b\bar{b}H}$	1.00 ± 0.10 (stat)$^{+0.12}_{-0.10}$ (syst)	0.99 ± 0.09 (stat)$^{+0.11}_{-0.09}$ (syst)
μ	$1.00^{+0.08}_{-0.07}$ (stat)$^{+0.10}_{-0.08}$ (syst)	0.94 ± 0.07 (stat)$^{+0.09}_{-0.08}$ (syst)

In order to probe the structure of the fermion and vector-boson induced contributions to the SM expected cross sections, a third set of signal strength modifiers is measured, merging together all the production mechanisms sensitive to either the fermion or vector-boson couplings of the H boson:

$$\mathcal{L} \propto \left(\mu_f \cdot (s_{ggH,t\bar{t}H,tH,b\bar{b}H}) + \mu_V \cdot (s_{VBF,VH}) + b\right),$$

where, $\mu_f \equiv \mu_{ggH,\,t\bar{t}H,b\bar{b}H,tH}$ and $\mu_V \equiv \mu_{VBF,VH}$. The results are extracted with a two-dimensional scan over a grid of μ_f and μ_V values, leading to:

$$\mu_f = 0.96^{+0.14}_{-0.12},\ \ \mu_V = 0.82^{+0.36}_{-0.31}$$

while the corresponding expected values are found to be:

$$\mu_f = 1.00\ ^{+0.15}_{-0.13},\ \ \mu_V = 1.00\ ^{+0.39}_{-0.33}$$

Figure 7.11 depicts the contour plot showing the 68 and 95% CL exclusion regions in the (μ_f, μ_V) plane. Similarly to all the other results presented above, no tension with the SM predictions is observed.

Dependence of the Results on the H Boson Mass

As mentioned above, all the results are obtained with a fixed value of the H boson mass of $m_H = 125.38$ GeV, corresponding to the most precise value measured to date for this observable. Hence, all the expected cross sections and branching ratios that enter the analysis are extrapolated at $m_H = 125.38$ GeV using the Yellow Report 4 [2] calculations as input. To study the dependence of the results on the actual value of

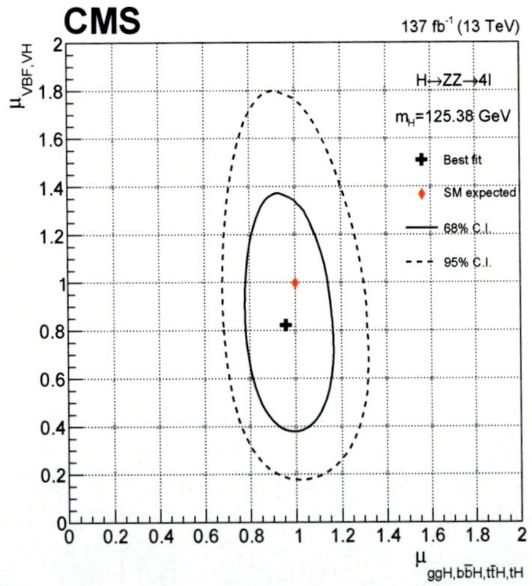

Fig. 7.11 Measurement of the signal strength modifiers for the fermion ($\mu_{\rm f} \equiv \mu_{\rm ggH,\,t\bar{t}H,b\bar{b}H,tH}$) and vector-boson ($\mu_{\rm V} \equiv \mu_{\rm VBF,VH}$) induced contributions to the SM expected cross sections. The solid and dashed lines show the 2D contours of the profile likelihood scan at 68% and 95% CL, respectively. The black cross indicates the best fit value, while the diamond represents the expected value for a SM Higgs boson

$m_{\rm H}$, the measurements of the signal strength modifiers presented above are repeated, leaving this parameter *profiled* in the fits.

Figure 7.12 shows the impact plots for the measurement of the inclusive $\rm H \to 4\ell$ signal strength modifier when $m_{\rm H}$ is fixed at 125.38 GeV (top) and when it is profiled in the fit (bottom). The ultimate precision of the measurement is limited by the same systematic uncertainties in the two cases. More precisely, the nuisance parameters with the largest impact on the results are found to be the ones on the ggH cross section, on the lepton identification efficiencies, and luminosity measurement in both cases. The unconstrained value of $m_{\rm H}$ is not found to pull the results significantly and it yields to a difference of at most 4% across all the signal strength modifiers with respect to the results obtained fixing it to 125.38 GeV. For example, the inclusive signal strength modifier is found to be:

$$\mu_{m_{\rm H}\ \rm fixed} = 0.94^{+0.12}_{-0.11}, \quad \mu_{m_{\rm H}\ \rm prof.} = 0.97^{+0.12}_{-0.11},$$

and the profiled value of the H boson mass is found to be $m_{\rm H} = 125.09^{+0.15}_{-0.14}\,(\rm stat)$ GeV. It is worth stressing that a precise determination of $m_{\rm H}$, along with a detailed estimate of the associated systematic uncertainties, goes beyond the scope of this analysis. Nevertheless, it is compelling to observe that the statistical uncertainty on the profiled value of $m_{\rm H}$ agrees with the current best measurement by the CMS Collaboration [1], reflecting that the $\rm H \to ZZ \to 4\ell$ decay channel drives the final precision of the dedicated $m_{\rm H}$ measurement.

A comparison between the signal strength modifiers for the five main production mechanisms of the H boson in the two scenarios considered in this study is presented in Fig. 7.13. All the results obtained with $m_{\rm H}$ unconstrained in the fits are found to be in agreement with the corresponding signal strengths obtained for $m_{\rm H} = 125.38$ GeV.

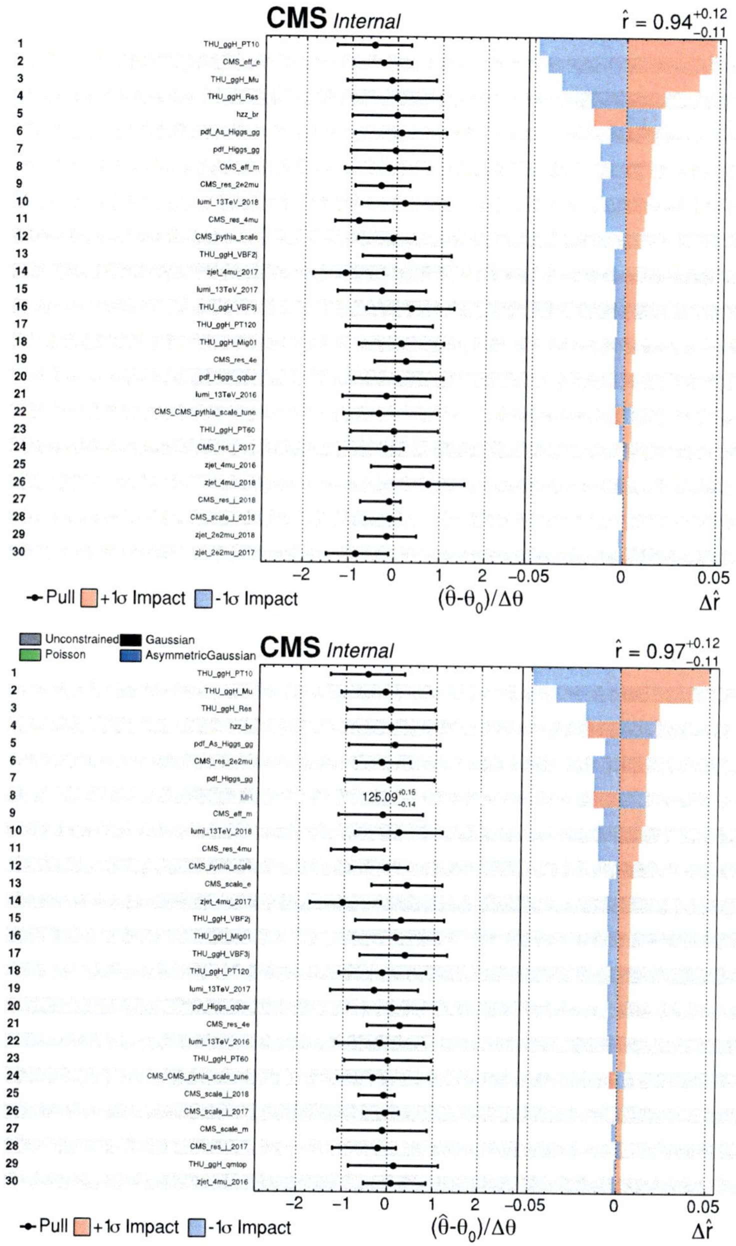

Fig. 7.12 Difference between the pre-fit and post-fit values of the nuisance parameters $\left(\left(\hat{\theta}-\theta_0\right)/\Delta\theta\right)$ considered in the analysis and their impact on the inclusive signal strength modifier μ ($\hat{r}$ in the plots) at a fixed value of $m_{\mathrm{H}} = 125.38$ GeV (top) and leaving m_{H} profiled in the fits (bottom)

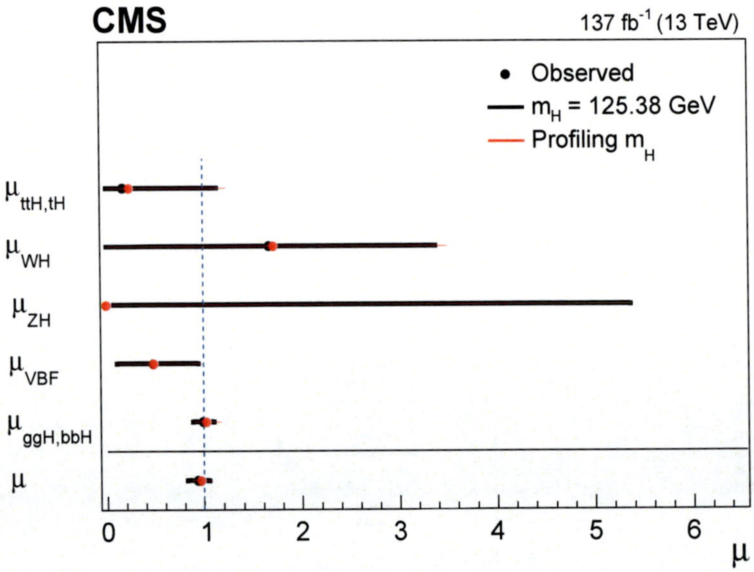

Fig. 7.13 Results of likelihood scans for the signal strength modifiers corresponding to the five main SM H boson production mechanisms measured at a fixed value of $m_{\mathrm{H}} = 125.38$ GeV (black lines) compared with the results obtained leaving m_{H} profiled in the fits (red lines). A discrepancy of no more than 4% is observed for all the measurements

Discrepancies of the order of 3% and 4% are observed in the central values of the WH and $\mathrm{t\bar{t}H}$ signal strengths, respectively, while the relative precision of the two measurement approaches is found to be equivalent. All the other signal strength measured are found to have a sub-percent dependence on m_{H}, as shown in the Figure.

7.2.2 *Simplified Template Cross Section*

The likelihood function of Eq. (6.7) is designed to extract the 19 cross sections of the merged STXS Stage 1.2 bins. As discussed in Sect. 6.5, these measurements are performed in the same way as the signal strength modifiers, with a slightly different treatment of some systematic uncertainties. More precisely, the set of THU NPs does not enter the measurement of the STXS cross sections. As discussed in Sect. 5.2, the goal of the STXS framework is to maximize the experimental sensitivity in each bin, while minimizing the dependence on the theory predictions. That is why THU NPs are associated with the SM prediction of the cross sections in each bin measured and can enter the analysis in a subsequent stage of interpretation of the results. For this reason it is important to quote the cross sections measured in the different STXS bins and also their correlations, to allow a posteriori re-interpretations of the measurements in physics scenarios different from the SM.

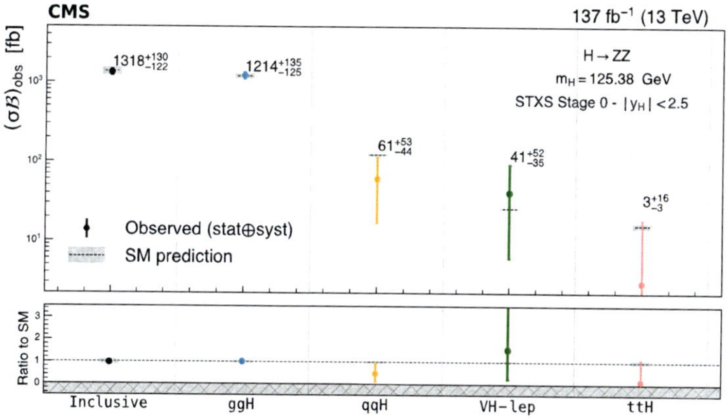

Fig. 7.14 The points with error bar represent the STXS Stage 0 measured cross sections $(\sigma\mathcal{B})_{\text{obs}}$, while the black dashed lines with gray uncertainty bands correspond to the SM predictions $(\sigma\mathcal{B})_{\text{SM}}$ at $m_{\text{H}} = 125.38$ GeV. The bottom panel shows the ratios of these two quantities, along with the corresponding uncertainty in each of the bins

As discussed in Sect. 5.2, two Stages of the STXS framework are measured in the context of the $\text{H} \rightarrow 4\ell$ analysis presented here: the Stage 0 and the Stage 1.2. The former targets the main production mechanisms of the SM H boson, while the latter allows to probe in fine-granular manner the phase space associated to each of them. The observed values of the product between the H boson cross section and the $H \rightarrow ZZ$ branching fraction ($(\sigma\mathcal{B})_{\text{obs}}$) and the corresponding SM expectations ($(\sigma\mathcal{B})_{\text{SM}}$) are presented in Figs. 7.14 and 7.15 for the STXS Stage 0 and the merged Stage 1.2, respectively. The corresponding numerical values are reported in Tables 7.3 and 7.4, and the correlation matrices for the different cross sections are presented in Fig. 7.16. Similarly to what observed for the signal strength modifiers, the dominant sources of experimental systematic uncertainties are the luminosity and the lepton efficiencies, while on the theoretical side the measurements are dominated by the NPs associated to category migration in the ggH sub-categories, which drive most of the sensitivity of the analysis.

7.2.3 *Fiducial Cross Section*

The measurement of cross sections in a fiducial volume designed to closely match the experimental acceptance provides a complementary approach to the STXS results presented in the previous section. More precisely, fiducial cross sections can be measured in differential bins of kinematic observables sensitive to the production or the decay side of the physical process under study, thus providing additional information for its complete characterization. The fiducial and differential measurements are designed to maximize the model independence of the results, and the price to pay is the fact that they remain agnostic to the production mechanisms of the H boson.

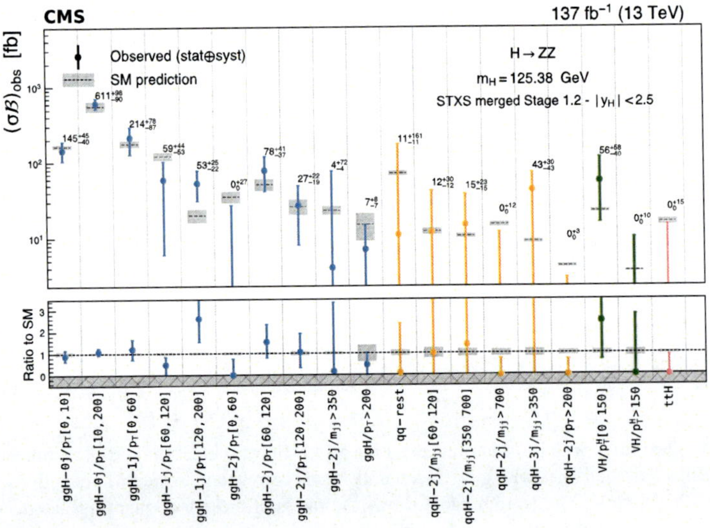

Fig. 7.15 The points with error bar represent the STXS merged Stage 1.2 measured cross sections $(\sigma\mathcal{B})_{\text{obs}}$, while the black dashed lines with gray uncertainty bands correspond to the SM predictions $(\sigma\mathcal{B})_{\text{SM}}$ at $m_{\text{H}} = 125.38$ GeV. The bottom panel shows the ratios of these two quantities, along with the corresponding uncertainty in each of the bins

Table 7.3 The STXS Stage 0 measured cross sections ($(\sigma\mathcal{B})_{\text{obs}}$) and the corresponding SM predictions ($(\sigma\mathcal{B})_{\text{SM}}$) are presented in the first two columns, along with their ± 1 standard deviation uncertainties. The last column shows the ratio between these two values. All the results are extracted at $m_{\text{H}} = 125.38$ GeV for $\text{H} \to \text{ZZ}$ decay

	$(\sigma\mathcal{B})_{\text{obs}}$ (fb)	$(\sigma\mathcal{B})_{\text{SM}}$ (fb)	$(\sigma\mathcal{B})_{\text{obs}}/(\sigma\mathcal{B})_{\text{SM}}$
ttH	3^{+16}_{-3}	15.9 ± 1.4	$0.16^{+0.98}_{-0.16}$
VH-lep	41^{+52}_{-35}	25.9 ± 0.8	$1.56^{+1.99}_{-1.34}$
qqH	61^{+53}_{-44}	122 ± 6	$0.50^{+0.44}_{-0.36}$
ggH	1214^{+135}_{-125}	1192 ± 95	$1.02^{+0.11}_{-0.10}$
Inclusive	1318^{+130}_{-122}	1369 ± 164	$0.96^{+0.10}_{-0.09}$

For all these reasons, in the analysis both STXS and fiducial cross sections are measured, thus providing a complete and sound characterization of the properties of the H boson in the $\text{H} \to 4\ell$ decay channel.

The measurement of fiducial and differential cross sections is performed by means of a one-dimensional maximum likelihood fit to the reconstructed 4ℓ invariant mass distribution ($N^{f,i}_{\text{exp}}(m_{4\ell})$) defined as of Eq. (6.11).

The values of the reconstruction efficiency (ε), as well as the number of reconstructed events considered in the signal region but not coming from the fiducial volume ($f^{f,i}_{\text{nonfid}}$) are computed for each production mechanism using dedicated

Table 7.4 The STXS merged Stage 1.2 measured cross sections ($(\sigma\mathcal{B})_{obs}$) and the corresponding SM predictions ($(\sigma\mathcal{B})_{SM}$) are presented in the first two columns, along with their ± 1 standard deviation uncertainties. The last column shows the ratio between these two values. All the results are extracted at $m_H = 125.38$ GeV for $H \to ZZ$ decay

	$(\sigma\mathcal{B})_{obs}$ (fb)	$(\sigma\mathcal{B})_{SM}$ (fb)	$(\sigma\mathcal{B})_{obs}/(\sigma\mathcal{B})_{SM}$
ggH-0j/p_T[0, 10]	145^{+45}_{-40}	164 ± 11	$0.89^{+0.28}_{-0.24}$
ggH-0j/p_T[10, 200]	611^{+98}_{-90}	561 ± 87	$1.09^{+0.17}_{-0.16}$
ggH-1j/p_T[0, 60]	214^{+78}_{-87}	177 ± 18	$1.21^{+0.44}_{-0.49}$
ggH-1j/p_T[60, 120]	59^{+44}_{-53}	121 ± 14	$0.48^{+0.37}_{-0.44}$
ggH-1j/p_T[120, 200]	53^{+25}_{-22}	20 ± 4	$2.62^{+1.24}_{-1.08}$
ggH-2j/p_T[0, 60]	0^{+27}_{-0}	35 ± 6	$0.00^{+0.76}_{-0.00}$
ggH-2j/p_T[60, 120]	78^{+41}_{-37}	51 ± 9	$1.53^{+0.81}_{-0.73}$
ggH-2j/p_T[120, 200]	27^{+22}_{-19}	26 ± 6	$1.06^{+0.87}_{-0.72}$
ggH-2j/$m_{jj} > 350$	4^{+72}_{-4}	23 ± 3	$0.17^{+3.2}_{-0.17}$
ggH/$p_T > 200$	7^{+8}_{-7}	15 ± 6	$0.47^{+0.56}_{-0.47}$
qqH-rest	11^{+161}_{-11}	71 ± 5	$0.15^{+2.27}_{-0.15}$
qqH-2j/m_{jj}[60, 120]	12^{+30}_{-12}	12.1 ± 1.2	$1.01^{+2.45}_{-1.01}$
qqH-2j/m_{jj}[350, 700]	15^{+23}_{-15}	10.5 ± 0.7	$1.41^{+2.21}_{-1.41}$
qqH-2j/$m_{jj} > 700$	0^{+12}_{-0}	15 ± 1	$0.00^{+0.77}_{-0.00}$
qqH-3j/$m_{jj} > 350$	43^{+30}_{-43}	8.9 ± 0.5	$4.84^{+3.38}_{-4.84}$
qqH-2j/$p_T > 200$	0^{+3}_{-0}	4.2 ± 0.2	$0.00^{+0.72}_{-0.00}$
VH-lep/p_T^H[0, 150]	56^{+58}_{-40}	22.3 ± 1.1	$2.49^{+2.60}_{-1.79}$
VH-lep/$p_T^H > 150$	0^{+10}_{-0}	3.6 ± 0.1	$0.00^{+2.79}_{-0.00}$
ttH	0^{+15}_{-0}	15.9 ± 1.4	$0.00^{+0.91}_{-0.00}$

simulation samples and they are presented in Table 7.5, where also the fraction of signal of signal events within the fiducial volume ($\mathcal{A}_{fid}$) is shown.

The $\mathcal{A}_{fid}$ and $(1 + f_{nonfid})\varepsilon$ columns highlight two fundamental features typical of fiducial cross section measurements. First of all, the number of events falling within the fiducial volume has a strong dependence on the H boson production mechanism: the values of $\mathcal{A}_{fid}$ for the ggH and VBF processes are up to 30% larger than the corresponding ones for the associated productions with vector bosons (VH) or top-quark

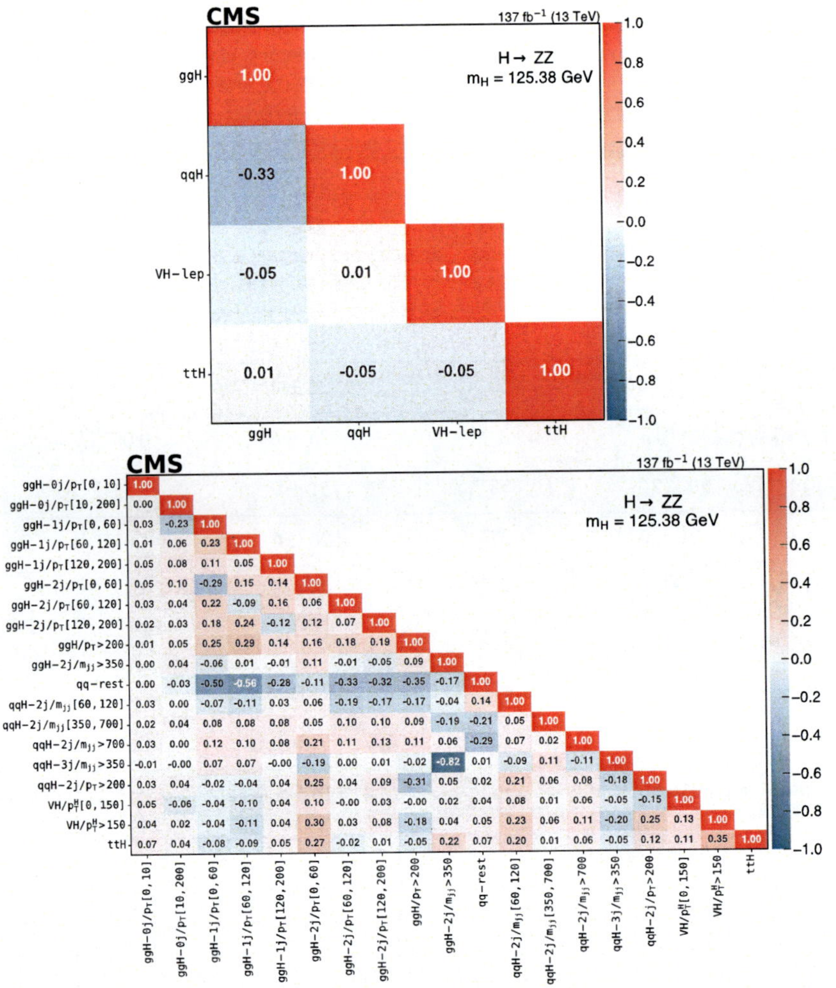

Fig. 7.16 Correlation matrices for the STXS Stage 0 (top) and merged Stage 1.2 (bottom) cross sections measured in the $H \to ZZ$ decay channel

pairs ($t\bar{t}H$). This effect is not surprising, as the fiducial volume is defined to closely match the events' selection of the analysis and to reduce the model dependence of the results. As a consequence, the acceptance of the events originating from associated vector bosons or top-quark pairs is lower, reflecting the possible presence leptons in the final states not produced by the H boson decay. The second feature worth highlighting is the actual model independence of the measurements. As discussed in Sect. 6.4, the value of $(1 + f_{\mathrm{nonfid}})\varepsilon$ regulates the contribution of each production mechanism to the fiducial cross section and is expected to agree on the different processes. As observed in Table 7.5, this is the case for most of the production modes but

Table 7.5 For each signal process (rows) the table shows the fraction of signal events within the fiducial phase space (acceptance $\mathcal{A}_{\text{fid}}$), the reconstruction efficiency (ε) for signal events in the fiducial phase space, as well as the ratio of the number of reconstructed events outside the fiducial phase space to that of the reconstructed events in the fiducial phase space (f_{nonfid}). All the values are reported at $m_{\text{H}} = 125$ GeV. The last column shows the model-dependence factor $(1 + f_{\text{nonfid}})\varepsilon$ which regulates the signal yield for a given fiducial cross section, as shown in Eq. (6.11). All the uncertainties listed are statistical only

Signal process	$\mathcal{A}_{\text{fid}}$	ε	f_{nonfid}	$(1 + f_{\text{nonfid}})\varepsilon$
ggH (POWHEG)	0.402 ± 0.001	0.598 ± 0.002	0.054 ± 0.001	0.631 ± 0.002
VBF	0.445 ± 0.002	0.615 ± 0.002	0.043 ± 0.001	0.641 ± 0.003
WH	0.329 ± 0.002	0.604 ± 0.003	0.078 ± 0.002	0.651 ± 0.004
ZH	0.340 ± 0.003	0.613 ± 0.005	0.082 ± 0.004	0.663 ± 0.006
$\text{t}\bar{\text{t}}\text{H}$	0.315 ± 0.004	0.588 ± 0.007	0.181 ± 0.009	0.694 ± 0.010

$\text{t}\bar{\text{t}}\text{H}$, where the low statistics available leads to a larger estimate of this contribution. Nevertheless, this process has a negligible contribution to the $\text{H} \rightarrow \text{ZZ} \rightarrow 4\ell$ decay (cf. Table 7.1), therefore the larger value of $(1 + f_{\text{nonfid}})\varepsilon$ has only a marginal impact on the final results.

The $\text{H} \rightarrow 4\ell$ integrated fiducial cross section measured in the analysis is presented in Fig. 7.17 as a function of the three final states (4e, 4μ, and 2e2μ, left) and of the centre-of-mass energy $\sqrt{s}$ (right). The dashed green and blue areas in the right plot of the figure correspond to the SM predictions at $\sqrt{s} = 7$, 8 TeV and $\sqrt{s} = 13$ TeV, respectively. A different color code has been chosen to avoid confusion between the two different set of results, obtained using slightly different approaches and MC generators. The former was computed at a fixed mass of $m_{\text{H}} = 125.00$ GeV using the fiducial volume definition of Ref. [3], while the latter was extrapolated at $m_{\text{H}} = 125.38$ GeV within the fiducial phase space defined as of Table 6.9. In addition, the acceptance is calculated using POWHEG at $\sqrt{s} = 13$ TeV and HRES [2, 4] at $\sqrt{s} = 7$ and 8 TeV.

The observed fiducial cross section for the $\text{pp} \rightarrow \text{H} \rightarrow 4\ell$ process is:

$$\sigma_{\text{fid}}^{\text{obs}} = 2.84^{+0.23}_{-0.22}\,(\text{stat})^{+0.26}_{-0.21}\,(\text{syst})\,\text{fb}$$

in agreement with the SM expectation of:

$$\sigma_{\text{fid}}^{\text{exp}} = 2.84 \pm 0.15\,\text{fb}$$

The integrated fiducial cross sections for each data taking period and for the full Run-II are presented in Table 7.6, along with the breakdown of the uncertainties into their statistical and systematic components. Similar to what observed for the inclusive and ggH signal strength modifiers, also these results point towards the fact that the analysis is entering the *precision physics era*, where the measurements start to be dominated by the systematic uncertainties, especially in their theoretical component.

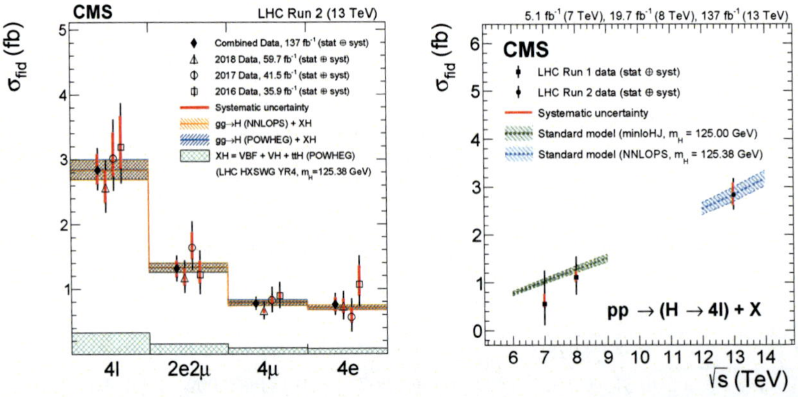

Fig. 7.17 *Left*: Inclusive fiducial cross section of the pp $\rightarrow$ H $\rightarrow 4\ell$ process and of the three final states ($4e, 4\mu, 2e2\mu$) considered in the analysis. *Right*: Inclusive fiducial cross section of the pp $\rightarrow$ H $\rightarrow 4\ell$ as a function of $\sqrt{s}$. The SM prediction for the region $\sqrt{s} = 6$–9 TeV uses the fiducial volume definition of Ref. [3] and the corresponding cross sections are computed at $m_{\rm H} = 125.0$ GeV, while for $\sqrt{s} = 12$–14 TeV the definition described in the text is used and SM predictions and measurements are calculated at $m_{\rm H} = 125.38$ GeV. The acceptance is calculated using POWHEG at $\sqrt{s} = 13$ TeV and HRES [2, 4] at $\sqrt{s} = 7$ and 8 TeV, and the total gluon fusion cross section and uncertainty are taken from Ref. [5]

Table 7.6 Inclusive fiducial cross section measured for the pp $\rightarrow$ H $\rightarrow 4\ell$ process at $m_{\rm H} = 125.38$ GeV and corresponding ± 1 standard deviation uncertainties. The results are shown separately for the three final states (first three columns) and data taking periods (first three rows) considered in the analysis. The last row shows the inclusive fiducial cross section measured for the Run-II of the LHC. The last column shows the inclusive 4ℓ measurement, for which the statistical and systematic uncertainties are given separately

	$2e2\mu$ (fb)	4μ (fb)	4e (fb)	Inclusive (fb)
2016	$1.22^{+0.38}_{-0.30}$	$0.89^{+0.22}_{-0.19}$	$1.07^{+0.44}_{-0.33}$	$3.19^{+0.68}_{-0.56} = 3.19^{+0.48}_{-0.45}\,(\text{stat})^{+0.48}_{-0.33}\,(\text{syst})$
2017	$1.64^{+0.41}_{-0.35}$	$0.82^{+0.21}_{-0.18}$	$0.56^{+0.29}_{-0.22}$	$3.01^{+0.60}_{-0.50} = 3.01^{+0.44}_{-0.41}\,(\text{stat})^{+0.41}_{-0.27}\,(\text{syst})$
2018	$1.17^{+0.27}_{-0.24}$	$0.66^{+0.15}_{-0.13}$	$0.73^{+0.24}_{-0.20}$	$2.57^{+0.42}_{-0.38} = 2.57^{+0.33}_{-0.31}\,(\text{stat})^{+0.27}_{-0.23}\,(\text{syst})$
2016–2018	$1.31^{+0.20}_{-0.19}$	$0.78^{+0.10}_{-0.10}$	$0.76^{+0.18}_{-0.16}$	$2.84^{+0.34}_{-0.31} = 2.84^{+0.23}_{-0.22}\,(\text{stat})^{+0.26}_{-0.21}\,(\text{syst})$

One of the complementary aspects of fiducial measurements with respect to the STXS approach is the possibility to measure cross sections in differential bins of particular kinematic observables. This aspect is compelling for two reasons: all the measurements are *unfolded* at generator level. Hence, significant discrepancies with respect to the SM predictions would hint to possible BSM physics and could not be attributed to detector effects; in addition, BSM effects are expected to be more pronounced in the tails of the distributions, and differential measurements are ideal for probing these regions as a function of different kinematic observables.

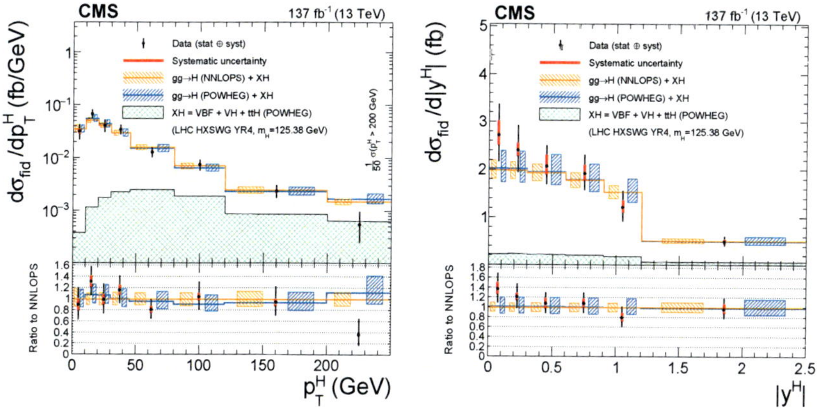

Fig. 7.18 Differential fiducial cross sections as a function of p_T^H (left), and $| y^H |$ (right). The dashed boxes represent the sum in quadrature of the acceptance and theoretical uncertainties on the POWHEG (blue) and NNLOPS (orange) predictions. The sub-dominant component of the signal (VBF + VH + $t\bar{t}$H) is denoted as XH and is represented by the green dashed histogram

Table 7.7 Differential fiducial cross sections and ±1 standard deviation uncertainties measured for the p_T^H observable at a fixed value of $m_H = 125.38$ GeV. The fourth and fifth column present the breakdown of the total uncertainty (unc.) into statistical and systematic components, respectively

Bin range (GeV)	$d\sigma_{fid}$ (fb)	unc.	stat.	syst.
0–10	0.32	$^{+0.11}_{-0.10}$	$^{+0.10}_{-0.09}$	$^{+0.04}_{-0.03}$
10–20	0.67	$^{+0.14}_{-0.13}$	$^{+0.13}_{-0.12}$	$^{+0.06}_{-0.05}$
20–30	0.41	$^{+0.12}_{-0.10}$	$^{+0.11}_{-0.10}$	$^{+0.04}_{-0.04}$
30–45	0.51	$^{+0.12}_{-0.10}$	$^{+0.11}_{-0.10}$	$^{+0.04}_{-0.04}$
45–80	0.45	$^{+0.10}_{-0.09}$	$^{+0.10}_{-0.09}$	$^{+0.04}_{-0.03}$
80–120	0.30	$^{+0.08}_{-0.07}$	$^{+0.07}_{-0.07}$	$^{+0.02}_{-0.02}$
120–200	0.19	$^{+0.06}_{-0.05}$	$^{+0.06}_{-0.05}$	$^{+0.01}_{-0.01}$
200–13000	0.03	$^{+0.02}_{-0.02}$	$^{+0.02}_{-0.01}$	$^{+0.00}_{-0.00}$

The differential cross sections are measured in bins of kinematic observables sensitive to both the production and the decay of the SM H boson in the 4ℓ channel, as presented below.

The differential cross sections in bins of the H boson transverse momentum (p_T^H) and rapidity ($| y^H |$) are shown in Fig. 7.18 and the corresponding numerical values are presented in Tables 7.7 and 7.8, respectively.

A second set of differential measurements is presented in Fig. 7.19, where the cross sections are measured in bins of the number of associated jets (N^j) and of the transverse momentum of the leading jet (p_T^j) in the event. These variables are

Table 7.8 Differential fiducial cross sections and ±1 standard deviation uncertainties measured for the | y^{H} | observable at a fixed value of m_{H} = 125.38 GeV. The fourth and fifth column present the breakdown of the total uncertainty (unc.) into statistical and systematic components, respectively

Bin range	$d\sigma_{fid}$ (fb)	unc.	stat.	syst.
0.0–0.15	0.41	$^{+0.10}_{-0.08}$	$^{+0.09}_{-0.08}$	$^{+0.05}_{-0.03}$
0.15–0.3	0.36	$^{+0.08}_{-0.07}$	$^{+0.07}_{-0.07}$	$^{+0.03}_{-0.02}$
0.3–0.6	0.62	$^{+0.13}_{-0.11}$	$^{+0.11}_{-0.10}$	$^{+0.07}_{-0.05}$
0.6–0.9	0.57	$^{+0.12}_{-0.10}$	$^{+0.10}_{-0.10}$	$^{+0.06}_{-0.04}$
0.9–1.2	0.36	$^{+0.10}_{-0.09}$	$^{+0.09}_{-0.08}$	$^{+0.05}_{-0.03}$
1.2–2.5	0.64	$^{+0.15}_{-0.13}$	$^{+0.13}_{-0.12}$	$^{+0.08}_{-0.05}$

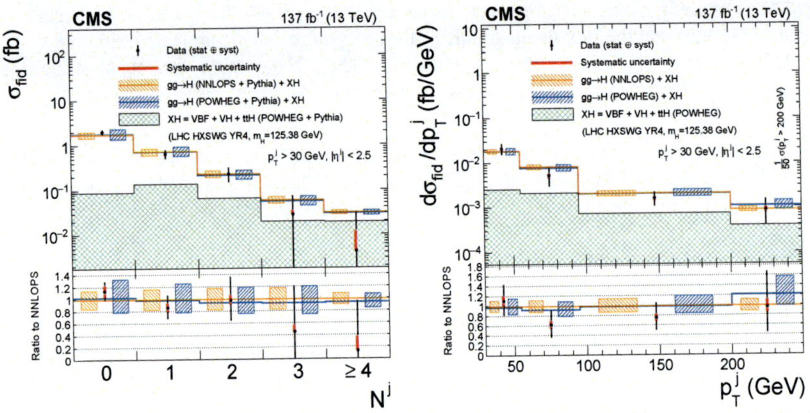

Fig. 7.19 Differential fiducial cross sections as a function of the number of associated jets (left), and p_{T} of the leading jet (right). The dashed boxes represent the sum in quadrature of the acceptance and theoretical uncertainties on the POWHEG (blue) and NNLOPS (orange) predictions. The sub-dominant component of the signal (VBF + VH + $t\bar{t}$H) is denoted as XH and is represented by the green dashed histogram

interesting as they allow to probe phase space regions more sensitive to specific production mechanisms, such as the VBF and VH processes. The observed values of the fiducial cross sections in each bin, along with a breakdown of the uncertainties into statistical and systematic components, are presented in Tables 7.9 and 7.10.

The response matrices ($\varepsilon_{i,j}$) that enter the unfolding procedure and quantify possible bin migrations between generator-level bins and reconstruction-level bins as of Eq. 6.11 are shown in Appendix B. These are found to be diagonal for all the observables considered in the analysis, confirming the well-suited definition of the fiducial phase space region in which these cross sections are measured. Small off-diagonal elements are present in the response matrices of the jet-related variables as a consequence of the low statistics available and of the intrinsic differences that

Table 7.9 Differential fiducial cross sections and ±1 standard deviation uncertainties measured for the N^{j} observable at a fixed value of $m_{\mathrm{H}} = 125.38$ GeV. The fourth and fifth column present the breakdown of the total uncertainty (unc.) into statistical and systematic components, respectively

Bin range	$d\sigma_{\mathrm{fid}}$ (fb)	unc.	stat.	syst.
0	2.00	$^{+0.29}_{-0.26}$	$^{+0.21}_{-0.20}$	$^{+0.20}_{-0.17}$
1	0.64	$^{+0.15}_{-0.14}$	$^{+0.14}_{-0.13}$	$^{+0.06}_{-0.04}$
2	0.23	$^{+0.09}_{-0.08}$	$^{+0.09}_{-0.08}$	$^{+0.02}_{-0.01}$
3	0.03	$^{+0.05}_{-0.03}$	$^{+0.05}_{-0.03}$	$^{+0.01}_{-0.00}$
≥4	0.00	$^{+0.03}_{-0.00}$	$^{+0.03}_{-0.00}$	$^{+0.01}_{-0.00}$

Table 7.10 Differential fiducial cross sections and ±1 standard deviation uncertainties measured for the $p_{\mathrm{T}}^{\mathrm{j}}$ observable at a fixed value of $m_{\mathrm{H}} = 125.38$ GeV. The fourth and fifth column present the breakdown of the total uncertainty (unc.) into statistical and systematic components, respectively

Bin range (GeV)	$d\sigma_{\mathrm{fid}}$ (fb)	unc.	stat.	syst.
30–55	0.52	$^{+0.16}_{-0.14}$	$^{+0.15}_{-0.13}$	$^{+0.05}_{-0.04}$
55–95	0.21	$^{+0.10}_{-0.09}$	$^{+0.10}_{-0.09}$	$^{+0.03}_{-0.02}$
95–200	0.16	$^{+0.07}_{-0.06}$	$^{+0.06}_{-0.05}$	$^{+0.02}_{-0.01}$
200–13000	0.04	$^{+0.03}_{-0.02}$	$^{+0.03}_{-0.02}$	$^{+0.01}_{-0.01}$

may be present between generator- and reconstruction-level jets. However, these off-diagonal effects are not sizable and do not significantly impact the results of the analysis.

The ultimate precision of the results is dominated by the systematic experimental uncertainties on the luminosity measurement and on the lepton efficiencies, while the impact of the systematic theoretical uncertainties is found to be almost negligible. On the other hand, the systematic contribution to the final uncertainty is marginal if compared to the statistical one. The residual model dependence of the results is studied by repeating the measurement using response matrices ($\varepsilon_{i,j}$) obtained from variations of the relative fraction of each SM production mechanism within its experimental constraint, similar to what was done for the Run-I version of this analysis [3]. The impact of the model dependence on the measurements is found to be of the order of few percents and in any case smaller than the experimental systematic uncertainties. This is a typical feature of fiducial and differential measurements, as the unfolding to generator-level observables substantially reduces any systematic effect arising from the reconstruction procedure or the detector imperfections. For this reason, this approach will play a fundamental role in the test of the SM predictions in the future of the LHC analyses, profiting from the larger statistics that will be available at the end of the Run-III data taking.

7.2.4 Fermionic and Bosonic Anomalous Couplings

The production and decay processes of the H boson are exploited to measure the anomalous contributions to the H boson couplings to vector bosons and fermions and to re-interpret these results in the context of higher-dimensional operators in the Effective Field Theory (EFT) [2]. Similarly, possible hints of CP-violation in the effective Hgg couplings are also sought, resulting in the first direct constraint of CP-violation using CP-sensitive observables.

The Lagrangian amplitude of the H boson couplings to fermions is:

$$A(\mathrm{Hff}) = -\frac{m_{\mathrm{f}}}{v}\overline{\psi}_{\mathrm{f}}\left(\kappa_{\mathrm{f}} + i\tilde{\kappa}_{\mathrm{f}}\gamma_5\right)\psi_{\mathrm{f}} \tag{7.1}$$

where $\overline{\psi}_{\mathrm{f}}$ and ψ_{f} are the Dirac spinors associated to the fermions, m_{f} is their mass, κ_{f} and $\tilde{\kappa}_{\mathrm{f}}$ are the corresponding coupling strengths, and v is the SM VEV of the H field. Eq. (7.1) can be used to study the anomalous couplings of the H boson to fermions, such as in $\mathrm{t\bar{t}H}$ and $\mathrm{b\bar{b}H}$ production mechanisms, but also in the less sensitive tHq and $gg \to$ ZH ones. More precisely, the SM predicts $\tilde{\kappa}_{\mathrm{f}} = 0$ and $\kappa_{\mathrm{f}} = 1$, and the presence of CP-even κ_{f} and CP-odd $\tilde{\kappa}_{\mathrm{f}}$ terms would lead to CP violation.

Additional anomalous contributions to the H boson couplings to vector bosons can be assessed from the study of the scattering amplitude describing the interaction between a spin-zero H boson and two spin-one gauge bosons $\mathrm{V_1V_2}$, such as ZZ, Zγ, $\gamma\gamma$, WW, and gg:

$$A(\mathrm{HV_1V_2}) = \frac{1}{v}\left[a_1^{\mathrm{VV}} + \frac{\kappa_1^{\mathrm{VV}} q_{\mathrm{V1}}^2 + \kappa_2^{\mathrm{VV}} q_{\mathrm{V2}}^2}{\left(\Lambda_1^{\mathrm{VV}}\right)^2} + \frac{\kappa_3^{\mathrm{VV}}\left(q_{\mathrm{V1}} + q_{\mathrm{V2}}\right)^2}{\left(\Lambda_Q^{\mathrm{VV}}\right)^2}\right] m_{\mathrm{V1}}^2 \varepsilon_{\mathrm{V1}}^* \varepsilon_{\mathrm{V2}}^*$$
$$+ \frac{1}{v} a_2^{\mathrm{VV}} f_{\mu\nu}^{*(1)} f^{*(2),\mu\nu} + \frac{1}{v} a_3^{\mathrm{VV}} f_{\mu\nu}^{*(1)} \tilde{f}^{*(2),\mu\nu}, \tag{7.2}$$

which parametrizes the electroweak H boson production modes (VBF, ZH, and WH), as well as the ggH production and the H $\to$ VV decay.
The $f^{(i)\mu\nu} = \varepsilon_{\mathrm{V}i}^{\mu} q_{\mathrm{V}i}^{\nu} - \varepsilon_{\mathrm{V}i}^{\nu} q_{\mathrm{V}i}^{\mu}$, $\tilde{f}_{\mu\nu}^{(i)} = \frac{1}{2}\varepsilon_{\mu\nu\rho\sigma} f^{(i),\rho\sigma}$, and $\varepsilon_{\mathrm{V}i}$, $q_{\mathrm{V}i}$, and $m_{\mathrm{V}i}$ terms in Eq. (7.2) represent the polarization vector, four-momentum, and pole mass of a gauge boson $i = 1, 2$, and Λ_1 and Λ_Q are the scales of BSM physics. The CP-even a_2^{gg} and CP-odd a_3^{gg} parameters define all the SM and BSM loop contributions to the Hgg coupling, as it is not possible to disentangle the top- and bottom-quark contributions to the gluon fusion loop from the ones originating from any other heavy BSM particle.

The only tree-level contributions in the SM that appear in Eq. (7.2) are $a_1^{\mathrm{ZZ}} \neq 0$ and $a_1^{\mathrm{WW}} \neq 0$. Assuming custodial symmetry $a_1^{\mathrm{ZZ}} = a_1^{\mathrm{WW}}$, and in the SM, $a_1^{\mathrm{ZZ}} = a_1^{\mathrm{WW}} = 2$. The rest of the ZZ and WW couplings are considered anomalous contributions, which are either small contributions arising in the SM due to loop effects or new BSM contributions. Among the anomalous contributions, considerations of symmetry and gauge invariance require $\kappa_1^{\mathrm{ZZ}} = \kappa_2^{\mathrm{ZZ}}$ and $a_1^{\mathrm{Z\gamma}} = a_1^{\gamma\gamma} = a_1^{gg} = \kappa_1^{\gamma\gamma} =$

$\kappa_2^{\gamma\gamma} = \kappa_1^{gg} = \kappa_2^{gg} = \kappa_1^{Z\gamma} = \kappa_3^{VV} = 0$. Hence, Eq. (7.2) includes a total of 13 independent parameters for the H boson couplings to EW gauge bosons and two parameters for the couplings to gluons. The CP-odd a_3^{VV} term is expected to be null in the SM and any deviation measured would result in CP violation in a given process.

The total of 13+2 independent parameters describing the HVV+Hgg couplings can be further reduced to 4+2 assuming custodial and SU(2)×U(1) symmetry, along with fixing $\sin^2\theta_w = 0.23119$. The four independent HVV parameters are a_1, a_2, a_3, and $\kappa_1/(\Lambda_1)^2$. A one-to-one relationship exists between the HVV+Hgg amplitude couplings described above and an EFT Lagrangian parameterization with dimension-4 and dimension-6 operators using the Higgs basis [2, 6]:

$$\delta c_z = \frac{1}{2}a_1 - 1\,, \tag{7.3}$$

$$c_{z\Box} = \frac{m_Z^2 s_w^2}{e^2}\frac{\kappa_1}{(\Lambda_1)^2}\,, \tag{7.4}$$

$$c_{zz} = -\frac{2s_w^2 c_w^2}{e^2}a_2\,, \tag{7.5}$$

$$\tilde{c}_{zz} = -\frac{2s_w^2 c_w^2}{e^2}a_3\,, \tag{7.6}$$

$$c_{gg} = -\frac{1}{2\pi\alpha_s}a_2^{gg}\,, \tag{7.7}$$

$$\tilde{c}_{gg} = -\frac{1}{2\pi\alpha_s}a_3^{gg}\,. \tag{7.8}$$

where in the SM, ignoring small loop-induced corrections, all EFT parameters related to the HVV amplitude (c_z, $c_{z\Box}$, c_{zz}, and $\tilde{c}_{zz}$) are expected to be zero, while the EFT parameters related to the Hgg amplitude are expected to be $c_{gg} = 1/(12\pi^2)$ and $\tilde{c}_{gg} = 0$.

The results are extracted starting from the events selected as of Sect. 5.1. Conversely to the signal-strength modifiers and STXS measurements presented above, a less granular events categorization is employed to set constraints on the HVV and Hgg anomalous couplings. More precisely, two independent categorization schemes, hereafter referred to as Scheme 1 and Scheme 2, are employed: the former targets the Htt and Hgg anomalous couplings, whereas the latter targets HVV anomalous couplings. The categorization relies on the same matrix element kinematic discriminants ($\mathcal{D}_{\text{1jet}}^{\text{VBF}}$, $\mathcal{D}_{\text{2jet}}^{\text{VBF}}$, $\mathcal{D}_{\text{2jet}}^{\text{WH}}$, $\mathcal{D}_{\text{2jet}}^{\text{ZH}}$) sensitive to the H boson production defined in Sect. 5.3.1. Scheme 1 comprises of the seven categories introduced in Sect. 5.3.1, of which the relative purity is shown in Fig. 5.9. Scheme 2 uses a total of six categories defined to enhance the sensitivity to the HVV coupling. More precisely, the VBF-2jet category is defined asking for the same number of leptons and jets as of in Scheme 1, but with a cut on max $\mathcal{D}_{\text{2jet}}^{\text{VBF},i} > 0.5$, where the index i refers to the fact that the discriminant is computed for all the BSM signal hypotheses considered in the analysis. The same strategy is employed for the definition of the VH-hadronic category. The $t\bar{t}H$ categories are not considered in Scheme 2, but rather they are merged together with the Untagged category. A Boosted category, not present in Scheme 1, is introduced to cope with the presence of events where not all associated particles are fully reconstructed.

The a_i real couplings of the Hff, Hgg, and HVV interactions introduced in Eqs. (7.2), (7.1) are used to parametrize the differential cross section of the on-shell H boson production mechanism i and its consequent decay to the final state f as:

$$\sigma(i \rightarrow \mathrm{H} \rightarrow j) \propto \frac{\left(\sum_{il} \alpha^{(j)} a_i a_l\right)\left(\sum_{mn} \alpha^{(f)}_{mn} a_m a_n\right)}{\Gamma_{\mathrm{H}}}, \tag{7.9}$$

where the $\alpha_{il}^{(k)}$ terms correspond to functions of the kinematic observables and are described with simulation. In principle, the total decay width of the H boson Γ_{H} depends on the couplings a_i and on the partial widths of the decays to unobserved or invisible final states, even though certain assumptions are made in the extraction of the results of the analysis as discussed in the following.

All the results are extracted in terms of signal strength modifiers and fractional contributions f_{ai} of the a_i couplings to the cross section of a given process. In such a way, the total width Γ_{H} and the SM-like couplings squared are absorbed into the signal strength definitions. More precisely, the effective fractional cross section for Hff couplings is:

$$f_{\mathrm{CP}}^{\mathrm{Hff}} = \frac{|\tilde{\kappa}_{\mathrm{f}}|^2}{|\kappa_{\mathrm{f}}|^2 + |\tilde{\kappa}_{\mathrm{f}}|^2} \operatorname{sign}\left(\frac{\tilde{\kappa}_{\mathrm{f}}}{\kappa_{\mathrm{f}}}\right) \tag{7.10}$$

The effective fractional cross section for Hgg couplings is:

$$f_{a3}^{ggH} = \frac{|a_3^{gg}|^2}{|a_2^{gg}|^2 + |a_3^{gg}|^2} \operatorname{sign}\left(\frac{a_3^{gg}}{a_2^{gg}}\right) \tag{7.11}$$

The fractional cross sections for HVV couplings can be written as:

$$f_{ai}^{\mathrm{VV}} \propto \frac{|a_i^{\mathrm{VV}}|^2}{\sum_j |a_j^{\mathrm{VV}}|^2} \operatorname{sign}\left(\frac{a_i^{\mathrm{VV}}}{a_1}\right). \tag{7.12}$$

Similarly to the signal strength modifiers and STXS cross section measurements, all the results are extracted by means of an unbinned extended maximum likelihood fit, with the Poisson marked model slightly modified to take into account the presence of fractional cross section for anomalous couplings:

$$\mathcal{P}_k(\vec{x}) = \sum_j \mu_j \mathcal{P}_{jk}^{\mathrm{sig}}\left(\vec{x}; \vec{\theta}_{jk}, \vec{f}_j\right) + \sum_i \mathcal{P}_{ik}^{\mathrm{bkg}}\left(\vec{x}; \vec{\theta}_{ik}\right). \tag{7.13}$$

The index k runs over all the reconstructed categories considered in the analysis, for all the signal processes j and background processes i. The $\vec{\theta}$ terms correspond to the constrained nuisance parameters, implemented in the likelihood function as described in Sect. 6.5, and f are the effective fractional cross sections defined above. The fit is performed simultaneously across all the seven (six) categories of Scheme 1 (Scheme 2), for all the three final states ($4e, 4\mu$, and $2e2\mu$), and the three data taking periods (2016, 2017, and 2018) considered in the analysis, resulting in a total of 63 (54) "categories" being fitted[3] in Scheme 1 (Scheme 2). The sensitivity to different BSM scenarios and anomalous couplings is enhanced by including in the fit the kinematic observables $\vec{x}$. These correspond to matrix element discriminants introduced in the analysis to separate signal and background hypotheses ($\mathcal{D}_{\text{bkg}}^{\text{kin}}$), as well as SM and BSM contributions.[4] These discriminants are defined as:

$$\mathcal{D}_{\text{alt}}\left(\mathbf{\Omega}\right) = \frac{\mathcal{P}_{\text{sig}}\left(\mathbf{\Omega}\right)}{\mathcal{P}_{\text{sig}}\left(\mathbf{\Omega}\right) + \mathcal{P}_{\text{alt}}\left(\mathbf{\Omega}\right)}, \tag{7.14}$$

$$\mathcal{D}_{\text{int}}\left(\mathbf{\Omega}\right) = \frac{\mathcal{P}_{\text{int}}\left(\mathbf{\Omega}\right)}{2\sqrt{\mathcal{P}_{\text{sig}}\left(\mathbf{\Omega}\right)\,\mathcal{P}_{\text{alt}}\left(\mathbf{\Omega}\right)}}, \tag{7.15}$$

where the probabilities $\mathcal{P}$ and the vectors $\vec{\Omega}$ refer to the same quantities introduced in Sect. 5.1.1. The subfix "sig" and "alt" identify a signal model and an alternative hypothesis, respectively. The label "int" identifies the interference between the two model contributions. A complete list of these observables is given in Table 7.11, while a more detailed description of each of them can be found in Ref. [7].

The probability distributions $\mathcal{P}_{jk}^{\text{sig}}$ and $\mathcal{P}_{ik}^{\text{bkg}}$ for signal and background are implemented in the analysis as binned multidimensional templates of the observables $\vec{x}$ reported in Table 7.11. A continuous dependence of these probability distributions on $\vec{\theta}_{jk}$ and $\vec{f}_j$ is achieved from analytical interpolations of the signal and background templates as further detailed in Sect. 5.2 of Ref. [7]. The measurements of μ_j and $\vec{f}_j$ are reinterpreted in terms of H boson anomalous couplings exploiting Eq. (7.9) to relate μ_j to the couplings and Eqs. (7.11)–(7.12) to relate $\vec{f}_j$ to coupling ratios. In order to get significant constraints from the fit, the H boson width in Eq. (7.9) is assumed to be the SM decay width,[5] without the inclusion of unobserved nor invisible decays.

A selected set of results is presented in the following, while a more extensive set of anomalous couplings measurements and their corresponding EFT interpretations is presented in Ref. [7].

[3] The term "categories" here should not be misunderstood with the reconstructed categories described in the text. Here it refers to the total number of channels that enter the fitting procedure.

[4] I.e. matrix element discriminants sensitive to the a_i terms in Eq. (7.1)–(7.2).

[5] Hence Γ_H is expressed as the sum of the nine decay modes dominant in the SM (cf. Chap. 1).

Table 7.11 List of the kinematic observables ($\vec{x}$) used in the fitting of the categories in Schemes 1 and 2 to extract constraints on the HVV and Hff anomalous couplings

Category	Observables $\vec{x}$ for fitting
Scheme 1	
Untagged	$\mathcal{D}_{\rm bkg}^{\rm kin}$
VBF-1jet	$\mathcal{D}_{\rm bkg}^{\rm kin}$
VBF-2jet	$\mathcal{D}_{\rm bkg}^{\rm kin}$, $\mathcal{D}_{\rm 2jet}^{\rm VBF}$, $\mathcal{D}_{0-}^{\rm ggH}$, $\mathcal{D}_{\rm CP}^{\rm ggH}$
VH-leptonic	$\mathcal{D}_{\rm bkg}^{\rm kin}$
VH-hadronic	$\mathcal{D}_{\rm bkg}^{\rm kin}$
$\rm t\bar{t}H$-leptonic	$\mathcal{D}_{\rm bkg}^{\rm kin}$, $\mathcal{D}_{0-}^{\rm t\bar{t}H}$
$\rm t\bar{t}H$-hadronic	$\mathcal{D}_{\rm bkg}^{\rm kin}$, $\mathcal{D}_{0-}^{\rm t\bar{t}H}$
Scheme 2	
Untagged	$\mathcal{D}_{\rm bkg}^{\rm kin}$, $\mathcal{D}_{\rm 0h+}^{\rm dec}$, $\mathcal{D}_{0-}^{\rm dec}$, $\mathcal{D}_{\Lambda 1}^{\rm dec}$, $\mathcal{D}_{\Lambda 1}^{\rm Z\gamma,dec}$, $\mathcal{D}_{\rm int}^{\rm dec}$, $\mathcal{D}_{\rm CP}^{\rm dec}$
Boosted	$\mathcal{D}_{\rm bkg}^{\rm kin}$, $p_{\rm T}^{4\ell}$
VBF-1jet	$\mathcal{D}_{\rm bkg}^{\rm kin}$, $p_{\rm T}^{4\ell}$
VBF-2jet	$\mathcal{D}_{\rm bkg}^{\rm kin}$, $\mathcal{D}_{\rm 0h+}^{\rm VBF+dec}$, $\mathcal{D}_{0-}^{\rm VBF+dec}$, $\mathcal{D}_{\Lambda 1}^{\rm VBF+dec}$, $\mathcal{D}_{\Lambda 1}^{\rm Z\gamma,VBF+dec}$, $\mathcal{D}_{\rm int}^{\rm VBF}$, $\mathcal{D}_{\rm CP}^{\rm VBF}$
VH-leptonic	$\mathcal{D}_{\rm bkg}^{\rm kin}$, $p_{\rm T}^{4\ell}$
VH-hadronic	$\mathcal{D}_{\rm bkg}^{\rm kin}$, $\mathcal{D}_{\rm 0h+}^{\rm VH+dec}$, $\mathcal{D}_{0-}^{\rm VH+dec}$, $\mathcal{D}_{\Lambda 1}^{\rm VH+dec}$, $\mathcal{D}_{\Lambda 1}^{\rm Z\gamma,VH+dec}$, $\mathcal{D}_{\rm int}^{\rm VH}$, $\mathcal{D}_{\rm CP}^{\rm VH}$

The measurement of anomalous couplings of the H boson to gluons is presented in the left plot of Fig. 7.20. Here, the μ_{ggH} and f_{a3}^{ggH} terms in Eq. (7.13) parametrize both the direct couplings of the H boson to SM fermions in the ggH loop and the presence of new BSM particles appearing in the loop.

The best fit value of f_{a3}^{ggH} is found to be consistent with zero, in agreement with the SM expectation, with a profiled value of $\mu_{\rm ggH} = 0.86^{+0.13}_{-0.11}$, in statistical agreement with the value obtained for the dedicated measurement of this signal strength modifier presented in Sect. 7.2.1. The measurement of μ_{ggH} and f_{a3}^{ggH} can be reinterpreted as limits on the CP-even and CP-odd couplings defined in Eqs. (7.7) and (7.8), as shown in the right plot of Fig. 7.20. In such a measurement the cross section and total width depend on c_{gg}, $\tilde{c}_{gg}$, κ_t, $\tilde{\kappa}_t$, and $\tilde{\kappa}_b$, but since all these couplings cannot be separated with a single process certain assumptions are needed. More precisely, only c_{gg} and $\tilde{c}_{gg}$ are left floating in the fit, while all the other couplings are fixed to their SM expectation.[6] A constraint on $(c_{gg}^2 + \tilde{c}_{gg}^2)$, corresponding to a ring on a two-parameter phase space, can be extracted from the measurement of μ_{ggH} and it

[6] I.e. $\kappa_t = \kappa_b = 1$ and $\tilde{\kappa}_t = \tilde{\kappa}_b = 0$. In addition, the contribution of the $\rm H \to \gamma\gamma$ and $\rm Z\gamma$ decays to the total width is fixed to the SM.

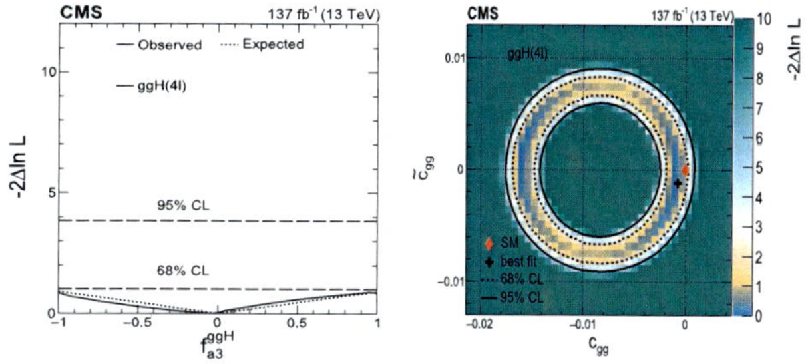

Fig. 7.20 Constraints on the anomalous couplings of the H boson to gluons in the ggH process using the $H \to 4\ell$ decay. *Left*: Observed (solid) and expected (dashed) likelihood scans of the CP-sensitive parameter f_{a3}^{ggH} obtained with μ_{ggH} profiled in the fit. The dashed horizontal lines show 68 and 95% CL. *Right*: Observed confidence level intervals on the c_{gg} and $\tilde{c}_{gg}$ EFT couplings extracted from the re-interpretation of the f_{a3}^{ggH} and μ_{ggH} measurement. The dashed and solid lines show the 68 and 95% CL exclusion regions in two dimensions, respectively. The black cross corresponds to the best fit value, resulting in the $(c_{gg}, \tilde{c}_{gg})$ and $(-c_{gg}, \tilde{c}_{gg})$ degeneracy explained in the text

gets resolved with the additional measurement of f_{a3}^{ggH}. The results are consistent with the SM expectation of $(c_{gg}, \tilde{c}_{gg}) = (0.0084, 0)$ at $1.1\,\sigma$. A degeneracy between any two points $(c_{gg}, \tilde{c}_{gg})$ and $(-c_{gg}, \tilde{c}_{gg})$ is observed, as none of the observables considered in the analysis is sensitive to the absolute sign of these two parameters.

The constraints on the anomalous H boson couplings to top quarks in the $\mathrm{t\bar{t}H}$ and tH process are shown in Fig. 7.21.

The left plot of the figure shows the results of the measurement of f_{CP}^{Htt} with μ_{ttH} profiled in the fit. The small sensitivity of the $H \to ZZ \to 4\ell$ channel to the $\mathrm{t\bar{t}H}$ production mechanism results in extremely loose constraints on this parameter, as shown by the red lines in the plot. Even though only two expected signal events are available for the measurement, the very clean signature of the 4ℓ channel allows to resolve an observed best-fit value. It corresponds to the pure CP-odd Yukawa coupling, but the result is statistically consistent, at $0.35\,\sigma$, with the pure CP-even Yukawa coupling expected in the SM. The $\mathrm{t\bar{t}H}$ signal strength modifier profiled in the fit is found to be $\mu_{\mathrm{t\bar{t}H}} = 0.17^{+0.70}_{-0.17}$, in statistical agreement with the corresponding parameter measured in Sect. 7.2.1. The figure shows the additional measurement of f_{CP}^{Htt} obtained from the combination with the $H \to \gamma\gamma$ decay channel [8]. Only the common systematic uncertainties are correlated between the two channels, as new particles appearing in the $H \to \gamma\gamma$ loops could have a different impact on the signal strengths of the two processes. The combined result excludes the pure pseudoscalar hypothesis at $3.2\,\sigma$, with an expected exclusion of $2.7\,\sigma$. The right plot of Fig. 7.21 shows the results of the EFT coupling interpretation in the $(\kappa_t, \tilde{\kappa}_t)$ plane, obtained under the assumption that no BSM particles contribute to the loop in the $H \to \gamma\gamma$ decay. In order to perform this interpretation, it is assumed that no BSM particle contributes to the loop in the $H \to \gamma\gamma$ decay, thus resolving the ambiguity that otherwise would be intrinsic in the interpretation of the $H \to \gamma\gamma$ decay. In addition, anomalous HVV couplings are not allowed in this interpretation, and the κ_b and $\tilde{\kappa}_b$

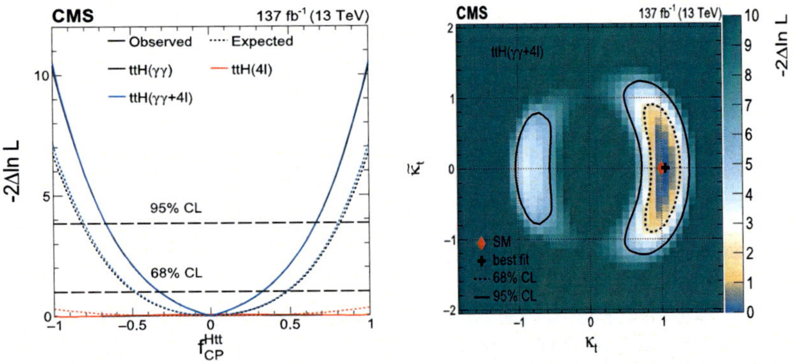

Fig. 7.21 Constraints on the anomalous couplings of the H boson to top quarks in the ttH process using the H→ 4ℓ and H→ $\gamma\gamma$ [8] decays. *Left*: Observed (solid) and expected (dashed) likelihood scans of $f_{\rm CP}^{Htt}$ in the ttH process in the H→ 4ℓ (red) and $\gamma\gamma$ (black) decay channels. Their combination (blue), obtained without correlation on the signal strengths of the two processes, is also shown. The results are obtained with the value of μ_{ttH} profiled in the fits. The dashed horizontal lines show 68 and 95% CL. *Right*: Observed confidence level intervals on the $\kappa_{\rm t}$ and $\tilde{\kappa}_{\rm t}$ EFT couplings extracted from the re-interpretations of the $f_{\rm CP}^{Htt}$ and μ_{ttH} measurements in the combined fit of the H→ 4ℓ and $\gamma\gamma$ channels. The signal strength modifier μ_{ttH} in the two channels is uncorrelated in the fit, as discussed in text. The dashed and solid lines show the 68 and 95% CL exclusion regions in two dimensions, respectively

are fixed to their SM expectations of 1 and 0, respectively. A pure yield measurement would resolve only a two-dimensional contour in the $(\kappa_t, \tilde{\kappa}_t)$ plane, similarly to the one obtained for $(-c_{gg}, \tilde{c}_{gg})$. However, values away from $\tilde{\kappa}_t$ are disfavoured by the introduction of $f_{\rm CP}^{Htt}$, and the inclusion of the tH process allows to resolve the ambiguity between the κ_t and $-\kappa_t$ values.

The measurement of the HVV anomalous couplings, under the $SU(2) \times U(1)$ custodial symmetry, are shown in Fig. 7.22 for the f_{a3}, f_{a2}, and $f_{\Lambda 1}$ fractional cross sections, obtained with the signal-strength modifiers $\mu_{\rm V}$, $\mu_{\rm ggH}$, and $\mu_{\rm t\bar{t}H}$ profiled in the fits. The steep trend of the likelihood function around the minima close to $f_{ai} = 0$ is a result of the production information used in the fits. More precisely, the q_i^2 terms that multiply the anomalous couplings in Eq. (7.2) are larger in the VBF and VH production modes than in the H $\rightarrow 4\ell$ decay. Hence, the VBF and VH cross sections increase quickly with f_{ai}, giving rise to the narrow minima observed. On the other hand, the constraints above $f_{ai} \sim 0.02$ are dominated by the decay information from the decay.

Exploiting Eqs. (7.3)–(7.6), the signal strength μ_V and the f_{a3}, f_{a2}, and $f_{\Lambda 1}$ parameters can be re-interpreted in terms of the δc_z, c_{zz}, $c_{z\Box}$, and $\tilde{c}_{zz}$ coupling strength parameters of the Higgs basis of the standard model effective field theory (SMEFT) formulation. Observed one- and two-dimensional constraints from a simultaneous fit of these EFT parameters are shown in Figs. 7.23 and 7.24, respectively. The c_{gg} and $\tilde{c}_{gg}$ couplings are left unconstrained.

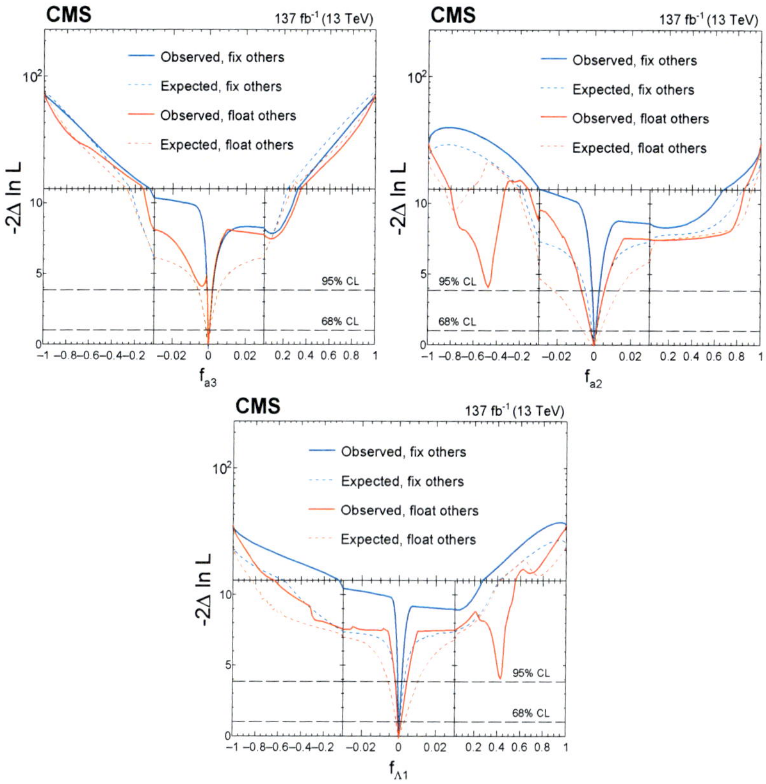

Fig. 7.22 Observed (solid) and expected (dashed) likelihood scans of f_{a3} (top left), f_{a2} (top right), and $f_{\Lambda 1}$ (bottom). The fits is performed separately for each coupling, with the other anomalous coupling fractions either set to zero or left unconstrained in the fit. The signal-strength modifiers μ_{V}, μ_{ggH}, and $\mu_{\mathrm{t\bar{t}H}}$ are profiled in the fits. The dashed horizontal lines show the 68 and 95% CL regions

The results presented in this section provide for the first time in a CMS analysis a simultaneous measurement of five independent HVV, two Hgg, and two Htt anomalous couplings of the H boson, as well as their re-interpretation in the context of dimension-six operators of an EFT Lagrangian. Similarly, the direct measurement of CP-violation effects in the Hgg coupling is performed for the first time, by employing CP-sensitive observables. All the results are still limited by the statistical contribution to their uncertainty, but are expected to improve further in future runs of the LHC, possibly unveiling the presence of BSM physics.

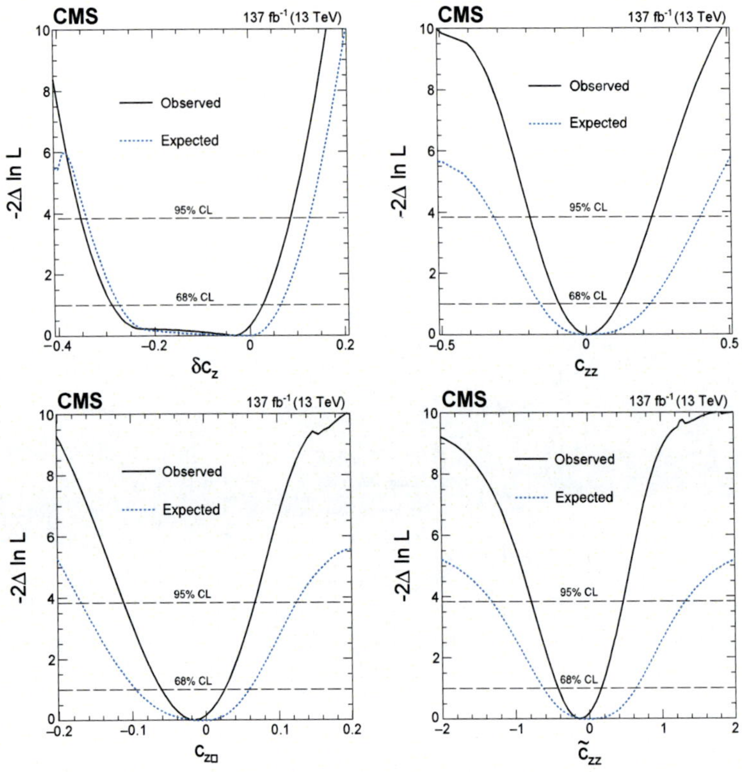

Fig. 7.23 Observed (solid) and expected (dashed) constraints from a simultaneous fit of EFT parameters δc_z (top left), c_{zz} (top right), $c_{z\Box}$ (bottom left), and $\tilde{c}_{zz}$ (bottom right) with the c_{gg} and $\tilde{c}_{gg}$ couplings left unconstrained

7.2.5 The H Boson Self-coupling

As mentioned in Chap. 1, the trilinear self-coupling (λ_{HHH}) of the Higgs boson is one of the free parameters of the SM, and a precise measurement of its value would represent a stringent test of the EWSB mechanism. Along with the Higgs boson couplings to the fermions of second and third generations, λ_{HHH} is still one of the more difficult SM observables to probe.

Double Higgs boson production is the most common process employed to study the Higgs boson self-coupling, as its cross section depends directly on λ_{HHH} at LO. However, its production cross section is three orders of magnitude lower than the single Higgs one, making the final precision achievable at HL-LHC on λ_{HHH} still unclear. The current best available constraints on λ_{HHH} are obtained from the combination of measurements of H boson pair production performed with the Run-II data-set. The ATLAS and CMS Collaborations current limits correspond to observed (expected) constraints at 95% CL of $-0.6\lambda_{\mathrm{HHH}}^{\mathrm{SM}} < \lambda_{\mathrm{HHH}} < 6.6\lambda_{\mathrm{HHH}}^{\mathrm{SM}}$ ($-2.1\lambda_{\mathrm{HHH}}^{\mathrm{SM}} <$

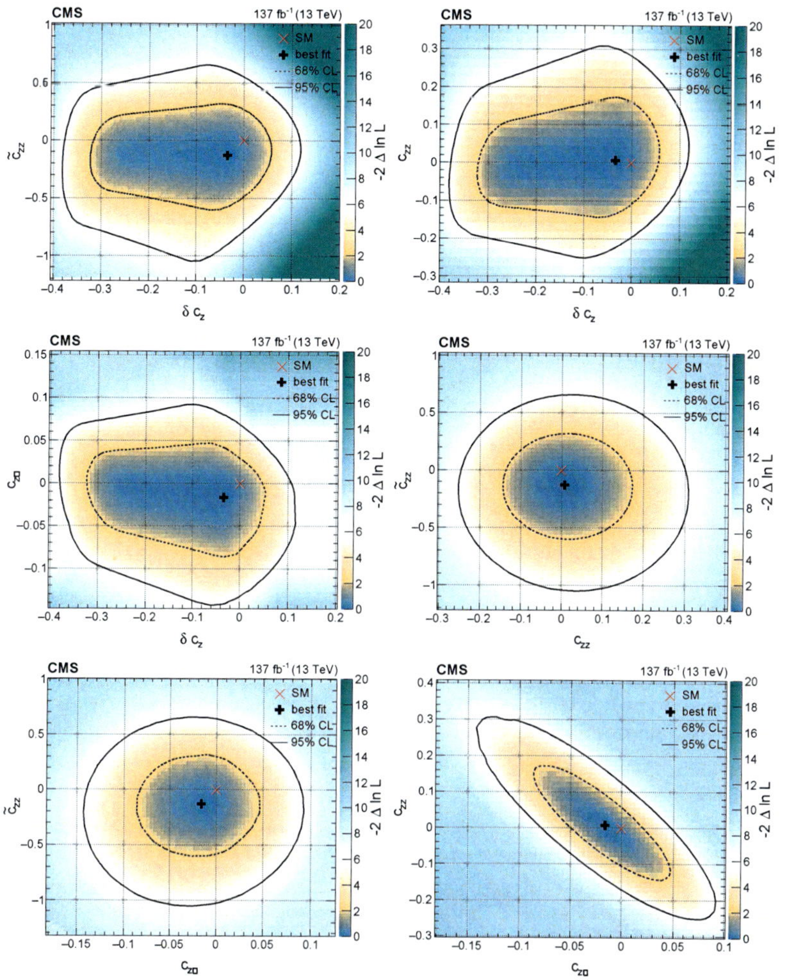

Fig. 7.24 Observed two-dimensional constraints from a simultaneous fit of EFT parameters δc_z, c_{zz}, $c_{z\Box}$, and $\tilde{c}_{zz}$ with the c_{gg} and $\tilde{c}_{gg}$ couplings left unconstrained

$\lambda_{\rm HHH} < 7.8\lambda_{\rm HHH}^{\rm SM}$) [9] and $-1.24\lambda_{\rm HHH}^{\rm SM} < \lambda_{\rm HHH} < 6.89\lambda_{\rm HHH}^{\rm SM}$ ($-0.89\lambda_{\rm HHH}^{\rm SM} < \lambda_{\rm HHH} < 7.12\lambda_{\rm HHH}^{\rm SM}$) [10], respectively. More stringent limits are expected from the double Higgs boson production analyses at the HL-LHC. In particular, projections at 3000 fb^{-1}, obtained from the combination ATLAS and CMS di-Higgs analyses, foresee limits of $0.57\lambda_{\rm HHH}^{\rm SM} < \lambda_{\rm HHH} < 1.5\lambda_{\rm HHH}^{\rm SM}$ at 68% CL [11].

A relatively recent proposal of measuring $\lambda_{\rm HHH}$ via the NLO electroweak radiative corrections in single-Higgs differential measurement [12, 13] is expected to give indirect constraints on $\lambda_{\rm HHH}$ competitive to those obtained from di-Higgs channels. This complementary approach would profit from a much larger statistics and

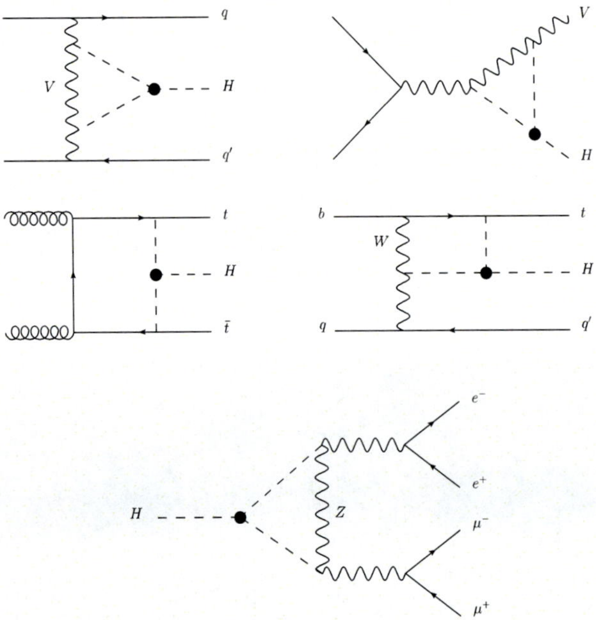

Fig. 7.25 Representative Feynman diagrams contributing to NLO electroweak radiative corrections in single-Higgs processes with anomalous trilinear couplings, indicated by the black dot. The top four diagrams represent anomalous trilinear couplings in production, while the bottom diagram depicts the NLO electroweak corrections in the $H \rightarrow ZZ \rightarrow 4\ell$ decay. Figure taken from [13]

it would set the ground for a complete understanding of the EWSB mechanism. Fig. 7.25 shows representative Feynman diagrams contributing to NLO electroweak radiative corrections in single-Higgs processes with anomalous trilinear couplings. The following presents the first model-independent measurement of λ_{HHH} performed in a CMS analysis exploiting the HZZ single-Higgs channel.

Similarly to what presented in the previous section for the HVV and Hff anomalous couplings, modifications to the Higgs boson self-coupling can be parametrized via the introduction of an anomalous coupling term directly in the SM Lagrangian:

$$\lambda_{\mathrm{HHH}} = \kappa_\lambda \lambda_{\mathrm{HHH}}^{\mathrm{SM}}, \tag{7.16}$$

which is proven to be equivalent to an EFT parametrization with the introduction of higher dimensional operators that affect only the Higgs self-coupling [13]. More precisely, the NLO EW corrections to a generic observable Σ in presence of anomalous trilinear coupling become:

$$\Sigma_{\lambda_{\mathrm{HHH}}}^{\mathrm{BSM}} = Z_{\mathrm{H}}^{\mathrm{BSM}} \Sigma_{\mathrm{LO}} \left(1 + \kappa_\lambda C_1 + \delta Z_{\mathrm{H}}\right), \tag{7.17}$$

where Σ_{LO} represents the LO prediction including higher-order corrections, $\delta Z_{\mathrm{H}} = -1.536 \times 10^{-3}$, and:

$$Z_{\mathrm{H}}^{\mathrm{BSM}} = \frac{1}{1 - (\kappa_\lambda^2 - 1)\delta Z_{\mathrm{H}}} \tag{7.18}$$

encloses all the new-physics contributions at one loop. While the δZ_{H} factor is universal, Eq. (7.17) has a dependence on the process and kinematics via the C_1 term and it is defined as the ratio of contribution to λ_{HHH} and the SM LO cross section. Additional sensitivity to κ_λ can be achieved by measuring differential cross sections, which have a non-flat dependence on this parameter. The values of C_1 in differential bins of a given observable can be computed using the reweighing tool provided in Ref. [13]. More precisely, one produces signal samples with the MadGraph5 aMC@NLO generator at LO in the SM, which get reweighed to include NLO electroweak loops responsible of $\mathcal{O}(\lambda_{\mathrm{HHH}})$ effects. The ratio between the two cross sections in bins of a given kinematic distribution at generator-level correspond to the $C_{1,i}$ values that enter Eq. (7.17).

As discussed in Ref. [13], the transverse momentum of the Higgs boson ($p_{\mathrm{T}}(\mathrm{H})$) is the most sensitive observable to probe anomalous contributions to the trilinear coupling and therefore it is used to infer exclusion limits on κ_λ. The values of $C_{1,i}$ in the different $p_{\mathrm{T}}(\mathrm{H})$ bins considered for the analysis are shown in Fig. 7.26 for the $\mathrm{t\bar{t}H}$ production mechanism and similar values are obtained for all the other production modes of the H boson. A coarser binning with respect to the one used to measure the $p_{\mathrm{T}}(\mathrm{H})$ differential cross section presented in Fig. 7.18 is used to facilitate the interpretation of the results and to cope with the limited statistics available for the computation of the $C_{1,i}$ values. More precisely, five bins with the following boundaries are chosen:

$$p_{\mathrm{T}}(\mathrm{H}) \in [0, 45)\ \mathrm{GeV},\ [45, 80)\ \mathrm{GeV},\ [80, 120)\ \mathrm{GeV},\ [120, 200)\ \mathrm{GeV},\ [200\ \mathrm{GeV}, +\infty)$$

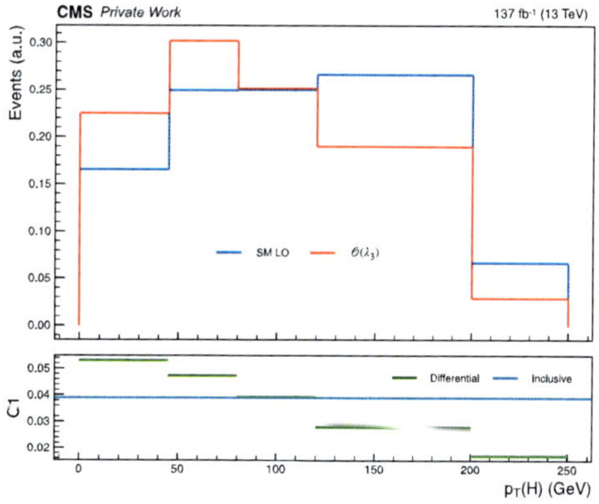

Fig. 7.26 Normalized $p_{\mathrm{T}}(\mathrm{H})$ spectra for $\mathcal{O}(\lambda_{\mathrm{HHH}})$ (red) and SM LO (blue) contributions in $\mathrm{t\bar{t}H}$. The bottom panel shows the inclusive C_1 value (blue line), obtained as the ratio of the two cross sections, as well as the $C_{1,i}$ values (green line) in the different bins considered for the analysis

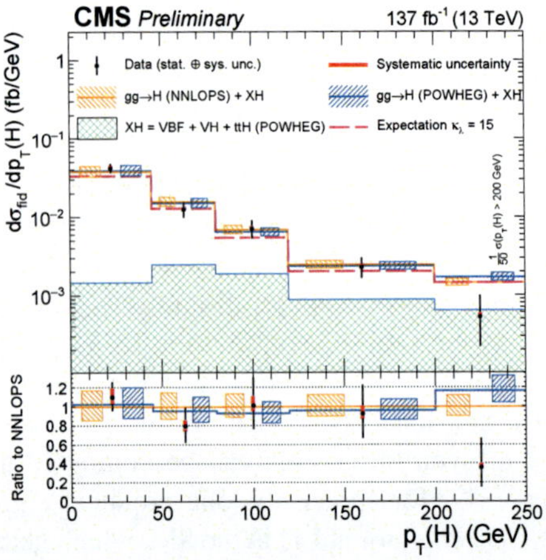

Fig. 7.27 Differential cross sections as a function of p_T^H in the bins used for the κ_λ limits extraction. The dashed boxes represent the sum in quadrature of the acceptance and theoretical uncertainties on the POWHEG (blue) and NNLOPS (orange) predictions. The sub-dominant component of the signal (VBF + VH + $t\bar{t}$H) is denoted as XH and is represented by the green dashed histogram. The purple line represents the differential cross section in bins of p_T^H expected for $\kappa_\lambda = 15$

The p_T(H) differential cross section is measured following the same strategy outlined in Sect. 7.2.3 and the result is presented in Fig. 7.27. Limits on κ_λ are extracted from a one-dimensional maximum likelihood fit with the signal model of Eq. (6.4) modified as follows:

$$S_k^{i,j}\left(\mu_{i,j}, \vec{\theta}_s\right) \longmapsto S_k^{i,j}\left(\mu_{i,j}\left(\kappa_\lambda\right), \vec{\theta}_s\right) = \mu_{i,j}^{\text{prod.}}(\kappa_\lambda) \times \mu^{\text{dec.}}(\kappa_\lambda) \times S_k^{i,j}\left(\vec{\theta}_s\right)\Big|_{\kappa_\lambda=1} \tag{7.19}$$

where $\vec{\theta}_s$ are the nuisance parameters of the model, and the $\mu_{i,j}^{\text{prod.}}$ and $\mu^{\text{dec.}}$ terms represent the κ_λ-dependent scaling of the H boson production mechanisms and branching fractions, respectively. The production side is parametrized as follows:

$$\mu_{i,j}^{\text{prod.}}\left(\kappa_\lambda, C_{1,i}\right) = \frac{1 + \kappa_\lambda C_{1,i} + \delta Z_H}{\left(1 - \left(\kappa_\lambda^2 - 1\right)\delta Z_H\right)\left(1 + C_{1,i} + \delta Z_H\right)}, \tag{7.20}$$

where the $C_{1,i}$ terms are computed per each production mechanism j in each bin i as described above. The scaling of the HZZ branching fraction[7] is parametrized as function of κ_λ as:

[7] The ZZ → 4ℓ branching ratio is fixed to the SM as it is found to have a negligible dependence on κ_λ, as detailed in Ref. [13].

$$\mu^{\text{dec.}}\left(\kappa_\lambda, C_1^{\Gamma_{ZZ}}\right) = 1 + \frac{(\kappa_\lambda - 1)\left(C_1^{\Gamma_{ZZ}} - C_1^{\Gamma_{tot}}\right)}{1 + (\kappa_\lambda - 1)\, C_1^{\Gamma_{tot}}} \tag{7.21}$$

where $C_1^{\Gamma_{ZZ}} = 0.0082$ and $C_1^{\Gamma_{tot}} = 2.5 \times 10^{-3}$ are taken from Ref. [13]. The distribution of the $\mu_{i,j}^{\text{prod.}}$ and $\mu^{\text{dec.}}$ scaling functions are shown in Fig. 7.28 for all the production mechanisms considered in the analysis. One can observe the relatively flat trend of $\mu^{\text{dec.}}$, which reflects the absence of kinematic dependence on the decay and that most of the constraining power comes from the production side. A more pronounced sensitivity to κ_λ is present in the $\mu_{i,j}^{\text{prod.}}$ scaling functions of the production side, shown for the different processes in the bins considered for the analysis, except for ggH, as only the inclusive calculation of C_1 is available in the current state-of-art literature [13]. It is interesting to observe the different sensitivity to κ_λ

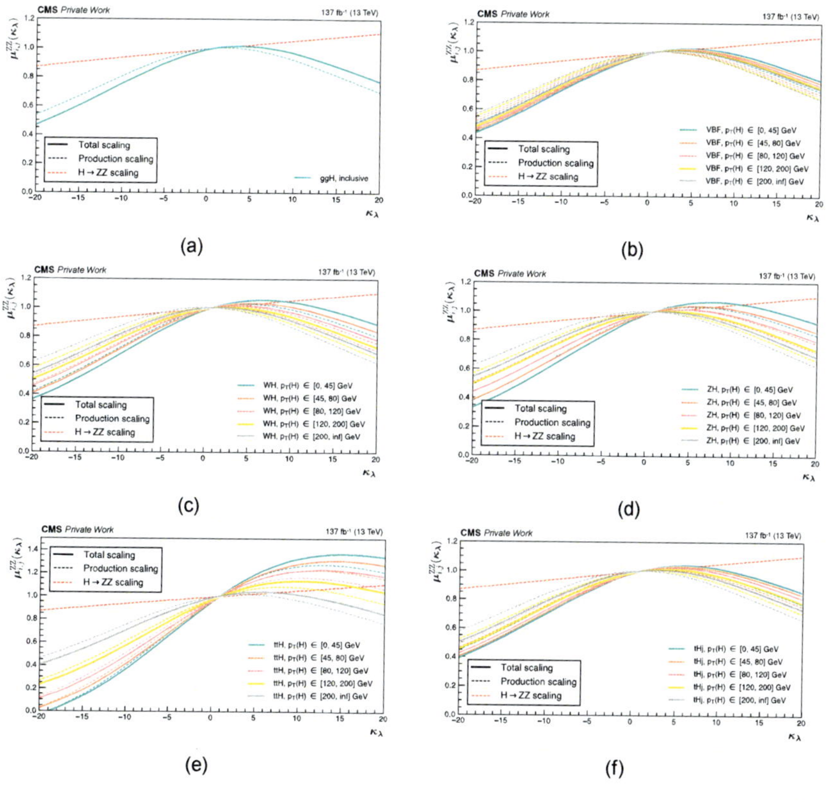

Fig. 7.28 Scaling function $\mu_{i,j}(\kappa_\lambda)$ as a function of κ_λ for the different H boson production mechanisms considered in the analysis: ggH (**a**), VBF (**b**), WH (**c**), ZH (**d**), $\text{t}\bar{\text{t}}$H (**e**), and tHq (**f**). The colored dashed lines represent the scaling functions of the production side in the differential bins of $p_{\text{T}}(\text{H})$ considered. The red dashed lines represent the scaling function of the H → ZZ decay, which is independent on the differential bin

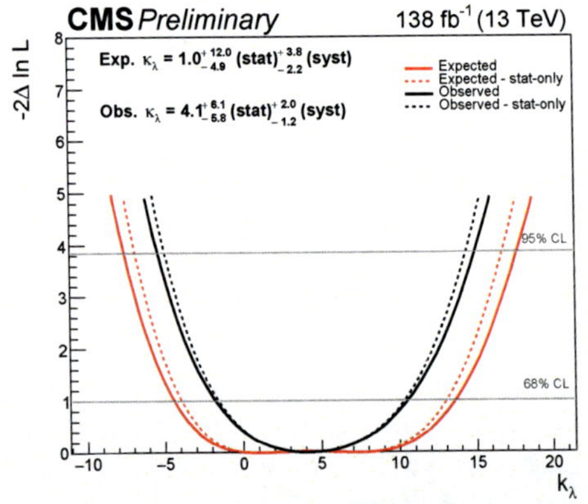

Fig. 7.29 Profile likelihood scan as a function of κ_λ for the observed data (black) and expected scan for a SM Higgs boson (red) derived from an Asimov dataset with $\kappa_\lambda = 1$. The statistical component of the profile likelihood is also shown, represented by the dashed lines

of each production mode: while ggH, VBF, and the production in association with a vector boson VH have a mild dependence on the trilinear coupling, the $t\bar{t}H$ process have a very strong dependence on κ_λ, with effects that can arrive up to 40 to 20% for $\kappa_\lambda \sim 10$ in the low $p_T(H)$ bins.

In the signal model, only anomalous contributions to the H boson self-coupling are considered. This assumption can be relaxed by performing fits where also the HVV and the Hff couplings are scaled by κ_V and κ_f. While this approach would result in more general limits on κ_λ, it is not expected to change significantly the constraints obtained with the parametrization given in Eq. (7.19), as discussed in Ref. [13]. Hence, given the loose sensitivity of the $H \to ZZ \to 4\ell$ decay to κ_λ, κ_V and κ_f are fixed to 1.0 in the following, in order not to introduce additional degrees of freedom that would make the fit unstable. This assumption does not introduce substantial loss of generality[8] to the results presented here and it can be relaxed in future analyses, where the $H \to ZZ$ will be combined with other decay channels. The profile likelihood ratio as a function of κ_λ is shown in Fig. 7.29. The measured value is found to be:

$$\kappa_\lambda^{\text{obs.}} = 4.1^{+6.4}_{-5.9} = 4.1^{+6.1}_{-5.8}(\text{stat})\ ^{+2.0}_{-1.2}(\text{sys})\ (68\%\ \text{CL.}) \tag{7.22}$$

The expected result assuming a SM Higgs boson is obtained from an Asimov dataset with $\kappa_\lambda = 1.0$ and it is found to be:

$$\kappa_\lambda^{\text{exp.}} = 1.0^{+12.6}_{-5.4} = 1.0^{+12}_{-4.9}(\text{stat})\ ^{+3.8}_{-2.2}(\text{sys})\ (68\%\ \text{CL}). \tag{7.23}$$

The shape of the expected profile likelihood ratio is completely dominated by the ggH production mechanism, which is the least sensitive to κ_λ, reflecting the observations from Fig. 7.28. Notwithstanding the little sensitivity of the $H \to ZZ \to 4\ell$

[8] Quoting from Ref. [13]: "We have found that in a global fit, [...], two additional degrees of freedom such as those considered here (anomalous couplings with vector bosons and/or the top quark) [...] have a tiny impact on the upper bound for positive κ_λ values."

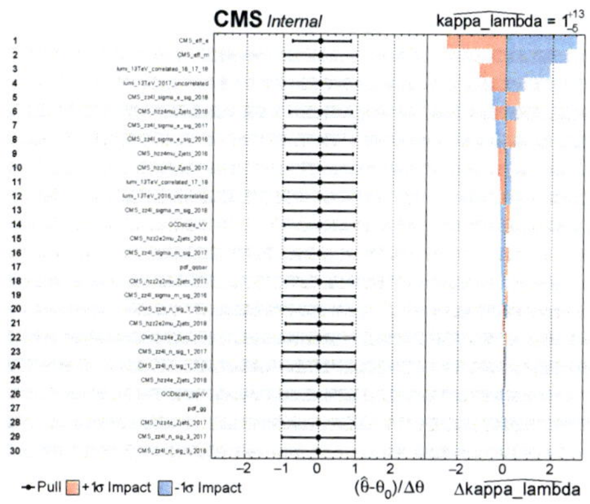

Fig. 7.30 Difference between the pre-fit and post-fit values of the nuisance parameters $\left(\left(\hat{\theta} - \theta_0\right) / \Delta\theta\right)$ considered in the analysis and their impact on the trilinear coupling of the H boson (κ_λ)

channel to the $t\bar{t}$H process, it is enough to resolve the second minimum at $\kappa_\lambda \simeq 9$, thus allowing the extraction of an exclusion limit on κ_λ and validating the possibility of constraining the Higgs boson self-coupling via single-Higgs decay. The likelihood function used for the measurement of κ_λ comprises the same systematic uncertainties considered in the rest of the analysis and described in Sect. 6.5. The impact of the NPs on the measured value of κ_λ is shown in Fig. 7.30. The main sources of systematic uncertainties are the ones on the lepton identification efficiencies and the luminosity measurements, while the theoretical uncertainties on the predicted cross sections is found to be smaller. Similar to the measurement of the HVV and Hff anomalous couplings presented in the previous section, also this result is limited by the statistical precision and is expected to improve in future runs of the LHC. However, the exclusion limits obtained on κ_λ can be improved when the H $\rightarrow$ ZZ decay is combined together with the other single-Higgs decay channels, following the same analysis strategy presented above. The CMS Collaboration measured exclusion limits on κ_λ exploiting the 137 fb^{-1} data collected during the Run-II and combining the main single-Higgs decay channels, for a 95% CL observed interval of $-3.5 < \kappa_\lambda < 12.7$ [10]. Similarly, the ATLAS Collaboration obtained an observed (expected) 95% CL of $-4.0 < \kappa_\lambda < 10.3$ from the combination of all the main single-Higgs decay channels exploiting a partial data-set collected during Run-II [9].

It is compelling to see that the results obtained with this indirect method yield to exclusion limits comparable to the ones obtained from direct di-Higgs searches even though this approach only has NLO sensitivity to κ_λ. It will be interesting to follow the developments of the single-Higgs constraints on κ_λ during the HL-LHC physics programme, as current projections obtained by the CMS Collaboration from the analysis of the $t\bar{t}$H and tH production mechanisms and the H $\rightarrow \gamma\gamma$ decay foresee constraints of $-4.1 < \kappa_\lambda < 14.1$ at 95% CL [11], expected to improve significantly once that all the production mechanisms and decay channels will be combined.

7.3 Summary

A complete characterization of the properties of the H boson in the $H \rightarrow ZZ \rightarrow 4\ell$ decay channel is performed exploiting proton-proton collisions at $\sqrt{s} = 13$ TeV, corresponding to 137 fb^{-1} of integrated luminosity collected by the CMS experiment during the full Run-II of the LHC.

The SM predictions are tested measuring the inclusive signal strength modifier, which is found to be $\mu = 0.94 \pm 0.07\,(\text{stat})^{+0.09}_{-0.08}\,(\text{syst})$, as well as the signal strength modifiers for the five main production mechanisms of the H boson. The WH and ZH signal strengths are measured separately for the first time in the $H \rightarrow ZZ \rightarrow 4\ell$ decay channel in the context of this analysis. All the measurements are performed at a fixed value of $m_H = 125.38$ GeV and the impact of leaving this parameter unconstrained in the fit is found to be negligible, with effects of less than 4% on all the results. The profiled value of m_H is found to be $m_H = 125.09\,^{+0.15}_{-0.14}$ GeV. Even though this analysis is not optimized for the measurement of the H boson mass, it is remarkable to observe that the precision on the profiled value of m_H is comparable to the official CMS result obtained combining the $H \rightarrow \gamma\gamma$ and $H \rightarrow ZZ$ decay channels [1], reflecting the fact that the high resolution available in the “golden channel” drives the ultimate sensitivity of this measurement.

A set of model-independent results is obtained measuring the cross section of the $H \rightarrow 4\ell$ process in a fiducial phase space volume that closely matches the reconstruction level selection. The inclusive fiducial cross section is found to be $2.84^{+0.23}_{-0.22}\,(\text{stat})^{+0.26}_{-0.21}\,(\text{syst})$ fb, in agreement with the SM prediction of 2.84 ± 0.15 fb. Fiducial cross sections are measured in differential bins of the transverse momentum and rapidity of the Higgs boson, the number of associated jets, and the transverse momentum of the leading associated jet.

The analysis also provides the measurement of cross sections in the STXS framework to enhance the experimental sensitivity while simultaneously reducing the theoretical dependence of the results. The cross sections of the main production mechanisms of the H boson are measured in the STXS Stage 0. To probe in much more detail each of these processes, cross sections in the fine granular bins of the STXS Stage 1.2 are also measured for the first time in a CMS analysis.

Measurements of CP-violation and anomalous couplings of the H boson to vector bosons (HVV) and fermions (Hff) are also performed. For the first time in a CMS analysis five HVV, two Hgg, and two Htt couplings are measured simultaneously. All the results are extracted imposing custodial and $SU(2)\times U(1)$ symmetries and are also interpreted in the context of EFT operators in the Higgs basis.

This analysis also presents the measurement of the H boson trilinear self-coupling (λ_{HHH}) using a single-Higgs decay channel for the first time in a Run-II analysis. The result is extracted in the form of a constraint on the anomalous coupling term (κ_λ) introduced in the SM Lagrangian to parametrize possible BSM effects that would scale λ_{HHH}. In order to enhance the sensitivity to κ_λ, the measurement is performed in differential bins of the transverse momentum of the H boson.
The measured value is found to be $\kappa_\lambda^{\text{obs.}} = 4.1^{+6.4}_{-5.9} = 4.1^{+6.1}_{-5.8}(\text{stat})\,^{+2.0}_{-1.2}(\text{sys})$ (68% CL.).

All the results are found to be in agreement with the expectations for a SM H boson, stressing the strength of the SM and the difficulty of unveiling new BSM physics.

References

1. Collaboration CMS (2020) A measurement of the Higgs boson mass in the diphoton decay channel. Phys Lett B 805:135425
2. de Florian D, et al (2016) Handbook of LHC higgs cross sections: 4. Deciphering the nature of the higgs sector. 2/2017, 10 2016
3. Collaboration CMS (2016) Measurement of differential and integrated fiducial cross sections for Higgs boson production in the four-lepton decay channel in pp collisions at $\sqrt{s} = 7$ and 8 TeV. JHEP 04:005
4. Grazzini M, Sargsyan H (2013) Heavy-quark mass effects in Higgs boson production at the LHC. JHEP 09:129
5. Anastasiou C, Duhr C, Dulat F, Furlan E, Gehrmann T, Herzog F, Lazopoulos A, Mistlberger B (2016) High precision determination of the gluon fusion Higgs boson cross-section at the LHC. JHEP 05:058
6. Gritsan AV, Roskes J, Sarica U, Schulze M, Xiao M, Zhou Y (2020) New features in the JHU generator framework: constraining Higgs boson properties from on-shell and off-shell production. Phys Rev D 102(5):056022
7. Collaboration CMS (2021) Constraints on anomalous Higgs boson couplings to vector bosons and fermions in its production and decay using the four-lepton final state. Phys Rev D 104(5):052004
8. Collaboration CMS (2020) Measurements of $\mathrm{t\bar{t}H}$ Production and the CP Structure of the Yukawa Interaction between the Higgs Boson and Top Quark in the Diphoton Decay Channel. Phys Rev Lett 125(6):061801
9. ATLAS Collaboration (2022) Constraining the Higgs boson self-coupling from single- and double-Higgs production with the ATLAS detector using pp collisions at $\sqrt{s} = 13$ TeV. Submitted to Phys Lett B
10. Collaboration CMS (2022) A portrait of the Higgs boson by the CMS experiment ten years after the discovery. Nature 607(7917):60–68
11. Cepeda M et al (2019) Report from working group 2: Higgs physics at the HL-LHC and HE-LHC. CERN Yellow Rep Monogr 7:221–584
12. Degrassi G, Giardino PP, Maltoni F, Pagani D (2016) Probing the Higgs self coupling via single Higgs production at the LHC. JHEP 12:080
13. Maltoni F, Pagani D, Shivaji A, Zhao X (2017) Trilinear Higgs coupling determination via single-Higgs differential measurements at the LHC. Eur Phys J C 77(12):887

Chapter 8
Conclusions

In June 2012 the ATLAS and CMS Collaborations at CERN announced the discovery of a scalar particle compatible with the Higgs boson predicted by the Standard Model (SM) of particle physics. The discovery was possible exploiting data from proton-proton collisions at a centre-of-mass energy of $\sqrt{s} = 7$ and 8 TeV, collected during the Run-I (2009–2013) phase of the LHC. The Higgs boson was detected as a statistically relevant fluctuation at around 125 GeV, with respect to a background-only hypothesis, from the combination of the results obtained with several decay channels.

The Higgs boson decay into two Z vector bosons that consequently decay into four leptons has been one of the most sensitive channels for the discovery. The four-lepton decay channel is also known as the "golden channel" because of its many virtues that greatly contributed to the success of the Run-I analyses: it features a large signal-to-background ratio, a completely resolved final state, and an excellent mass resolution. After a two years long shutdown of the machine, the LHC operations resumed in 2015 with the Run-II (2015–2018), at the unprecedented centre-of-mass energy of $\sqrt{s} = 13$ TeV, targeting the precision measurement of the properties of the Higgs boson as well as the detection of possible signs of physics beyond the Standard Model (BSM).

This thesis presented a study of the Higgs boson properties in the four-lepton decay channel exploiting the full Run-II statistics, for a total of 137 fb^{-1} analyzed. The predictions of the SM have been tested measuring the production cross sections of the Higgs boson in the "golden channel" and the results have been presented in the form of signal strength modifiers, defined as the ratio of the observed cross section to its SM expectation. The inclusive $\text{H} \rightarrow \text{ZZ} \rightarrow 4\ell$ signal strength modifier is found to be $\mu = 0.94 + 0.07\,(\text{stat})^{+0.09}_{-0.08}\,(\text{syst})$ and even in the signal strength modifiers of the main production mechanisms no deviation from the SM was observed.

M. Bonanomi, *Response of the High Granularity Calorimeter HGCAL and Characterisation of the Higgs Boson*, Springer Theses,
https://doi.org/10.1007/978-3-031-26833-5_8

The $H \to 4\ell$ cross section has also been measured in a fiducial phase space region, defined to closely match the events' selection, after the unfolding of detector and reconstruction effects and its observed value is found to be $\sigma_{\text{fid}} = 2.84^{+0.23}_{-0.22}$ (stat)$^{+0.26}_{-0.21}$ (syst) fb, in agreement with the SM prediction of 2.84 ± 0.15 fb. In order to probe in more detail the SM and to possibly unveil hints of physics BSM, fiducial cross sections have also been measured in differential bins of the transverse momentum and rapidity of the Higgs boson, the number of associated jets, and the transverse momentum of the leading associated jet. All the measurements on data have been found to be in agreement with the predictions from the theory.

While fiducial cross sections have several advantages, they come at the price of an inclusive description of the production modes of the Higgs boson. The Simplified Template Cross Sections (STXS) framework was introduced at the beginning of Run-II with the goal of providing a common ground for the analyses to perform production cross sections measurements, enhancing the experimental sensitivity, whilst at the same time reducing the dependence on the theoretical predictions. The STXS framework defines a set of mutually exclusive bins that have been evolving from the beginning of Run-II to cope with the increasingly large statistics collected by the experiments. In addition to Stage 0, which corresponds to the inclusive cross section of each production mechanism, the cross sections of Stage 1.2 have been measured in the analysis, for a total of 19 cross sections that provide a fine granular description of the entire production phase space. All the results have been found to be in agreement with the SM predictions and neither hints of new physics have been detected in the STXS bins designed to provide enhanced sensitivity to BSM scenarios, even though most of these bins are still limited by the small statistics available.

An even more stringent test of the SM has been performed measuring the anomalous contributions of the Higgs boson couplings to vector bosons (HVV) and fermions (Hff). For the first time in a CMS analysis a total of nine anomalous contributions to the Higgs boson couplings have been measured simultaneously: five HVV, two anomalous couplings to gluons (Hgg), and two Htt. Additional measurements of CP-violation have also been performed, but all the results have been found to be consistent with the SM within their uncertainties.

One of the most compelling tests of the SM consists in the measurement of the trilinear self-coupling of the Higgs boson, as it is directly related to the value of the scalar potential and it is a free parameter of the theory. Hence, precision measurements of this observable would ultimately disclose the secrets of the SM and for this reason many analyses are, and will be even more in the near future, working on this topic.

Modifications induced by the presence of new physics can be parametrized from the introduction of an anomalous coupling term (κ_λ) directly in the SM Lagrangian, defined as the ratio of the observed Higgs boson self-coupling to its SM prediction. The value of κ_λ is often constrained from the study of double Higgs boson production, as its cross section depends on this parameter at LO. More recently, complementary strategies have been sought in order to overcome the limited statistics typical of these channels. Modifications of the Higgs boson self-coupling contribute to NLO electroweak corrections to the single Higgs boson production cross sections, which are three orders of magnitude larger than the typical di-Higgs ones. Hence, these processes can be used to infer constraints on κ_λ complementary to the ones obtained from the study of the double Higgs boson channels. Exclusion limits on κ_λ have been extracted for the first time in a single-Higgs analysis with CMS, exploiting the cross section measurement in differential bins of the transverse momentum of the Higgs boson and resulting in an observed value of $\kappa_\lambda^{\mathrm{obs.}} = 4.1^{+6.4}_{-5.9} = 4.1^{+6.1}_{-5.8}(\mathrm{stat})\ ^{+2.0}_{-1.2}(\mathrm{sys})$ (68%CL.).

The complete characterization of the H boson properties in the "golden channel" did not unveil any sign of BSM physics and all the results are found to be in agreement with the SM predictions within their uncertainties. The full Run-II data set opened the doors to the precision physics domain, with the error on many of the results commencing to be limited by the systematic component of the uncertainties. However, in some cases, the measurements are still limited by the statistics available. Given the remarkable predictive power of the SM when including the H boson, possible new physics will require even more extensive studies at higher statistics. To further stress the SM theory and probe its detailed predictions, CERN intends to commence the operations of the High-Luminosity LHC (HL-LHC) project by the end of 2027. The HL-LHC is expected to integrate 10 times the final LHC luminosity, thus extending the physics reach of the experiments but also leading to a higher pile-up rate and to unprecedented radiation levels. A massive upgrade of the LHC infrastructure and the detectors is necessary to maintain the current physics performance in such a harsh environment. The CMS Collaboration will replace the current endcap calorimeters with a High Granularity Calorimeter (HGCAL), designed to provide improved discrimination power and radiation hardness in the busy environment of the HL-LHC. The HGCAL will be the first large-scale silicon-based imaging calorimeter ever employed in a high energy physics experiment. The analysis of the test beam data collected in October 2018 with the first large-scale prototype of the HGCAL presented in this thesis provides a comprehensive assessment of the physics performance of the detector and its design, with a particular focus on the electromagnetic compartment of the prototype. In addition to the measurement of the energy resolution and linearity, the high granularity of the prototype is exploited to study the longitudinal and transverse containment of electromagnetic cascades, as well as to measure the position and angular resolution of the detector.

The results of the data analysis and the straightforward operation of the HGCAL prototype in the October 2018 test beam validate the detector design and suggest that it will meet the physics requirements expected in the HL-LHC operations.

This thesis presented two topics strictly related to the immediate past and future of the LHC. On one side, the "golden channel" was used to probe in detail the properties of the Higgs boson, exploiting the full Run-II data collected by the CMS experiment. On the other side, the "silicon road" towards the ultimate design of the HGCAL was explored in view of the installation of this innovative detector for the HL-LHC operations.

The "golden channel" is a longstanding character of the Higgs physics at the LHC, and it will continue to mark important milestones even in the years to come. If the Run-II statistics enables detailed studies of the production cross sections of the Higgs boson, the next round of analyses will set significant breakthroughs in understanding the EWSB mechanism and the SM predictions. In addition to the measurements of the Higgs boson's mass and width, future versions of this analysis envisage the use of innovative techniques and improved strategies to enhance the sensitivity to BSM scenarios and the precision of the results.

The path of the "silicon road" will lead the CMS experiment to the installation of the HGCAL, thus improving the detector's performance in the very forward region. The HGCAL, along with all the other upgrades of the CMS detector, will ensure detailed studies of the Higgs boson and SM processes, measurements of BSM physics, and access to rare phenomena, such as vector boson-initiated processes.

It will be interesting to follow the "golden channel" path along the "silicon road" in the near future, as a large plethora of possible studies exploiting this final state appears on the horizon. What is sure is that some of the questions raised at the beginning of the Run-II are still open, and the four-lepton decay channel and the HGCAL will be two of the pillars in the quest for answers in the years to come.

Appendix A
Dataset Used in the Analysis

The datasets used for 2016, 2017 and 2018 data taking are listed in Tables A.1, A.2, and A.3, respectively, along with the integrated luminosity.

Table A.1 Datasets used in the 2016 analysis

Run-range	Dataset	Integrated luminosity
272007-275376	/DoubleMuon/Run2016B-17Jul2018-v1/MINIAOD	5.892 fb^{-1}
	/DoubleEG/Run2016B-17Jul2018-v1/MINIAOD	
	/MuonEG/Run2016B-17Jul2018-v1/MINIAOD	
	/SingleElectron/Run2016B-17Jul2018-v1/MINIAOD	
	/SingleMuon/Run2016B-17Jul2018-v1/MINIAOD	
275657-276283	/DoubleMuon/Run2016C-17Jul2018-v1/MINIAOD	2.646 fb^{-1}
	/DoubleEG/Run2016C-17Jul2018-v1/MINIAOD	
	/MuonEG/Run2016C-17Jul2018-v1/MINIAOD	
	/SingleElectron/Run2016C-17Jul2018-v1/MINIAOD	
	/SingleMuon/Run2016C-17Jul2018-v1/MINIAOD	
276315-276811	/DoubleMuon/Run2016D-17Jul2018-v1/MINIAOD	4.353 fb^{-1}
	/DoubleEG/Run2016D-17Jul2018-v1/MINIAOD	
	/MuonEG/Run2016D-17Jul2018-v1/MINIAOD	

(continued)

M. Bonanomi, *Response of the High Granularity Calorimeter HGCAL and Characterisation of the Higgs Boson*, Springer Theses,
https://doi.org/10.1007/978-3-031-26833-5

Table A.1 (continued)

Run-range	Dataset	Integrated luminosity
	/SingleElectron/Run2016D-17Jul2018-v1/MINIAOD	
	/SingleMuon/Run2016D-17Jul2018-v1/MINIAOD	
276831-277420	/DoubleMuon/Run2016E-17Jul2018-v1/MINIAOD	4.117 fb^{-1}
	/DoubleEG/Run2016E-17Jul2018-v1/MINIAOD	
	/MuonEG/Run2016E-17Jul2018-v1/MINIAOD	
	/SingleElectron/Run2016E-17Jul2018-v1/MINIAOD	
	/SingleMuon/Run2016E-17Jul2018-v1/MINIAOD	
277772-278808	/DoubleMuon/Run2016F-17Jul2018-v1/MINIAOD	3.186 fb^{-1}
	/DoubleEG/Run2016F-17Jul2018-v1/MINIAOD	
	/MuonEG/Run2016F-17Jul2018-v1/MINIAOD	
	/SingleElectron/Run2016F-17Jul2018-v1/MINIAOD	
	/SingleMuon/Run2016F-17Jul2018-v1/MINIAOD	
278820-280385	/DoubleMuon/Run2016G-17Jul2018-v1/MINIAOD	7.721 fb^{-1}
	/DoubleEG/Run2016G-17Jul2018-v1/MINIAOD	
	/MuonEG/Run2016G-17Jul2018-v1/MINIAOD	
	/SingleElectron/Run2016G-17Jul2018-v1/MINIAOD	
	/SingleMuon/Run2016G-17Jul2018-v1/MINIAOD	
280919-284044	/DoubleMuon/Run2016H-17Jul2018-v1/MINIAOD	8.857 fb^{-1}
	/DoubleEG/Run2016H-17Jul2018-v1/MINIAOD	
	/MuonEG/Run2016H-17Jul2018-v1/MINIAOD	
	/SingleElectron/Run2016H-17Jul2018-v1/MINIAOD	
	/SingleMuon/Run2016H-17Jul2018-v1/MINIAOD	
	/DoubleMuon/Run2016H-17Jul2018-v1/MINIAOD	
	/DoubleEG/Run2016H-17Jul2018-v1/MINIAOD	
	/MuonEG/Run2016H-17Jul2018-v1/MINIAOD	
	/SingleElectron/Run2016H-17Jul2018-v1/MINIAOD	
	/SingleMuon/Run2016H-17Jul2018-v1/MINIAOD	
	/DoubleMuon/Run2016H-17Jul2018-v1/MINIAOD	
	/DoubleEG/Run2016H-17Jul2018-v1/MINIAOD	
	/MuonEG/Run2016H-17Jul2018-v1/MINIAOD	
	/SingleElectron/Run2016H-17Jul2018-v1/MINIAOD	
	/SingleMuon/Run2016H-17Jul2018-v1/MINIAOD	

Table A.2 Datasets used in the 2017 analysis

Run-range	Dataset	Integrated luminosity
297046-299329	/DoubleMuon/Run2017B-17Nov2017-v1/MINIAOD	4.792 fb^{-1}
	/DoubleEG/Run2017B-17Nov2017-v1/MINIAOD	
	/MuonEG/Run2017B-17Nov2017-v1/MINIAOD	
	/SingleElectron/Run2017B-17Nov2017-v1/MINIAOD	
	/SingleMuon/Run2017B-17Nov2017-v1/MINIAOD	
299368-300676	/DoubleMuon/Run2017C-17Nov2017-v1/MINIAOD	9.755 fb^{-1}
	/DoubleEG/Run2017C-17Nov2017-v1/MINIAOD	
	/MuonEG/Run2017C-17Nov2017-v1/MINIAOD	
	/SingleElectron/Run2017C-17Nov2017-v1/MINIAOD	
	/SingleMuon/Run2017C-17Nov2017-v1/MINIAOD	
302030-303434	/DoubleMuon/Run2017D-17Nov2017-v1/MINIAOD	4.319 fb^{-1}
	/DoubleEG/Run2017D-17Nov2017-v1/MINIAOD	
	/MuonEG/Run2017D-17Nov2017-v1/MINIAOD	
	/SingleElectron/Run2017D-17Nov2017-v1/MINIAOD	
	/SingleMuon/Run2017D-17Nov2017-v1/MINIAOD	
303824-304797	/DoubleMuon/Run2017E-17Nov2017-v1/MINIAOD	9.424 fb^{-1}
	/DoubleEG/Run2017E-17Nov2017-v1/MINIAOD	
	/MuonEG/Run2017E-17Nov2017-v1/MINIAOD	
	/SingleElectron/Run2017E-17Nov2017-v1/MINIAOD	
	/SingleMuon/Run2017E-17Nov2017-v1/MINIAOD	
305040-306462	/DoubleMuon/Run2017F-17Nov2017-v1/MINIAOD	13.50 fb^{-1}
	/DoubleEG/Run2017F-17Nov2017-v1/MINIAOD	
	/MuonEG/Run2017F-17Nov2017-v1/MINIAOD	
	/SingleElectron/Run2017F-17Nov2017-v1/MINIAOD	
	/SingleMuon/Run2017F-17Nov2017-v1/MINIAOD	
278820-280385	/DoubleMuon/Run2017G-PromptReco-v1/MINIAOD	7.721 fb^{-1}
	/DoubleEG/Run2017G-PromptReco-v1/MINIAOD	
	/MuonEG/Run2017G-PromptReco-v1/MINIAOD	
	/SingleElectron/Run2017G-PromptReco-v1/MINIAOD	
	/SingleMuon/Run2017G-PromptReco-v1/MINIAOD	
281207-284068	/DoubleMuon/Run2017H-PromptReco-v1/MINIAOD	8.857 fb^{-1}
	/DoubleEG/Run2017H-PromptReco-v1/MINIAOD	
	/MuonEG/Run2017H-PromptReco-v1/MINIAOD	
	/SingleElectron/Run2017H-PromptReco-v1/MINIAOD	
	/SingleMuon/Run2017H-PromptReco-v1/MINIAOD	
	/DoubleMuon/Run2017H-PromptReco-v2/MINIAOD	
	/DoubleEG/Run2017H-PromptReco-v2/MINIAOD	
	/MuonEG/Run2017H-PromptReco-v2/MINIAOD	
	/SingleElectron/Run2017H-PromptReco-v2/MINIAOD	
	/SingleMuon/Run2017H-PromptReco-v2/MINIAOD	
	/DoubleMuon/Run2017H-PromptReco-v3/MINIAOD	
	/DoubleEG/Run2017H-PromptReco-v3/MINIAOD	
	/MuonEG/Run2017H-PromptReco-v3/MINIAOD	
	/SingleElectron/Run2017H-PromptReco-v3/MINIAOD	
	/SingleMuon/Run2017H-PromptReco-v3/MINIAOD	

Table A.3 Datasets used in the 2018 analysis

Run-range	Dataset	Integrated luminosity
315252-316995	/DoubleMuon/Run2018A-17Sep2018-v2/MINIAOD	14.02 fb^{-1}
	/MuonEG/Run2018A-17Sep2018-v1/MINIAOD	
	/EGamma/Run2018A-17Sep2018-v2/MINIAOD	
	/SingleMuon/Run2018A-17Sep2018-v2/MINIAOD	
317080-319310	/DoubleMuon/Run2018B-17Sep2018-v1/MINIAOD	7.06 fb^{-1}
	/MuonEG/Run2018B-17Sep2018-v1/MINIAOD	
	/EGamma/Run2018B-17Sep2018-v1/MINIAOD	
	/SingleMuon/Run2018B-17Sep2018-v1/MINIAOD	
319337-320065	/DoubleMuon/Run2018C-17Sep2018-v1/MINIAOD	6.90 fb^{-1}
	/MuonEG/Run2018C-17Sep2018-v1/MINIAOD	
	/EGamma/Run2018C-17Sep2018-v1/MINIAOD	
	/SingleMuon/Run2018C-17Sep2018-v1/MINIAOD	
320673-325175	/DoubleMuon/Run2018D-PromptReco-v2/MINIAOD	31.75 fb^{-1}
	/MuonEG/Run2018D-PromptReco-v2/MINIAOD	
	/EGamma/Run2018D-PromptReco-v2/MINIAOD	
	/SingleMuon/Run2018D-PromptReco-v2/MINIAOD	

Appendix B
Fiducial Cross Sections: Response Matrices

The differential cross sections are measured in bins of the H boson transverse momentum ($p_{\mathrm{T}}^{\mathrm{H}}$) and rapidity ($|\ y^{\mathrm{H}}\ |$) as well as in bins of the number of associated jets (N^{j}) and of the transverse momentum of the leading jet ($p_{\mathrm{T}}^{\mathrm{j}}$) in the event. The response matrices $\epsilon_{i,j}^{f}$ that enter Eq. (6.11) for the observables considered in the analysis are depicted in Figs. B.1, B.2, B.3, B.4 and B.5 for the various production mechanisms of the H boson.

The matrices showing the correlations among the fitted fiducial cross sections measured in the analysis are shown in Fig. B.6.

M. Bonanomi, *Response of the High Granularity Calorimeter HGCAL and Characterisation of the Higgs Boson*, Springer Theses,
https://doi.org/10.1007/978-3-031-26833-5

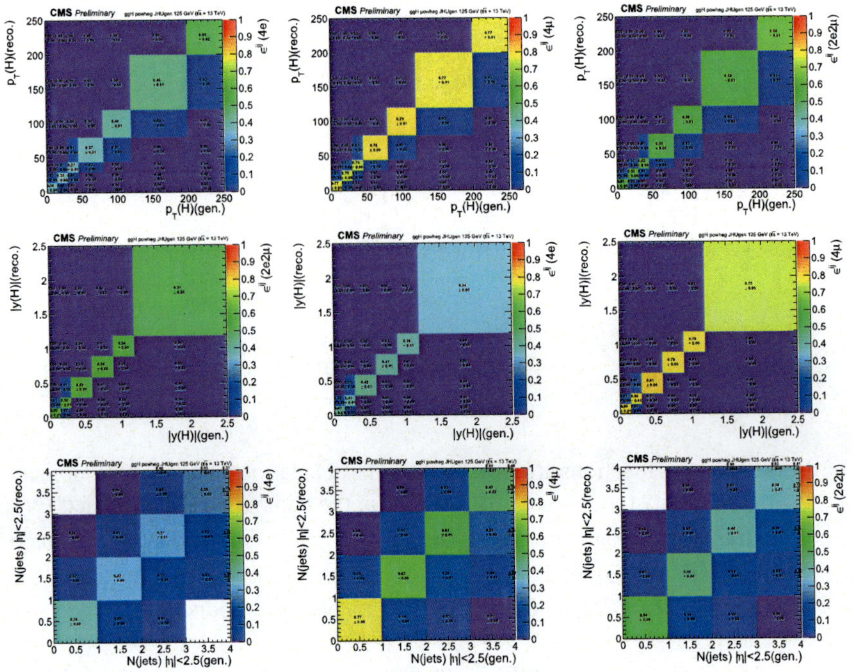

Fig. B.1 Efficiency matrices for the p_T(H) (top) and | yH | (middle) and N(jets) (bottom) observables for gluon fusion production mode in the 4e (left) 4μ (middle), and 4μ (right), $2e2\mu$ final states

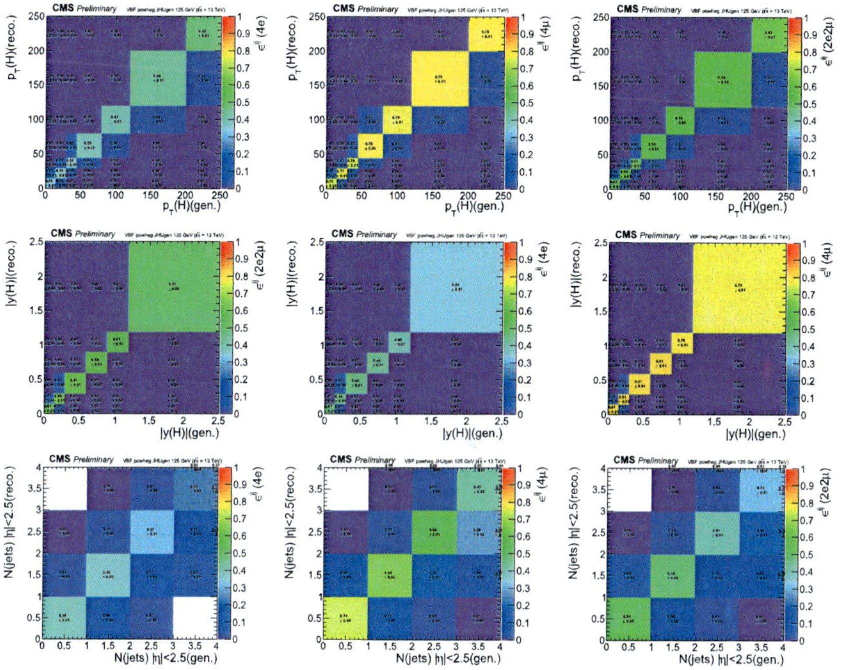

Fig. B.2 Efficiency matrices for the p_T(H) (top) and | yH | (middle) and N(jets) (bottom) observables for vector boson fusion production mode in the 4e (left) 4μ (middle), and 4μ (right), $2e2\mu$ final states

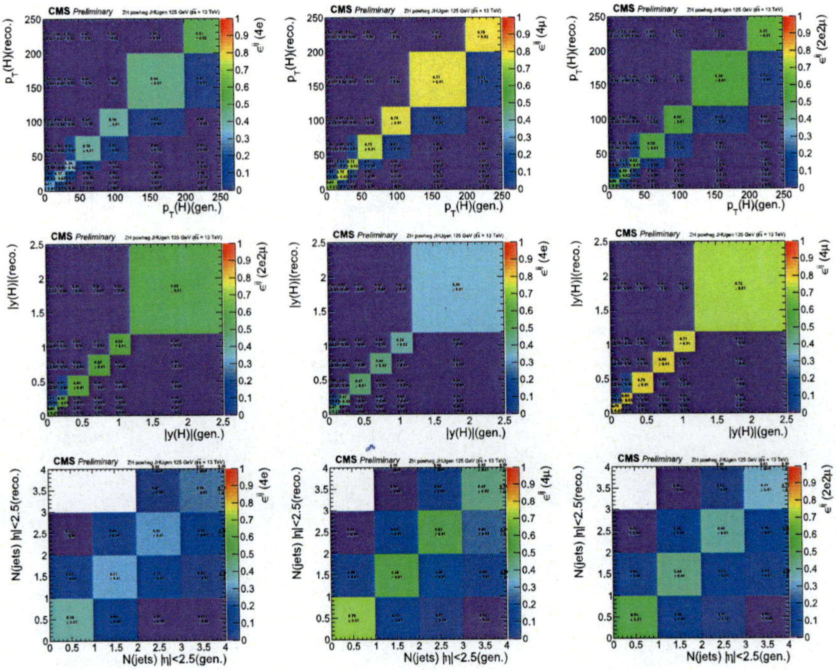

Fig. B.3 Efficiency matrices for the p_T(H) (top) and | yH | (middle) and N(jets) (bottom) observables for ZH production mode in the 4e (left) 4μ (middle), and 4μ (right), $2e2\mu$ final states

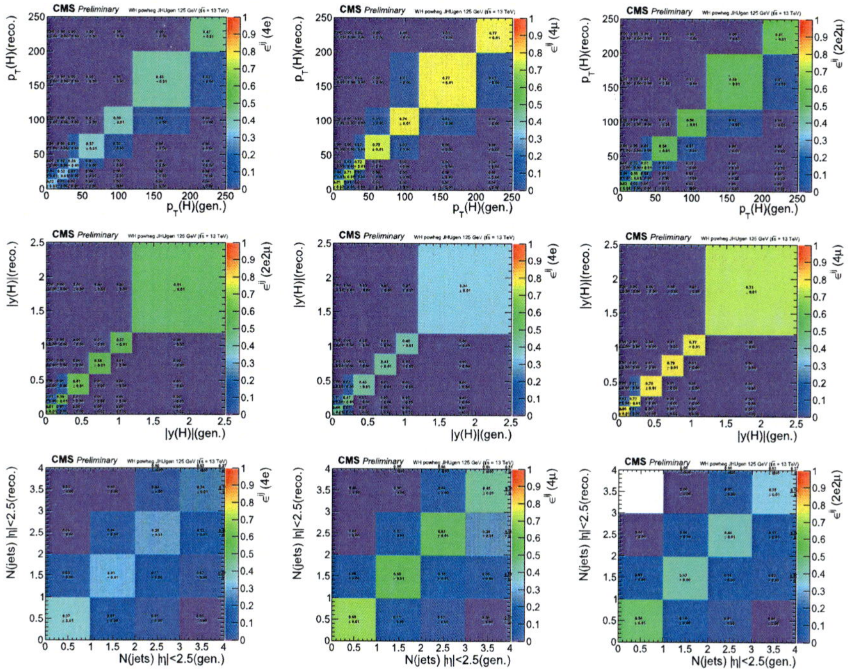

Fig. B.4 Efficiency matrices for the p_T(H) (top) and | yH | (middle) and N(jets) (bottom) observables for WH production mode in the 4e (left) 4μ (middle), and 4μ (right), $2e2\mu$ final states

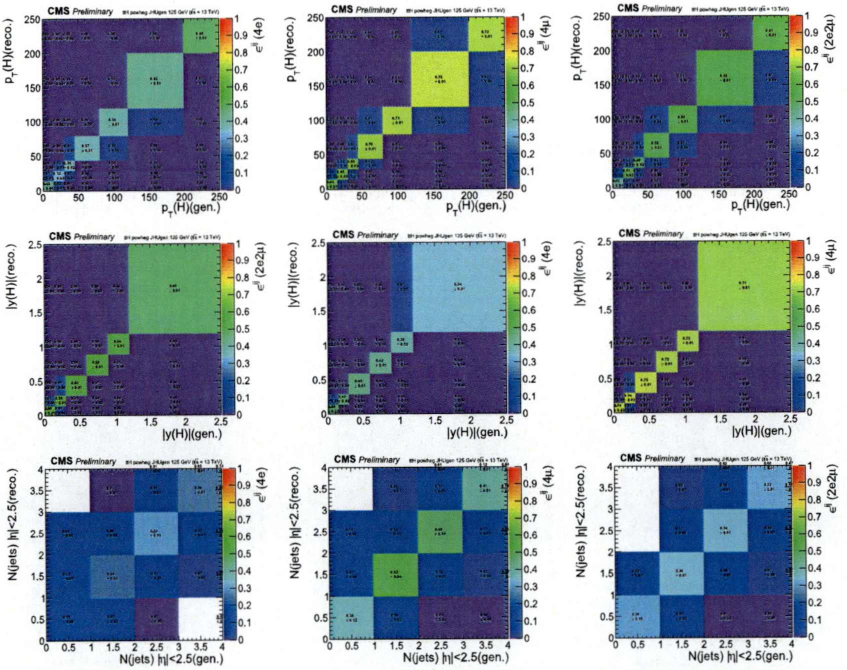

Fig. B.5 Efficiency matrices for the p_T(H) (top) and | yH | (middle) and N(jets) (bottom) observables for tt̄H production mode in the 4e (left) 4μ (middle), and 4μ (right), $2e2\mu$ final states

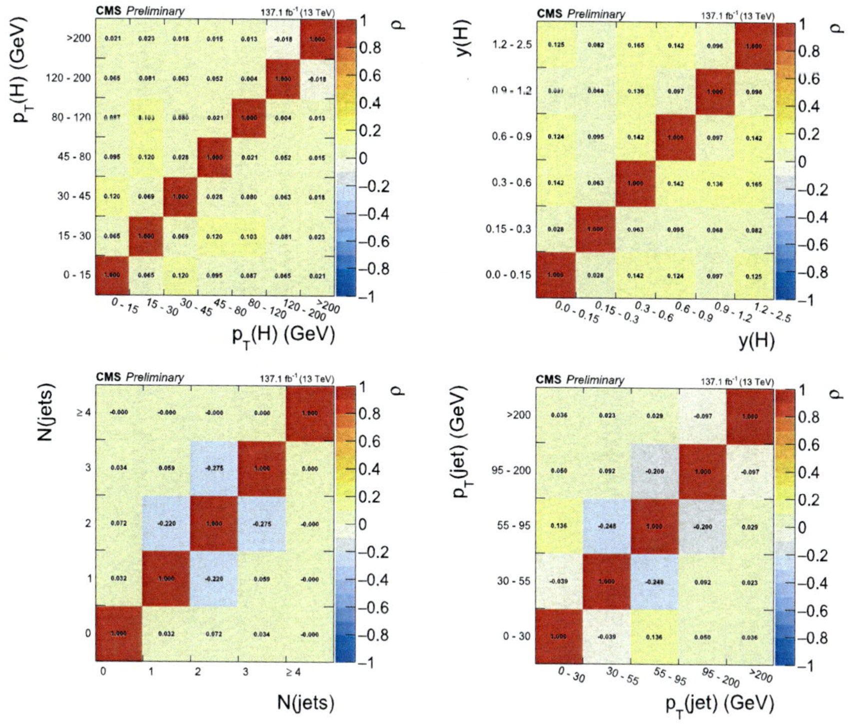

Fig. B.6 Correlation matrices for the measurement of the differential cross sections in bins of the H boson transverse momentum (p_T^H, top left) and rapidity (abs y^H, top right) as well as in bins of the number of associated jets (N^j, bottom left) and of the transverse momentum of the leading jet (p_T^j, bottom right) in the event

Matteo Bonanomi | Curriculum Vitae

matteo.bonanomi@cern.ch

Experience

University of Hamburg **Hamburg**
Postdoctoral Researcher *2022–Present*
CMS group of Prof. Johannes Haller. Research topics: precision measurement of Higgs boson properties, searches for heavy Higgs bosons, Higgs boson pair production, statistical methods in data analysis. Supervision of Ph.D. and M.Sc. students.

M. Bonanomi, *Response of the High Granularity Calorimeter HGCAL and Characterisation of the Higgs Boson*, Springer Theses,
https://doi.org/10.1007/978-3-031-26833-5

Matteo Bonanomi | Curriculum Vitae

matteo.bonanomi@cern.ch

Experience

University of Hamburg **Hamburg**
Postdoctoral Researcher *2022–Present*
CMS group of Prof. Johannes Haller. Research topics: precision measurement of Higgs boson properties, searches for heavy Higgs bosons, Higgs boson pair production, statistical methods in data analysis. Supervision of Ph.D. and M.Sc. students.

M. Bonanomi, *Response of the High Granularity Calorimeter HGCAL and Characterisation of the Higgs Boson*, Springer Theses,
https://doi.org/10.1007/978-3-031-26833-5

Laboratoire Leprince-Ringuet (École Polytechnique) **Palaiseau**
Ph.D. in Particle Physics *2018–2021*
Study of the properties of the Higgs boson in the four leptons decay channel and on the validation of the High Granularity Calorimeter (HGCAL) design. Supervisors: Prof. Roberto Salerno, Prof. Yves Sirois

École Polytechnique **Palaiseau**
M.Sc. thesis co-supervisor *2019–2021*
I co-supervised three M.Sc. thesis projects during my Ph.D.

École Polytechnique **Palaiseau**
Teaching Assistant *2019–2021*
Frontal and practical lectures: experimental aspects of particle physics. Referent: Prof. Alexandre Zabi

University of Milano-Bicocca **Milan**
Physics I Tutor *2017– 2018*
Exercises on classical mechanics, thermodynamics and special relativity. Referent: Prof. Marco Paganoni

Research Awards

CMS Thesis Award: Ph.D. Thesis awarded by CMS, "rewarding the most outstanding theses for the CMS experiment, providing inspiration and motivation for the new generations of students in CMS and elsewhere."
Best Poster Award: Best Poster Award at the 15th Topical Seminar on Innovative Particle and Radiation Detectors, Siena 2019. Title: "Beam-tests of CMS High Granularity Calorimeter prototypes at CERN."
INFN Grant winner: INFN Grant winner: "Particle physics to understand the Universe", three months of work at CERN.

Selected List of Publications

1. CMS, Collaboration: "Differential cross section measurements in the Higgs boson to four-lepton decay channel in proton-proton collisions at $\sqrt{s} = 13$ TeV". CMS–PAS–HIG–21–009.
2. CMS, Collaboration: "Measurements of production cross sections of the Higgs boson in the four-lepton final state in proton-proton collisions at $\sqrt{s} = 13$ TeV". Eur. Phys. J. C 81 (2021), Nr. 6, S. 488.

Laboratoire Leprince-Ringuet (École Polytechnique) **Palaiseau**
Ph.D. in Particle Physics *2018–2021*
Study of the properties of the Higgs boson in the four leptons decay channel and on the validation of the High Granularity Calorimeter (HGCAL) design.
Supervisors: Prof. Roberto Salerno, Prof. Yves Sirois

École Polytechnique **Palaiseau**
M.Sc. thesis co-supervisor *2019–2021*
I co-supervised three M.Sc. thesis projects during my Ph.D.

École Polytechnique **Palaiseau**
Teaching Assistant *2019–2021*
Frontal and practical lectures: experimental aspects of particle physics.
Referent: Prof. Alexandre Zabi

University of Milano-Bicocca **Milan**
Physics I Tutor *2017– 2018*
Exercises on classical mechanics, thermodynamics and special relativity.
Referent: Prof. Marco Paganoni

Research Awards

CMS Thesis Award: Ph.D. Thesis awarded by CMS, "rewarding the most outstanding theses for the CMS experiment, providing inspiration and motivation for the new generations of students in CMS and elsewhere."
Best Poster Award: Best Poster Award at the 15th Topical Seminar on Innovative Particle and Radiation Detectors, Siena 2019. Title: "Beam-tests of CMS High Granularity Calorimeter prototypes at CERN."
INFN Grant winner: INFN Grant winner: "Particle physics to understand the Universe", three months of work at CERN.

Selected List of Publications

1. CMS, Collaboration: "Differential cross section measurements in the Higgs boson to four-lepton decay channel in proton-proton collisions at $\sqrt{s} = 13$ TeV". CMS–PAS–HIG–21–009.
2. CMS, Collaboration: "Measurements of production cross sections of the Higgs boson in the four-lepton final state in proton-proton collisions at $\sqrt{s} = 13$ TeV". Eur. Phys. J. C 81 (2021), Nr. 6, S. 488.

3. CMS, Collaboration: "Constraints on anomalous Higgs boson couplings to vector bosons and fermions in its production and decay using the four-lepton final state". Phys. Rev. D 104 (2021), Nr. 5, S. 052004.
4. Acar, B. u.a.: "Construction and commissioning of CMS CE prototype silicon modules".JINST 16 (2021), Nr. 04, S. T04002.
5. Acar, B. u.a.: "Response of a CMS HGCAL silicon-pad electromagnetic calorimeter prototype to 20-300 GeV positrons".JINST 17 (2022) Nr. 05, P05022.
6. Bonanomi, M.: "Higgs boson measurements in the WW and ZZ final states with the CMS Detector". PoS ICHEP2020 (2020), S. 049.
7. Bonanomi, M.: "Beam-tests of CMS High Granularity Calorimeter prototypes at CERN". JINST 15 (2020), Nr. 04, C04001.
8. Abbiendi, G. u.a.: "A study of muon-electron elastic scattering in a test beam". JINST 16 (2021), Nr. 06, P06005.
9. Abbiendi, G. u.a.: "Results on multiple Coulomb scattering from 12 and 20 GeV electrons on carbon targets". JINST 15 (2020), Nr. 01, S. 01.